Principles and Practices of
Plant Genomics

——— Volume 1: Genome Mapping ———

Editors

Chittaranjan Kole
Albert G. Abbott
Department of Genetics and Biochemistry
Clemson University
South Carolina
USA

Science Publishers

Enfield (NH)   Jersey   Plymouth

SCIENCE PUBLISHERS
An imprint of Edenbridge Ltd., British Isles.
Post Office Box 699
Enfield, New Hampshire 03748
United States of America

Website: *http://www.scipub.net*

sales@scipub.net (marketing department)
editor@scipub.net (editorial department)
info@scipub.net (for all other enquiries)

Library of Congress Cataloging-in-Publication Data

Principles and practices of plant genomics/editors, Chittaranjan Kole,
  Albert G. Abbott.
    p. cm.
  Includes bibliographical references and index.
  ISBN 978-1-57808-525-5 (hardcover)
    1. Plant genomes. 2. Plant genetics. I. Kole, Chittaranjan. II.
  Abbott, Albert G.
  QK981.P75 2008
  572.8'62—dc22

                                                      2007047428

ISBN 978-1-57808-525-5

© 2008, Copyright reserved

Published by Science Publishers, Enfield, NH, USA
An imprint of Edenbridge Ltd.

Dedicated to

Prof. Blannie Bowen

Vice Provost for Academic Affairs
The Pennsylvania State University

The Prologue

Dear Prof. Kole

Having read the preface and contents of your series of volumes on plant genetics, I wish to thank you to have undertaken together with Professor Abbott the heavy work to produce an updated encyclopedia on molecular plant genetics. I am sure that your series on "Principles and Practices of Plant Genomics" will have a tremendous, worldwide impact on plant sciences and their applications.

Our human society has benefited for thousands of years from plants found in their natural environments. Soon people learned to bring about improvements by breeding, as well as by choosing most appropriate soils to grow the plants serving for their nutrition, as well as for medical and ornamental uses.

A true paradigmatic change has started a few decades ago with the progress of scientific investigations at the molecular level. Novel opportunities have become available by a combination of genetic engineering and classical breeding to obtain even more appropriate plant variants for the service to the human population. With novel research strategies and with steadily improving knowledge on gene structure and functions, we can expect that already in the near future many cultivated plants will benefit themselves of an improved health, be more resistant to pests and, what will be of great importance, will provide improved nutritional values to our daily diets. Agriculture will enter a new, more sustainable era.

Your publication series can greatly help the scientists working on this field to reach this goal and to carry out their work, with the required care and responsibility, to the benefit both of the human society and of its environment.

Please take these words as an encouragement to complete and also to steadily update your important project.

With my best regards.

Yours Sincerely,

January 23, 2008

Prof. Werner Arber
Professor Emeritus for Molecular Microbiology
University of Basel
Nobel Laureate in Medicine 1978

Foreword

The elucidation of the double-helix structure of the Deoxyribose Nucleic Acid (DNA) molecule in 1953 by Drs. James Watson, Francis Crick, Maurice Wilkins and Rosalind Franklin marked the beginning of what is now known as *the new genetics*. Research during the last 54 years in the fields of molecular genetics and recombinant-DNA technology has opened up new opportunities in agriculture, medicine, industry, and environment protection.

Availability of DNA-based molecular markers since 1980s has facilitated construction of complete genetic maps in several plants of academic and economic importance. Several computer software also have paved the way for handling huge numbers of markers and segregating individuals to frame these maps, also for detection of positions of genes and gene clusters, known as quantitative trait loci, on them.

Molecular marker-based genetic mapping has provided not only comprehensive depictions of the genomes but also facilitated elucidation of phylogenetic relationships and evolutionary pathways, map-based cloning of genes and above all use of molecular markers in various schemes of crop improvement, including germplasm characterization, marker-aided selection and introgression of genes through marker-assisted backcross breeding.

The strategies of construction of BAC and YAC libraries and advent of high-throughput sequencing of DNA have built the platform for physical mapping of chromosomal regions and even whole genomes. The new millennium has already witnessed complete sequencing of the genomes of the model plant Arabidopsis and the leading crop plant rice followed by the genomes of a forest tree poplar and a fruit tree peach. Genome initiatives in many more crop plants individually and families comprising important crop plants are indeed progressing fast.

One of the spectacular happening during this period is the merging of information technology and biological science leading to a new discipline called as bioinformatics. This subject is now playing a pivotal role in warehousing biological data on sequences and functions of genes, transcripts, expressed sequence tags and proteins, and their searching to utilize for various purposes. We have now several new branches with the 'omics' sciences such as genomics, transcriptomics, proteomics and metabolomics. However, all of them relates to the genomes and reside under the common roof of genomics.

There is little doubt that the genomics has opened up uncommon opportunities for enhancing the productivity, profitability, sustainability and stability of major cropping systems. It has also created scope for developing crop varieties resistant/tolerant to biotic and abiotic stresses through an appropriate blend of Mendelian and molecular breeding techniques. It has led to the possibility of undertaking anticipatory breeding to meet potential changes in temperature, precipitation and sea level as a result of global warming. There are new opportunities for fostering pre-breeding and farmer-participatory breeding methods in order to continue the merits of genetic efficiency with genetic diversity.

In the coming decades, farm women and men in population-rich but land-hungry countries like India and China will have to produce more food and other agricultural commodities to meet home needs and to take advantage of export opportunities, under conditions of diminishing per capita availability of arable land and irrigation water and expanding abiotic and biotic stresses. The enlargement of the gene pool with which breeders work will be necessary to meet these challenges. Genomics supplemented by transgenics provide breeders with a powerful tool for enlarging the genetic base of crop varieties and to pyramid genes for a wide range of economically important traits.

Genetics can boast of being the subject that progressed with the most rapid speed within the shortest period of time and contributed stupendously in unraveling the nature and function of genes and also towards the genetic improvement of useful microbes, animals and plants in the last century. In fact, the *green revolution* of the sixties was the contribution of classical genetics and conventional breeding. Another revolution in agriculture, medicine and environment is in the wing and we are ready to welcome it as *gene revolution* to be driven by genomics-aided breeding. However, for a meaningful and sustainable application of plant genomics, importance of *orphan* crops which have immense potential for improving food and nutrition security should be taken into account.

The book series "Principles and Practices of Plant Genomics" is thus a timely contribution. It provides an authoritative account of recent progress in plant genomics including structural and functional genomics and their use in molecular- and genomics-assisted breeding. It also provides a road map for genomic research in the present century. I express our gratitude to Profs. Chittaranjan Kole and Albert Abbott for this labor of love in spreading knowledge among both students and scientists on the opportunities as well as challenges opened up by the genomics era.

M S Swaminathan
Chairman
M.S. Swaminathan Research Foundation

Preface to the Series

It is an unequivocal fact that genomics has emerged as the leading discipline in plant sciences. The enormous global efforts in the last two and a half decades have led to a very rapid development of this subject. Currently genetic linkage mapping employing molecular markers has become routine for development of chromosome maps in genetically refractory species, mapping of economic genes and gene clusters, studies on evolution and phylogenetic relationships, map-based cloning of useful genes, and marker-aided breeding. Physical mapping of plant genomes using artificial chromosomes and integration of physical and genetic maps have provided more power of resolution to depict the structural basis of genes and also to elucidate genome colinearity and evolution facilitating exploration of homologous desired genes employing comparative genomics. Whole-genome sequencing of the model plant Arabidopsis followed by rice and poplar has already demonstrated the power of structural genomics in elucidating focused understanding of genome organization using sequence data that represents the ultimate level of genetic information. The success stories of genome initiatives in microbes, animals and plants have inspired 'initiatives' at national and international levels to sequence genomes of an array of other plant genomes. Information on genetic and physical maps, colinearity and whole-genome sequencing serve as the starting point for assignment of biological meaning to putative genes with no known phenotype. For this precise purpose, functional genomics emerged to address the function of genes discovered through sequencing efforts. Employment of concepts and strategies including expressed sequence tags, reverse genetics and transcriptional profiling have facilitated identification and discovery of genes and their expressions in specific tissues and stages.

Plant genomics has already found its deserved place in various courses under agriculture, medicine, and environment sciences. Application of molecular markers for several purposes in crop improvement has fostered the development of another discipline, molecular breeding. Recent developments in functional genomics, transgenics and bioinformatics have vastly enriched the resources of today's plant breeding. Information on a large number of genes handled simultaneously by genomics and thereby genetic integration of diverse processes, tissues and organisms will highly benefit future plant improvement endeavors. Therefore, future molecular breeding will encompass marker-assisted breeding, transgenic breeding and genomics-aided breeding. Plant genomics and molecular breeding are going to play the pivotal role in the fields of agriculture, medicine, environment and ecology in this and hopefully in the coming centuries.

However, through academic and research related interactions with fellow research workers on plant genomics, we recognized the need of a handbook for ready reference. Textbooks explaining the fundamentals and applications of plant genomics are of significant value to students, teachers and scientists of the public and private sectors. Plant genomics is basically the frontier branch of genetics but is complemented strongly by biochemistry, microbiology, computational biology, and bioinformatics; with supplementation from various other disciplines including crop production, statistics, physiology, pathology, entomology, horticulture, just to name a few. Most interestingly in today's world, people from all these fields learn, teach and/or practice plant genomics. Thus, it is essential that academics in these broad disciplines have available the lucid deliberations of the basic concepts and strategies of plant genomics to enhance their own studies, training and expertise. This was the main driving force for taking up this endeavor to popularize the science and art of plant genomics.

The subject of plant genomics has grown really rapidly within a short period of two and a half decades but with distinct phases. Since the mid-eighties, basic structural genomics emerged comprising construction of genetic linkage maps using molecular markers and mapping of genes and quantitative trait loci. The following phase included application of molecular markers for various purposes of crop improvement, popularly known as molecular breeding. And the third phase involved physical genome mapping with artificial chromosomes culminating in advanced structural genomics including whole-genome mapping, and functional genomics. We have tried to follow almost the same trend while organizing the chapters on different topics under the three volumes of this text book series.

In the first volume we briefly introduce the historical background and overview on genome mapping in the first chapter. Subsequently, we present 10 chapters deliberating on different types of molecular markers, their detection, relative merits, shortcomings and applications; types of mapping populations, methods of their generation, applications; basic concepts and schematic depiction of construction of genetic linkage maps; concepts and strategies of mapping genes controlling qualitative and quantitative traits on framework genetic linkage maps; rationale, methodologies and implications of comparative mapping; principles, strategies, and outcome of map-based cloning; overviews on the recent advances on plant genomics and genome initiatives; and finally computer strategies and software employed in plant molecular mapping and breeding.

In the second volume, we emphasize areas of application of molecular mapping. We present a general deliberation on the fundamentals of molecular breeding and applications of molecular markers for germplasm characterization; concepts and application of molecular mapping and breeding for yield, quality and their related component traits; biotic and abiotic stresses; and physiological traits of economic importance. Some plant groups, for example fruit crops, forest trees, fodder, forage and turf grasses, and polyploid crop plants exhibit unique breeding problems and necessarily require special breeding strategies. We have dedicated independent chapters for them. We include a chapter on transgenic breeding considering its growing popularity and contribution in crop improvement. We also include a chapter on intellectual property rights (IPR) considering the legal implications in using, developing and commercializing plant varieties and genotypes.

In the third volume on advanced genomics, we present an overview on the advances of plant genomics made in the last century; deliberations on the genomics resources; concepts, tools, strategies and achievements of comparative, evolutionary and functional genomics and whole-genome sequencing. We also present critical reviews on the already completed genome initiatives and glimpses on the currently progressing genome initiatives. We also have a deliberation on application of the genomics information in genetic improvement of crop plants. We include a chapter on facilitation of the progress of plant genomics researches considering the requirements of generous funding, state-of-the-art facilities and well-trained human resources. Finally, we present depiction of a map of the road for plant genomics to be traveled by us and scientists of our next generations in the twenty-first century.

Articles and reviews on all these concepts, strategies, tools and achievements in genome analysis and improvement abound but are scattered mainly on the pages of periodicals. Most of them deal with a focus on a particular crop plant or a group of related crop plants. Recently, a seven-volume series on 'Genome Mapping and Molecular Breeding in Plants' was published by Springer that encompasses 82 chapters on more than 100 plant species of economic importance.

Interested readers may enrich their knowledge from this and other resources as well. However, in these three books, we have paid particular attention to the basic principles, strategies, and requirements and applications of various facets of plant genomics useful for students to develop a comprehensive understanding of the subjects and to assist academicians and researchers in disseminating and using the knowledge.

We express our deep gratitude to Prof. Werner Arber, Nobel Laureate in Medicine 1978 for kindly contributing the Prologue for this series and his generous comments about this endeavor. We are grateful to Prof. M.S. Swaminathan for kindly scripting the foreword for this treatise. Views of these science missionaries and social visionaries will be immensely useful for the present and future travelers in the world of plant genomics and will remain as words of inspiration for the present and future generations of plant scientists.

We must mention here that working with the authors of the chapters of these three volumes has been highly enriching, enlightening and entertaining. We wish to express our thanks and gratitude to these devoted and reputed scientists for their excellent chapters and above all, their affection and cooperation during the pleasant and painful periods of gestation and delivery of these three book volumes. They appreciated the importance of these textbooks and donated their time to contribute these clear and concisely written chapters.

The authors of some chapters of our series have included a few previously published illustrations in original or modified versions. We believe they have obtained permission for such reproduction from the competent authors or publishers of the articles containing these illustrations. However, as editors we believe that it is our moral duty to keep in record our acknowledgement for all these authors, publishers and sources for the helps that facilitated the improvement of the contents and formats of the concerned deliberations.

We look forward to constructive criticisms and suggestions from all corners for future improvement of the contents and format of these three book volumes.

We must express our thanks and gratitude to the publishers for their constant co-operation extended to us and the authors. We remember that on many occasions we failed to stick to the dead lines of submission of manuscripts, proof corrections and other required inputs due mainly to the conflicts of schedules and other unforeseen assignments and commitments. However, the concerned staff did bear with us with laudable patience.

Chittaranjan Kole
Albert G. Abbott

Preface to the Volume

Genome of an organism is depicted by genetic linkage mapping and physical mapping. Genome mapping started with genetic linkage mapping and contributed enormously in genome analysis and its improvement. Physical mapping emerged later and is the prelude to structural and functional genomics. We have therefore accentuated on genetic linkage mapping in this volume with glimpses on physical mapping as it will be delineated emphatically in Volume 3 on advanced genomics.

Genetic linkage mapping practiced since the mid-eighties paved the way for emergence of the full-grown subject of plant genomics. The first chapter of this volume provides an overview on the historical background, basics and rationale of genetic mapping. It is well known that linkage mapping requires three pivotal components: a segregating population that experiences optimum recombinations spread over the chromosomes of the genome; an optimum number of polymorphic molecular markers to be used as landmarks to evenly cover the genome; and a computer program to frame the linkage groups preferably representing the haploid number of chromosomes of the genome based on recombination events of the markers among the individuals of the population used. Genetic linkage mapping evolved from first generation maps called preliminary linkage maps through second generation maps such as high-resolution, high-density and saturated maps, to the third generation maps like the consensus, integrated, comparative and reference maps. The evolution of the genetic maps obviously was contributed by the concomitant evolution of the mapping populations, from F_2s to pseudo-testcross; molecular markers from RFLP to SNPs; and mapping software from Linkage I to JoinMap. We have devoted four chapters with schematic depictions of the diverse molecular markers, mapping populations, mapping programs; and development of linkage maps using them. We have included four chapters on application of genetic linkage maps for positioning genes controlling simply inherited trait loci controlling qualitative characters; detection of putative genomic locations controlling quantitative traits; comparative mapping; and positional cloning of genes and QTLs.

The chapters on bioinformatics and its application on plant genetics, mapping and breeding; plant genome initiatives and computing strategies and software are equally related to and useful for understanding and practicing genetic linkage mapping, molecular breeding, and structural and functional genomics. We placed them in the first volume envisaging their implications in the basics of plant genome analysis and improvement.

The chapters of this volume have been contributed by 30 scientists from 9 countries and all of them have pioneering contributions in plant genomics and molecular breeding. We met and interacted with almost all of them at least once in San Diego during the Genome Meetings. It was really an enriching and entertaining experience to have 'linkage' with them again to produce this volume. We appreciate their earnest efforts to prepare their deliberations with class-room approach with lucid texts, good examples and illustrative figures mostly from their own works. We are thankful to all of them for their contributions, cooperation and *e-company*!

We must take the responsibly for any mistake in planning, editing and designing of this volume and will look forward to suggestions for its improvement from all corners. If this volume is considered to be useful to students, scientists and industries, the credit must go only to the authors and the publisher.

Chittaranjan Kole
Albert G. Abbott

Contents

List of Contributors

Albert G. Abbott

Department of Genetics and Biochemistry, Clemson University, 116 Jordan Hall, Clemson, SC 29634, USA; Phone: +1-864-656-3060; Fax: +1-864-656-6879; *e-mail:* aalbert@clemson.edu

Amy Frary

Department of Biological Sciences, Clapp Lab, Mount Holyoke College, South Hadley, MA 01075, USA; Phone: +1-413-538-3015; Fax: +1-413-538-2548; *e-mail:* afrary@mtholyoke.edu

Brian Yandell

University of Wisconsin, Animal Sciences Building, 1675 Observatory Drive (Room 146), Biometry Program, Madison, Wisconsin 53706, USA; Phone: +1-608-262-1157; Fax: +1-608-262-0032; *e-mail:* byandell@wisc.edu

Chenguang Wang

Department of Statistics, University of Florida, Gainesville, FL 32611, USA; Phone: +1-352-265-8035; Fax: +1-352-392-8162; *e-mail:* cgwang@cog.ufl.edu

Chittaranjan Kole

Department of Genetics and Biochemistry, 111 Jordan Hall, Clemson University, Clemson, SC 29634, USA; Phone: +1-864-656-3060; Fax: +1-864-656-6879; *e-mail:* ckole@clemson.edu

Christopher A. Cullis

Department of Biology, Case Western Reserve University, 10900 Euclid Avenue, Cleveland, OH 44106-7080, USA; Phone: +1-216-368-5362; Fax: +1-216-368-4672; *e-mail:* cac5@case.edu

David Edwards

Australian Centre for Plant Functional Genomics, Institute for Molecular Biosciences and School of Land, Crop and Food Sciences, University of Queensland, Brisbane, QLD 4072, Australia; Phone: +61-7-3346-2615; Fax: +61-7-3346-2101, *e-mail:* Dave.Edwards@ acpfg.com.au

David Lee

Centre for Plant Genetics, Breeding and Evaluation, NIAB, Huntingdon Road, Cambridge, CB3 0LE, UK; Phone: +44-1223-342241; Fax: +44-1223-342269; *e-mail:* david.lee@niab.com

Elisa Mihovilovich

International Potato Center, Apartado 1558 Lima 12, Peru; Phone: +51-1-349-6017; Fax: +51-1-349-5638; *e-mail:* e.mihovilovich@cgiar.org

Hanamareddy Biradar

Department of Genetics & Plant Breeding, University of Agricultural Sciences, GKVK, Bangalore 560065, India; Phone: +91-80-23624967; Fax: +91-80-23330267; *e-mail:* bghreddy@rediffmail.com

Hong-Bin Zhang

Department of Soil and Crop Sciences, Texas A&M University, 2474 TAMU, College Station, Texas 77843-2474, USA; Phone: +1-979-862-2244; Fax: +1-979-845-0456; *e-mail:* hbz7049@tamu.edu

Ilan Levin

Institute of Plant Sciences, Agricultural Research Organization, The Volcani Center, P.O. Box 6, Bet Dagan 50250, Israel; Phone: +972-3-9683477; Fax: +972-3-9669642; *e-mail:* vclevini@volcani.agri.gov.il

Ilan Paran

Department of Plant Genetics and Breeding, Agricultural Research Organization, The Volcani Center, P.O. Box 6, Bet Dagan 50250, Israel; Phone: +972-3-9683943; Fax: +972-3-9669642; *e-mail:* iparan@volcani.agri.gov.il

Jacqueline Batley

Australian Centre for Plant Functional Genomics, Centre for Integrated Legume Research and School of Land, Crop and Food Sciences, University of Queensland, Brisbane, QLD 4072, Australia; Phone: +61-7-3346-9948; Fax: +61-7-3346-3556; *e-mail:* J.Batley@uq.edu.au

Jiahan Li

Department of Statistics, University of Florida, Gainesville, FL 32611, USA; Phone: +1-352-392-3806; Fax: 352-392-8555; *e-mail:* jiahanli@ufl.edu

M. Maheswaran

Department of Plant Breeding and Genetics, Agricultural College and Research Institute, Tamil Nadu Agricultural University, Coimbatore 641003, India; Phone: +91-9443550818; *e-mail:* mahes@tnau.ac.in

Merideth Bonierbale

International Potato Center, Apartado 1558, Lima 12, Peru; Phone: +51-1-349-6017; Fax: +51-1-349-5638; *e-mail:* m.bonierbale@cgiar.org

Milind B. Ratnaparkhe

Division of Plant Sciences, University of Missouri-Columbia, 210 Waters Hall, MO 65211, USA; Phone: +1-573-884-0451; Fax: +1-573-884-1467; *e-mail:* ratnaparkhem@missouri.edu

Myron M. Chang

Department of Epidemiology and Health Policy Research, University of Florida, Gainesville, FL 32611, USA; Phone: +1-352-273-0553; Fax: +1-352-392-8162; *e-mail:* mchang@cog.ufl.edu

Peter Bradbury

Cornell University, USDA-ARS, 741 Rhodes Hall, USA; Phone: +1-607-255-5392; Fax: +1-607-255-0323; *e-mail:* pjb39@cornell.edu

Peter J. Maughan

285 WIDB, Brigham Young University, Provo, Utah 84602, USA; Phone: +1-801-422-8698; Fax: +1-801-422-0008; *e-mail:* Jeff_Maughan@byu.edu

Reinhard Simon

International Potato Center, Apartado 1558, Lima 12, Peru; Phone: +51-1-349-6017; Fax: +51-1-349-5638; *e-mail:* s.reinhard@cgiar.org

Rongling Wu

Department of Statistics, Institute of Food and Agricultural Sciences, Statistical Consulting Unit, 409 McCarty Hall C, P.O. Box 110339, University of Florida, Gainesville, FL 32611-0339, USA; Phone: +1-352-392-3806; Fax: +1-352-392-8555; *e-mail:* rwu@stat.ufl.edu

Runqing Yang

School of Agriculture and Biology, Shanghai Jiaotong University, Shanghai 200240, PRC; Phone: +1186-21-54745853; Fax: +1186-21-54745853; *e-mail:* runqingyang@sjtu.edu.cn

Sami Doganlar

Department of Biology, Izmir Institute of Technology, Gulbahcekoyu Campus, Urla 35430, Izmir, Turkey; Phone: +90-232-7507547; Fax: +90-232-7507509; *e-mail:* samidoganlar@iyte.edu.tr

Shailaja Hittalmani

Department of Genetics & Plant Breeding, University of Agricultural Sciences, GKVK, Bangalore 560065, India; Phone: +91-80-2362-4967; Fax: +91-80-2333-0277; *e-mail:* shailajah_maslab@rediffmail.com

Silvia Doveri

Department of Bioorganic Chemistry and Biopharmacy, University of Pisa, Via Bonanno 33, 56126 Pisa, Italy; Phone: +39-50-2219695; Fax: +39-50-2219660; *e-mail:* silvia.doveri@gmail.com

T. N. Girish

Department of Genetics & Plant Breeding, University of Agricultural Sciences, GKVK, Banglaore 560065, India; Phone: +91-80-23624967; Fax: +91-80-23330267; *e-mail:* girishchain@yahoo.co.in

Wayne Powell

Diversity Genomics Group, NIAB, Huntingdon Road, Cambridge, CB3 0LE, UK; Phone: +44-1223-342280; Fax: +44-1223-277602; *e-mail:* wayne.powell@niab.com

Yanru Zeng

School of Forestry and Biotechnology, Zhejiang Forestry University, Lin'an, Zhejiang 311300, PRC; Phone: +1186-571-63743858; Fax: +1186-571-63732738; *e-mail:* zengyr@hotmail.com

List of Abbreviations

AB-QTL	Advanced backcross-QTL
AFLP	Amplified fragment length polymorphism
ANOVA	Analysis of variance
AP-PCR	Arbitrarily primed-PCR
BAC	Bacterial artificial chromosome
BC	Backcross
BGI	Beijing Genome Institute (PRC)
BIBAC	Binary BAC
BIM	Bayesian interval mapping
BSA	Bulked segregant analysis
CAPS	Cleaved amplified polymorphic sequence
CBC	Clone-by-clone
CIM	Composite interval mapping
cM	CentiMorgan
DAF	DNA amplification fingerprinting
DDB	DNA Data Bank (Japan)
DH	Doubled haploid
DHPLC	Denaturing high-pressure liquid chromatography
DP	Donor parent
DUS	Distinctness, Uniformity and Stability
EBI	European Bioinformatics Institute (UK)
EM	Expectation-maximization
EMBL	European Molecular Biology Laboratory (Germany)
EMS	Ethyl-methane sulfonate
eQTL	Expression QTL
EST	Expressed sequence tag
FAO	Food and Agriculture Organization (Italy)
FAQ	Frequently asked questions

FISH   Fluorescent in situ hybridization
GC   Genomic control
GDPC   Genomic diversity and phenotype connection
GEO   Gene expression omnibus
GIS   Geographic information system
GISH   Genomic in situ hybridization
GLM   General linear model
GOA   Gene ontology annotation
GSS   Genome survey sequence
HICF   High information content fingerprint
HIF   Heterogeneous inbred family
HK   Haley-Knott (regression)
HTR   Haplotype trend regression
HWE   Hardy-Weinberg equilibrium
IBD   Identical by descent
IBI   International *Brachypodium* Initiative
ICARDA   International Center for Agricultural Research in the Dry Areas (Syria)
ICIS   International Crop Information System
ICRISAT   International Crop Research Institute for the Semi-Arid Tropics (India)
IF_2 Immortalized F_2
IM   Interval mapping
IMP   Multiple imputation
InDel   Insertion-deletion
INSDC   International Nucleotide Sequence Database Collaboration
IPCR   Inverse PCR
IPGRI   International Plant Genetic Research Institute (Italy)
IRAP   Inter-retrotransposon amplified polymorphism
IRRI   International Rice Research Institute (The Philippines)
ISSR   Inter-simple sequence repeat
IWGSC   International Wheat Genome Sequencing Consortium
JGI   Joint Genome Institute (USA)
KEGG   Kyoto Encyclopedia of Genes and Genomes (Japan)
LBC   Large-insert bacterial clone
LCR   Ligase chain reaction
LD   Linkage disequilibrium
LIS   Legume Information System
LRR   Leucine-rich repeat
LOD   Logarithm of odds

LPD	Log posterior density
LR	Log-likelihood ratio
LSD	Least-square difference
LTR	Long terminal repeat
MAS	Marker-assisted selection
MBL	Maximum bin length
MCMC	Markov chain Monte Carlo
MCPDs	Multi-crop passport descriptors
MF	Methylation filtration
MIAME	Minimum information about a microarray experiment
MIM	Multiple interval mapping
MIP	Molecular inversion probe
MIPS	Munich Institute for Protein Sequences (Germany)
ML	Maximum likelihood
MLM	Mixed linear model
MPSS	Massively parallel signature sequencing
MS	Mean squares
MTP	Minimal tiling path
NASC	Nottingham Arabidopsis Stock Center (UK)
NBS	Nucleotide binding site
NCGR	National Center for Genome Resources (USA)
NIL	Near(ly)-isogenic line
NR	Non-recombinant
ORF	Open reading frame
PAC	P1-derived artificial chromosome
PBC	Plasmid-based bacterial clone
PCR	Polymerase chain reaction
PFGE	Pulsed-field gel electrophoresis
PHYLIP	Phylogeny inference package
PLRV	Potato leafroll virus
QTL	Quantitative trait loci
RAPD	Random(ly) amplified polymorphic DNA
RBIP	Retrotransposon-based insertional polymorphism
REMAP	Retrotransposon-microsatellite amplified polymorphism
RFLP	Restriction fragment length polymorphism
RI	Recombinant inbred
RIL	Recombinant inbred line
RNAi	RNA (mediated) interference
RP	Recurrent parent
RT-PCR	Real-time reverse transcription PCR

SAGE	Serial analysis of gene expression
SALOD	Sum of adjacent LOD scores
SARF	Sum of adjacent recombination fraction
SBE	Single-based extension
SBS	Sequencing by synthesis
SCAR	Sequence characterized amplified region
SFP	Single feature polymorphism
SHOM	Sequencing by hybridization with oligonucleotide matrix
SIM	Simple interval mapping
SNP	Single nucleotide polymorphism
SNuPE	Single nucleotide extension
SRAP	Sequence related amplification polymorphism
S-SAP	Sequence-specific amplified polymorphism
SSBL	Sum of squares of bin length
SSCP	Single-strand conformational polymorphism
SSD	Single seed descent
SSR	Simple sequence repeat
STS	Sequence tagged site
TAC	Transformation-competent artificial chromosome
TAIR	The Arabidopsis Information Resource (USA)
TASSEL	Trait analysis by association, evolution and linkage
T-DNA	Transferred-DNA
TIGR	The Institute of Genomic Research (USA)
TILLING	Targeting induced local lesions in genomes
TRAP	Target region amplification polymorphism
TSP	Traveling salesman's problem
UHD	Ultra high-density (map)
USDA	United States Department of Agriculture
WGS	Whole-genome shotgun
YAC	Yeast artificial chromosome

1 | Fundamentals of Plant Genome Mapping

Chittaranjan Kole[1]* and Albert G. Abbott[2]

[1]Department of Horticulture, 316 Tyson Building,
The Pennsylvania State University, University Park, PA 16802,
USA; *e-mail*: cuk10@psu.edu

[2]Department of Genetics and Biochemistry, 116 Jordan Hall,
Clemson University, Clemson, SC 29634, USA

The rediscovery of the 'Laws of Inheritance' by Gregor Johann Mendel in the early months of 1900 was immediately followed by an array of elegant experiments. Many of them conducted on both plants and animals validated unequivocally the 'Law of Segregation' of the two alleles of a locus (factors or elements, coined as 'genes' later in 1909 by Wilhelm L. Johannsen). The term 'genetics', the study of inheritance was also first proposed during this period by William Bateson in his letter to Adam Sedgwick of April 18, 1905, which was formally made public at the Third International Conference on Genetics at London in 1906.

However, certain observations in these early genetic investigations exhibited some deviations from those noted by Mendel in his experiments with garden pea, *Pisum sativum*. The significant ones included intra-allelic and inter-allelic interactions; penetrance and expressivity; pleiotropism; multiple allelism; polygenic inheritance; extra-nuclear inheritance and linkage. All these phenomena had a tremendous impact on evolution of the concept of genome; its depiction through chromosome maps constructed based on genetic markers; and its improvement by using these markers in breeding.

*Present address: Department of Genetics and Biochemistry, 111 Jordan Hall, Clemson University, Clemson, SC 29634, USA

1 THE GENOME CONCEPTS

1.1 Types of Genomes

The 'Bovery-Sutton Chromosome Theory' or 'Chromosomal Theory of Heredity' proposed independently by Theodor Boveri and Walter S. Sutton in 1902 (Sutton 1903) stands as one of the most significant milestones in genetics as it enunciated for the first time that the genes are physically present in the chromosomes, and they are transmitted from the parents to their progeny following the segregation and crossing over of the homologous chromosomes contributed by each parent. The earlier experiments on inheritance focused on the chromosomes that are contained in the nucleus. However, Mendel himself carried out reciprocal crosses showing identical results and thereby ruling out cytoplasmic involvement. Thus, in the era of Mendelian genetics, the pattern of nuclear inheritance defined the genome. Later on, the haploid chromosome content of an organism was defined as its 'genome'. Figure 1 depicts the genomes of Arabidopsis, rice and human, the three leading model organisms for genome mapping.

The essence of all the above deliberations is that the chromosomes are the bearers of the genes that govern the characters of an organism. These characters include those attributes that serve as the basis for species distinction, described as 'specificity' and those attributes by which individuals of a species could be differentiated from one another, termed 'variation'.

In the post-Mendelian period, the transient influence of maternal parent on the offspring (denoted maternal effect) was observed. Involvement of cytoplasmic organelles with stable transmission of characters of the maternal parent to the offspring led to the concept of cytoplasmic inheritance, also known as extra-nuclear, extra-chromosomal or epigenetic inheritance. Well-organized cytoplasmic organelles like mitochondria and chloroplasts of higher plants and animals were found to carry genes (chondriogenes and plastogenes) and were deservedly recognized as genomes. However, their small size, limited gene repertoire, and non-Mendelian segregation excluded them from classical genetic mapping experiments that involved biparental crosses to enumerate parental and recombinant genotypes. Many other cytoplasmic entities including kappa particles of *Paramecium* and F-factors of *Escherichia coli*, etc. were also found to carry genes. They were also not included in early genome mapping efforts for the same reasons as mitochondria and chloroplast.

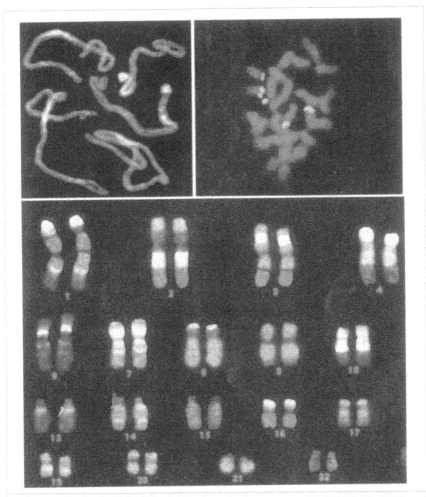

Fig. I Genomes A. *Arabidopsis thaliana* (2n = 10), B. *Oryza sativa* (2n = 24) and C. *Homo sapiens* (2n = 46).

Sources: http://www.esa.int/esaHS/SEMHQU2IU7E_business_0.html; ((c) European Communities 1995-2006)

N. Ohmido, Y. Akiyama and K. Fukui (1998) Plant Molecular Biology 38: 1043-1052 *http://www.mun.ca/biology/scarr/FISH_chromosome_painting.htm.* (with permission from Dr. Steven M. Carr and Genetix Ltd., UK)

Therefore at this period, the sum total of all the genes, nuclear and cytoplasmic, in a cell of an organism defines the genome of that organism and the blue print of its life controlling all its features, both structural and functional.

Subsequent to the discovery of chromosomal inheritance, was the discovery of DNA as the genetic material (Avery et al. 1944), except a few RNA-viruses, and their presence in chromosomes and the cytoplasmic organelles and particles detectable by chemical staining (Fuelgen 1924), led to a change in the genome concept. In the neo-Mendelian period of genetics, the sum total of all DNA contained in an organism or in the cells of an organism is known as the genome. However, much later it was proven that only a small portion of the DNA, at least in higher plants and animals, encoded heritable characters.

It is of worth mentioning here that the genomes of the eukaryotes contrast markedly to those of prokaryotes. The main genome of the eukaryotes is separated by a nuclear envelope from the cytoplasm. The constitution of the nuclear genome of different organisms exhibits variation in sequences, in DNA content (**C**-value) and the number of chromosomes. The genomes of prokaryotes and eukaryotes differ in many attributes as described in the Table 1 below.

Table 1   Major differences between prokaryotic and eukaryotic genomes.

Characteristics	Prokaryotes	Eukaryotes
Genome size	600 kb to 9.5 Mb	3 Mb to 140,000 Mb
Average gene size/number	950 bp/4300	2500 bp/19,000
Operon-like regulatory unit	General	No
Horizontal gene transfer	Significant	Negligible
Rate of non-coding sequences	Low	High
Intron	Rare	General
Redundant gene number	Rare	General
Ploidy level	Haploidy	Haploidy to Polyploidy
Chromosome number	One	More than one
Heterozygosity	No	Yes

The first prokaryotic genomes to be studied in depth at the molecular level were those of the bacterium *Escherichia coli* and its bacteriophage. In eukaryotes, the first genomes to be studied in great detail at the molecular level were those of mitochondria and chloroplasts. Finally, attention turned to detailed studies of the nuclear genomes of selected eukaryotes: an yeast (*Saccharomyces cerevisiae*), a nematode (*Caenorhabditis elegans*), a mammal (*Homo sapiens*) and two plants (dicot: *Arabidopsis thaliana*, monocot: *Oryza sativa*).

In this book, we will discuss only mapping of the genomes of higher plants of economic importance. Mapping of the genomes of model plants, animals, prokaryotes and cytoplasmic organelles is beyond the scope of this book.

1.2 Nuclear DNA Content in Plant Genomes

The total amount of DNA in the (haploid) genome is a characteristic of each living species known as its C-value. The fact that the DNA content of an organism is much greater than that required to code for and regulate the production of all necessary proteins has been termed the C-value paradox. In eukaryotes, the C-value is defined as the amount of DNA per haploid genome (1C = haploid nucleus, 2C = diploid nucleus and 4C = tetraploid nucleus or a diploid nucleus that is about to divide by mitosis). There is enormous variation in the range of C-values, from as little as a mere 10^6 bp for a mycoplasma to as much as 10^{11} bp for some plants and animals. Table 2 summarizes the range of C-values found in some plant species. We have included examples mainly from some plants of agricultural importance, except Arabidopsis, the model plant species. More detailed information can be obtained from Arumuganathan and Earle (1991) and the web site http://www.rbgkew.org.uk/cvalues.

1.3 Genome or Chromosome Mapping

Genome mapping, otherwise also known as chromosome mapping, establishes the road-map of a genome. Mapping of chromosomes helps to locate important genes and manipulate them. Chromosome mapping helps to identify the molecular environment of both coding and non-coding DNA sequences. The map of a chromosome can be of two types: genetic map and physical map (Figure 2).

Genetic Map

A genetic map identifies the linear arrangement of genes on a chromosome and is assembled from meiotic recombination data. These are theoretical maps which give the chromosomal order of genes and the distance of separation expressed as the percent of recombination between them. They cannot pinpoint the physical whereabouts of genes or determine how far apart they are in base pairs. Distances on this map are not directly equivalent to physical distances. The unit of measurement is the centiMorgan (cM), defined as 1% recombination between any two genetic loci.

Table 2 Variation in nuclear DNA content of some plant species.

Common Name	Species	Chrom. No. (2n)	Ploidy level	pg/1C	Mbp/1C
Arabidopsis	Arabidopsis thaliana	10	2	0.16	157
Rice	Oryza sativa	24	2	0.50	490
Maize	Zea mays	20	2	2.73	2671
Wheat	Triticum aestivum	42	6	17.33	16979
Barley	Hordeum vulgare	14	2	5.55	5439
Soybean	Glycine max	40	2	1.13	1103
Rapeseed	Brassica napus	38	4	1.15	1127
Sunflower	Helainthus annuus	34	2	2.43	2377
French bean	Phaselous vulgaris	22	2	0.60	588
Pea	Pisum sativum	14	2	4.88	4778
Potato	Solanum tuberosum	48	4	2.10	2058
Tomato	Lycopersicon esculentum	24	2	0.88	1005
Orange	Citrus sinensis	18	2	0.63	613
Grapes	Vitis vinifera	38	2	0.43	417
Peach	Prunus persica	16	2	0.28	270
Sweet cherry	Prunus avium	16	2	0.35	343
Sour cherry	Prunus ceraus	32	4	0.63	613
Banana	Musa acuminata	22	2	0.63	613
Radish	Raphanus sativus	18	2	0.55	539
Onion	Allium cepa	16	2	16.75	16415
Cotton	Gossypium hirsutum	52	4	3.23	3161
Tobacco	Nicotiana rustica	48	4	7.28	7130
Tobacco	Nicotiana tabacum	48	4	5.85	5733
Alfalfa	Medicago sativa	32	4	1.75	1715
Rose	Rosa x hybrida	14	2	0.59	578
Petunia	Petunia hybrida	14	2	1.68	1642
Antirrhinum	Antirrhinum majus	16	2	1.60	1568
Eucalyptus	Eucalyptus globolus	–	2	0.58	564

Source: http://www.rbgkew.org.uk/cvalues

Physical Map

Physical maps identify the actual physical position of genes on a chromosome. Distances are measured in base pairs (bp), kilobases (kb = 1,000 bases) or megabases (mb = 1 million bases).

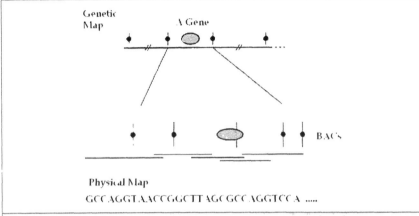

Fig. 2 Diagrammatic representation of a section of a genetic map with markers and position of a gene, five BACs and physical map with an arbitrary nucleotide sequence.

Both genetic and physical maps can be global or local, indicating coverage of the entire genome or a genome section, respectively, depending on the objectives of mapping and resources available.

Complete genetic maps are now available in almost all the plant species of academic and economic importance. Complete physical maps are available in only a few plant species (e.g., Arabidopsis, rice, poplar, peach). Information from genetic mapping is now used for crop improvement purposes. Most of the deliberations of this book will focus on genetic mapping. Details on genetic mapping and molecular breeding in many individual economic plants may be seen in the book series 'Genome Mapping and Molecular Breeding in Plants' (Kole 2006-2007).

2 GENETIC MAPPING AND MARKERS IN THE ERA OF CLASSICAL GENETICS

The idea of chromosomes emerged in 1870's. However, behavior of chromosomes during cell division, mitosis, meiosis and fertilization, was depicted later in the last two decades of the 19th century. As mentioned earlier, the importance of chromosomes in heredity was depicted only in 1902 by Theodor Boveri and Walter S. Sutton, however, the presence of genetic materials in chromosomes was unequivocally proven by Calvin B. Bridges in 1916.

The deviation of the 'Law of Independent Assortment' the so-called second law of inheritance of Mendel was first observed by William Bateson and his coworkers in 1906. They observed partial linkage between two characters in *Lathyrus odoratus*. This was validated by Thomas Hunt Morgan who confirmed the existence of linkage in segregating populations of fruit fly, *Drosophila melanogaster* (Morgan 1911a, b). Since the genes for the characters he was using were both sex-linked, Morgan concluded that there must be exchange of materials between homologous chromosomes, in accordance with an earlier hypothesis of chromosome exchange formulated by Hugo De Vries in 1903. Morgan and Cattell (1912) proposed the term 'crossing over' to describe this physical exchange, which we now accept as manifested cytologically by a chiasma, a cross shaped structure commonly observed between non-sister chromatids during meiosis (Figure 3). Morgan's student, Alfred H. Sturtevant, realized the potential of this observation by pointing out that the proportion of cross-overs could be used as an index of the distance between any two genes. He named this proportion centiMorgan (cM) after his mentor Thomas Hunt Morgan. One cM is defined as the genetic distance between two loci with a recombination frequency of 1% and is also often called a 'map unit'. The map of sex chromosomes in *Drosophila* was described by Sturtevant (1913) for the first time. It was followed by linkage mapping in many different species. By 1922 over 50 genes had been mapped on the 4 *Drosophila* chromosomes.

Construction of linkage maps in *Drosophila* elicited interest in plant scientists, who also used simple-inherited 'visible' or 'morphological'

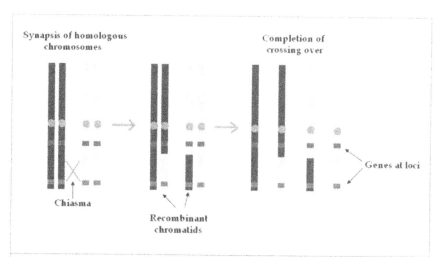

Fig. 3 Synapsis, chiasma formation and crossing over between two homologous chromosomes during meiosis producing two parental and two recombinant chromatids.

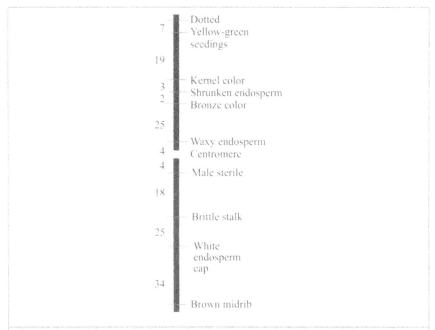

Fig. 4 A genetic map of chromosome 9 of maize with position and order of ten loci detected by morphological markers and the centromere. The names of the morphological characters are on the right. The marker intervals (in centiMorgan, not to the scale) are on the left.

traits to map genes in chromosome maps. Genetic maps based on such traits were developed in some plants, for example in tomato, pea and maize. The first linkage study in tomato, which ranks as one of the classical organisms for such studies, can be traced back to Jones (1917), who reinterpreted data of Hedrick and Booth (1907) on the co-segregation of dwarfness (*d*) and elongate (ovate) fruit shape (*o*) as the consequence of linkage between them. Ernest W. Lindstrom followed with an intensive study of linkage on chromosome 2, utilizing Jones' markers in addition to peach fruit (*p*) and compound inflorescence (*s*) (Rick 1991). Figure 4 shows a genetic map of chromosome 9 based on morphological markers in one of the model plants of classical genetics, maize.

Initially these traits were mapped by using segregation data of two phenotypes, one wild type and another abnormal or unusual phenotype of the so-called 'mutant' appearing spontaneously in populations. Successful demonstration of induction of mutation by X-rays in *Drosophila* by Hermann J. Muller in 1927 and barley and maize by

Lews J. Stadler in 1928 facilitated production of mutants by using various physical and chemical mutagens.

The landmarks on the genetic maps based on qualitative variation of the morphological or visible traits were later known as 'markers'. The concept of a marker originated with Karl Sax who observed, in as early as 1923, an association of seed coat pattern and pigment markers with size differences in the common bean, *Phaseolus vulgaris*. This is perhaps the cornerstone for the discipline of marker-assisted breeding.

The morphological traits used as markers had several advantages with regard to genetic mapping. Firstly, they are easily discernible and facilitate enumeration (e.g., eye color or wing size in Drosophila; seed color or height in garden pea). Figure 5 shows contrasting phenotypes with regard to leaf pubescence in *Brassica rapa* syn. *campestris*. Secondly, detection and use of morphological markers are simple and cheap.

However, these morphological markers have several shortcomings as well. Some of them are mentioned below.

1. The number of morphological markers that could be detected from a single cross involving two individuals is very few. Construction of a complete map requires a large number of crosses.

2. Morphological markers cover only a limited fraction of the total genome of an organism.

Fig. 5 Difference in leaf pubescence in two genotypes of *Brassica rapa* syn. *campestris*.

3. Morphological traits are influenced by environment and may exhibit penetrance and expressivity.

4. Many morphological traits exhibit dominant intra-allelic interactions. Therefore, homozygous dominant and heterozygous individuals cannot be distinguished.

5. Many morphological traits show inter-allelic interactions leading to deviated dihybrid segregation ratios that are not useful for linkage analysis.

6. Several mutant phenotypes have lethal or sublethal effects that could lead to distorted segregation ratios and thereby erroneous estimation of recombination fractions.

All these limitations of morphological markers in genetic linkage mapping invoked a search for alternative genetic markers. Thus, the Neo-Mendelian era in the period of molecular genetics saw an abundance of such new markers that have highly enriched, if not saturated, the arsenal of plant geneticists and breeders.

3 GENETIC MAPPING AND MARKERS IN THE ERA OF MOLECULAR GENETICS

3.1 Isoenzyme Markers

Protein markers, particularly isozyme (= isoenzyme) markers, emerged as a promising tool for genetic mapping (Markert and Moller 1959). These were used for development of partial genetic maps in quite a few organisms (Tanksley and Orton 1983). Isozymes have been used in genome analysis of higher plants both to determine evolutionary pathways and phylogenetic relationships, and in gene mapping. Isozymes exhibit allelism and for an independent locus alleles can be separated by electrophoresis (Tiselius 1959) on starch or polyacrylamide gels based on peptide size and charge at a set pH.

However, isozymes are also phenotypic markers, in that they can be affected by the tissue, growth stage and conditions of plant growth. Tissues need also to be fresh or properly treated before protein extractions or erroneous results may be generated. Protein systems lack adequate polymorphism, genome coverage was low and there was a requirement for the gene to be expressed at the phenotypic level to make detection possible. The number of polymorphic loci is very limited within a gene pool, and polymorphism may be very low.

3.2 Historical Background of DNA Markers

Several revolutionary concepts and techniques laid the foundation of molecular genetics. The most significant ones include elucidation of the structure of DNA (Watson and Crick 1953a) and its plausible mode of replication (Watson and Crick 1953b); experimental evidence of semi-conservative mode of replication (Meselson and Stahl 1958); the exact mode of gene function, the 'central dogma' elucidated in two steps, studies of transcription and translation (Brenner et al. 1961; Jacob and Monod 1961) and artificial synthesis of gene (Khorana 1968). However, discovery of certain concepts and techniques, such as specificity of restriction endonucleases (Smith and Wilcox 1970); development of recombinant DNA technology by Paul Berg and his associates in 1972 (Jackson et al. 1972); Southern blotting (Southern 1975); DNA sequencing (Maxam and Gilbert 1977; Sanger et al. 1977) and cloning of genes (Goeddel et al. 1979) had direct impact on the development of DNA-based molecular markers and their use in genetic linkage mapping and in breeding.

3.3 The DNA Markers

The first DNA-based genetic markers were the restriction fragment length polymorphism (RFLP) markers used by David Botstein and coworkers in 1980 in human genome mapping. This was immediately adapted to several animal and plant systems. The advent of these markers, accelerated our understanding of the structure and function of the genome. The discovery of polymerase chain reaction (PCR) (Mullis et al. 1986; Mullis and Faloona 1987) was another landmark and proved to be a unique process that brought about a new class of DNA marker, the random amplified polymorphic DNA (RAPD) marker (Williams et al. 1990; Welsh and McClelland et al. 1990). Several other DNA markers followed. However, they basically rely on the principle(s) of RFLP and/or PCR. An array of DNA markers are now available to choose from for mapping of a specific organisms genome, and genetic diversity studies in populations.

3.4 Molecular Genetic Mapping: New Concepts, Tools and Strategies

The advent of molecular markers dramatically changed the scenario of genetic linkage mapping in the last two decades of the 20th century. Owing to several advantages of DNA based markers over morphological or isoenzyme markers, complete genome mapping became a reality. The possibility to produce infinite numbers of such markers facilitating complete genome coverage, phenotypic neutrality, absence of inter-allelic interactions, co-dominant molecular phenotype, and independence of the age of the plants and the stage or type of tissue for detection made them invaluable tools for construction of genetic linkage maps and the concomitant mapping of genes.

The use of molecular markers to framework genetic linkage maps and to position genes on them in any organism came to be known as molecular genetic mapping. It is important to note however that molecular genetic mapping is a potential tool supplementary to classical genetic linkage mapping, not its substitute.

Molecular mapping was a highly promising and powerful tool because of the availability of molecular markers but the development of several other concepts and strategies had immense contribution as well. Some of them are briefed below.

3.5   Mapping Populations and Genetic Stocks

Classical genetic maps were constructed by using mainly F_2 and backcross populations as the mapping populations. Later on, several types of segregating populations were available for mapping including recombinant inbred (RI), doubled haploid (DH), and near-isogenic (NI) lines. The former two mapping populations facilitated growing of the lines over time and space under replicated designs while extricating the precise contribution of the genotypic component towards total phenotypic variation for a quantitative trait. This assisted in mapping of the putative chromosomal locations harboring the polygenic clusters controlling quantitative traits, known as quantitative trait loci (QTL). The NI lines were useful in saturation of a targeted chromosomal region and detection of markers tightly linked to a gene co-segregating with it. The concept of bulked segregant analysis, developed by Michelmore et al. (1991) for gene tagging, is basically similar to two NI lines. Use of aneuploid stocks, substitution lines or addition lines also helped in anchoring molecular linkage maps to the corresponding chromosomes. This strategy has been used effectively in rice, tomato, cotton and wheat.

3.6   Computer Programs for Molecular Mapping

Genetic mapping in classical genetics involved only a few genes as landmarks to develop chromosome maps. Calculations for two-point analysis to assort the genes into groups equivalent to the haploid number of chromosomes and three-point analysis to align the genes with their order and distance on a particular linkage group could be done easily. However, handling of hundreds of molecular markers and genes for mapping invoked the development of computer programs. The first computer program used for such a purpose was Linkage I devised by Suiter et al. (1983). The most widely used program for molecular map construction, MAPMAKER v 2.0, was proposed by Lander et al. (1987). This group improved their program and came up with a new version, MAPMAKER/EXP 3.0 in 1992 (Lincoln et al. 1992a). It was followed by several other programs for map construction, such as Cri-Map (Weaver et al. 1992), G-Mendel (Liu and Knapp 1992), JoinMap (Stam 1993) and MapManager (Manly 1995). The choice of a program, however, may depend on: the kind of computer available (Macintosh/IBM); the criteria for assembling marker loci and maps; the type of mapping populations; the specific use of the genetic map and other factors.

One of the unique contributions of molecular genetic mapping is the possibility of detecting putative genomic regions controlling quantitative traits. It is now possible to study the correlation of phenotypic trait data of segregating individuals of a population with their marker data to identify the putative genomic locations of the gene clusters controlling a character (Paterson et al. 1988). The concept of interval mapping (Lander and Botstein 1989) to detect such QTLs in marker intervals using complete framework maps became a popular strategy of mapping polygenes. A number of excellent computer programs are available for QTL mapping. These include programs proposed by Haseman and Elston (1972), Amos and Elston (1989), Goldgar (1990), Haley and Knott (1992), Kruglyak and Lander (1995), Olson (1995), Almasy and Blangero (1998), Ghosh and Majumder (2000a, b). Of these the first such program used most widely is the computer program MAPMAKER/QTL 1.1 (Lincoln et al. 1992b).

4 APPLICATIONS OF GENETIC MAPPING

4.1 Marker-Assisted Crop Improvement

The major contribution of molecular genetic mapping is undoubtedly marker-assisted crop improvement. The markers linked closely or flanking genes and QTLs in linkage maps can be used for: introgression of favorable genes and QTLs from a donor genotype into a recipient genetic background; marker-assisted selection of plants in breeding programs; germplasm screening for target traits; and studies on durability, uniformity and stability (DUS) of commercial varieties. Initially, markers at a distance of < 5 cM apart from the trait loci were thought to be useful in breeding. Development of second generation maps, for example saturated and high-density maps, increased the efficiency of their use where markers were tightly linked, mostly < 1 cM apart from the target trait loci. Pyramiding race/biotype-specific resistance genes to develop durable resistance in host plants against diseases and insects is a significant contribution of molecular mapping and breeding. Development of high-resolution or fine-scale map of a target chromosomal region containing QTLs through marker-assisted backcross breeding paved the way for Mendelization of QTLs, another significant achievement. Development of sequence converted markers such as sequence tagged site (STS) and sequence characterized amplified region (SCAR) also helped enormously in marker-assisted breeding. Application of molecular markers in crop improvement will be discussed in detail in Volume 2 dedicated to Molecular Breeding.

4.2   Map-Based Cloning of Genes and QTLs

Several strategies are available now for isolation of useful oligogenes and polygenes. However, map-based or positional cloning is still the method of choice. Most of the plant genes to-date have been isolated employing a map-based cloning strategy. Information on map position of the target genes and saturation of the chromosomal region flanked by markers have facilitated chromosome walking and more importantly chromosome landing. Recently, application of comparative mapping has also paved the way for isolation of desirable genes.

4.3   Elucidation of Phylogenetic Relationships and Evolutionary Pathways

Studies on origin, evolution and phylogenetic relationships among close taxa were traditionally based on morphotaxonomy and chemotaxonomy. Later on, the use of repetitive sequences of nuclear DNA, chloroplast DNA and mitochondrial DNA also reinforced such studies. However, development of comparative maps using a common set of probes and/or primers facilitated detection of colinearity of chromosomes of two or more species or genera and delineation of the most plausible evolutionary pathways in several species. In these studies, the most precise phylogenetic reconstructions for taxa were based on the homology of the positions of markers, and genes.

Additionally, detection of chromosomal regions in molecular maps with skewed or distorted segregation threw light of the evolutionary significance of lethal and sublethal alleles related to biological fitness. Molecular mapping significantly increased the power to detect duplicated chromosomal regions and chromosomal restructuring; all possible pathways of genome evolution. These levels of genome resolution elucidated the origin of today's so-called diploid and paleoploid plants through polyploidization followed by chromosomal restructuring. Comparison of molecular maps based on synthetic and natural polyploids also helped in elucidation of genome dynamism.

4.4 Construction of Third Generation Chromosome Maps

Development of so-called third generation maps, for example integrated, consensus, community and molecular cytogenetic maps were realized through the quick developments in: (1) several types of highly numerous DNA markers; (2) the clever use of diverse mapping populations and cytogenetic stocks; (3) the availability of appropriate computer software (e.g., JoinMap); (4) the use of innovative tools such as pseudo-testcross, paternal and maternal mapping, in situ hybridization (e.g., GISH, FISH); and above all (5) the concerted efforts and collaboration of scientists on a global basis particularly through sharing of permanent mapping populations and marker probes/primers. These maps are of academic interest and serve as the platform for national and international genome initiatives, providing information useful for crop improvement.

4.5 Role of Genetic Mapping in Genomics Arena

Formulation of routine techniques for cloning large segments of DNA in vectors (Schmidt et al. 1995) such as BACs (Shizuya et al. 1992) and YACs (Burke et al. 1987), cloned insert DNA fingerprinting (acrylamide, agarose or HICF; Wu et al. 2005) and high throughput genomic DNA sequencing methods (http://en.wikipedia.org/wiki/DNA_sequencing) facilitated the construction of physical maps comprising of contiguous overlapping mega base DNA clones which spanned large regions of chromosomes and represented a major step toward complete sequencing of entire genomes. While genetic mapping using DNA markers facilitated the completeness of coverage of an entire genome with closely spaced landmarks, physical mapping and genome sequencing have provided the complete nucleotide sequence providing the most extensive structural characterization of a genome.

However, physical and genetic mapping should not be parallel roads. Integration of the two resultant maps is important from an applied point of view. Placement of markers linked to genes of economic importance and positioning of genes and gene-related sequences on the physical maps provide the substrate for studies on gene structure, function, regulation, and expression. In addition, the use of sequence information to construct new probes and primers in generating new markers has great utility in molecular breeding. Fundamentals and applications of physical mapping will be thoroughly described in Volume 3.

5 CONTRIBUTORY DISCIPLINES AND BENEFICIARIES

Genome mapping stands on the interface of classical genetics and molecular biology with contributions from many diverse disciplines. These disciplines are genetics, breeding, biochemistry, microbiology and statistics; and supplementary disciplines such as computer science, bioinformatics, bacteriology, mycology, virology, entomology, nematology, physiology, agronomy, horticulture, ecology, taxonomy, cytology and sociology. The beneficiaries include the fields of agriculture, medicine and environment.

6 CONCERNS AND CAUTIONS

Construction of a molecular genetic linkage map of a new crop involves enormous cost, time, space and labor. These are amplified to a greater extent for development of whole-genome physical maps. In certain cases, identification of objectives for genome mapping programs is driven by available technologies, databases and physical facilities. This is particularly relevant to developing countries. The limited resources available should prioritize the mapping objectives of a region, state or nation.

6.1 Selection of Crop

Each and every agricultural belt has its own mandate crops thus, genome mapping in key species is more relevant. In certain cases, scientists adopt a crop on which s/he has the training and expertise, and access to the plant materials. A country where millets are grown on hundreds of hectares of land must focus its resources on them. This is also true for countries growing pulses, tea, rubber, banana, medicinal plants, etc. The list of orphan and beggar crops is too long, at least of Asia and Africa, to choose some for immediate attention.

6.2 Selection of Traits

The requirements of the industries processing outputs for the end users vary over space and time. The priority problems that relate to, for example, biotic and abiotic stresses; consumer and industry preferences; transportation and storage; cultural and religious mandates and taste habits; and laws on the substitution of crops must be addressed during selection of the target traits to be mapped and manipulated.

6.3 Selection of Base Materials

The genetic background of any landrace or indigenous variety has attained stability through thousands of years of natural selection, however, they require genetic tailoring of a few genes for increased yield and improved quality. They must be used as one of the parents in biparental crossings for development of mapping populations and be used as the recipient parent for gene introgression. Examples abound to substantiate that the some molecular geneticists and breeders have selected parental genotypes only for academic purposes and without any thought on the application value of their mapping endeavors.

6.4 Converging Approach

The field of plant genome mapping is a panmictic arena. Students and scientists of all the disciplines; people from the corporate sectors, regulatory agencies, media; and social activists should obtain some level of training in this area to attain a working knowledge of the basic tenets of this field. This will enable them to accurately promote the value of these scientific approaches to the public thus philosophically tasting and distributing the fruits produced by them.

References

Almasy L, Blangero J (1998) Multipoint quantitative trait linkage analysis in general pedigrees. Am J Hum Genet 62: 1198-1211

Amos Cl, Elston RC (1989) Robust methods for the detection of genetic linkage for quantitative data from pedigrees. Genet Epidemiol 6(2): 349-360

Arumuganathan K, Earle ED (1991) Nuclear DNA content of some important plant species. Plant Mol Biol Rep 9(3): 211-215

Avery OT, McLeod CM, McCarty M (1944) Studies on the chemical nature of the substance inducing transformation of pneumococcal types. J Exp Med 79: 137-158

Beadle GW (1932) A possible influence of the spindle fibre on crossing-over in *Drosophila.* Proc Natl Acad Sci USA 18 (2): 160-165

Botstein D, White RL, Skolnick M, Davis RW (1980) Construction of a genetic linkage map in man using restriction fragment length polymorphisms. Am J Hum Genet 32: 314-331

Brenner S, Jacob F, Meselson M (1961) An unstable intermediate carrying information from genes to ribosomes for protein synthesis. Nature 190: 576-581

Bridges CB (1916) Nondisjunction as proof of the chromosome theory of heredity. Genetics 1: 1-52, 107-163

Burke DT, Carle GF, Olson MV (1987) Cloning of large segments of exogenous DNA into yeast by means of artificial chromosome vectors. Science 236: 806-812

De Vries H (1903) Fertilization and hybridization. In: Intracellular Pangenesis including a paper on Fertilization and Hybridization. Open Court Publishing Co, Chicago, IL, USA, pp 217-263 [Translated from Befruchtung und Barstardierung

Fuelgen R, Rossenbeck H (1924) Mikroskopisch chemischer Nachweiss einer Nucleinsaure vom Typus der Thymonucleinsaure und die darauf besuehnde elektive Fiirbung von Zellkernen in mikroskopischen PrBparaten. 2 Physiol Chem 135: 203

Ghosh S, Majumder PP (2000a) Mapping a quantitative trait locus via the EM algorithm and Bayesian classification. Genet Epidemiol 19: 97-126

Ghosh S, Majumder PP (2000b) An improved procedure of mapping a quantitative trait locus via the EM algorithm using posterior probabilities. J Genet 79: 47-53

Goeddel DV, Kleid DG, Bolivar F, Heyneker HL, Yansura DG, Crea R, Hirose T, Kraszewski A, Itakura K, Riggs AD (1979) Expression in *Escherichia coli* of chemically synthesized genes for human insulin. Proc Natl Acad Sci USA 76(1): 106-110

Goldgar, DE (1990) Multipoint analysis of human quantitative genetic variation. Am J Hum Genet 47: 957-967

Haley CS, Knott SA (1992) A simple regression method for mapping quantitative trait loci in line crosses using flanking markers. Heredity 69: 315-324

Haseman JK, Elston RC (1972) The investigation of linkage between a quantitative trait and a marker locus. Behav Genet 2: 3-19

Hedrick UP, Booth NO (1907) Mendelian characters in tomatoes. Proc Am Soc Hort Sci 5: 19-24

Jackson DA, Symons RH, Berg P (1972) Biochemical method for inserting new genetic information into DNA of Simian Virus 40: circular SV40 DNA molecules containing lambda phage genes and the galactose operon of *Escherichia coli*. Proc Natl Acad Sci USA 69: 2904-2909

Jacob F, Monod J (1961) Genetic regulatory mechanisms in the synthesis of proteins. J Mol Biol 3: 318-356

Jones DF (1917) Linkage in *Lycopersicum*. Am Nat 51: 608-621

Khorana HG (1968) Synthesis in the study of nucleic acids. The Fourth Jubilee Lecture. Biochem J 109: 709-725

Kole C (ed) (2006-07) Genome Mapping and Molecular Breeding in Plants. 7 Volumes. Springer: Berlin, Heidelberg, New York

Kruglyak L, Lander ES (1995) A nonparametric approach for mapping quantitative trait loci. Genetics 139 (3): 1421-1428

Lander ES, Botstein D (1989) Mapping Mendelian factors underlying quantitative traits using RFLP linkage maps. Genetics 121(1): 185-199

Lander ES, Green P, Abrahamson J, Barlow A, Daly MJ, Lincoln SE, Newburg L (1987) MAPMAKER: an interactive computer package for constructing primary genetic linkage maps of experimental and natural populations. Genomics 1(2): 174-181

Lincoln S, Dally M, Lander E (1992a) Constructing genetic linkage maps with MAPMAKER/EXP 3.0. Whitehead Institute Technical Reports, edn 3

Lincoln S, Dally M, Lander E (1992b) Mapping genes controlling quantitative traits with MAPMAKER/QTL 1.1. Whitehead Institute Technical Reports, edn 2

Liu BH, Knapp SJ (1992) GMENDEL 2.0, a software for gene mapping. Oregon State University, Corvallis, Oregon, USA

Manly, KF (1995) New functions in Map Manager, a microcomputer program for genomic mapping. In: Plant Genome III Conf, San Diego, CA, USA

Markert CL, Moller F (1959) Multiple forms of enzymes. Tissue, ontogenetic, and species specific patterns 45: 753-763

Maxam AM, Gilbert W (1977) A new method for sequencing DNA. Proc Natl Acad Sci USA 74: 560-564

Meselson M, Stahl F (1958) The replication of DNA in *Escherichia coli*. Proc Natl Acad Sci USA 44: 671-682

Michelmore RW, Paran I, Kesseli RV (1991) Identification of markers linked to disease resistance genes by bulked segregant analysis: a rapid method to detect markers in specific genomic regions using segregating populations. Proc Natl Acad Sci USA 88: 9828-9832

Morgan TH (1911a) The application of the conception of pure lines to sex-limited inheritance and to sexual dimorphism. Am Nat 45: 65-78

Morgan TH (1911b) An attempt to analyze the constitution of the chromosomes on the basis of sex-limited inheritance in *Drosophila*. J Exp Zool 11: 365-414

Morgan TH, Cattell E (1912) Data for the study of sex-linked inheritance in *Drosophila*. J Exp Zool 13: 79-101

Muller HJ (1927) Artificial transmutation of the gene. Science 66: 84-87

Mullis K, Faloona F, Scharf S, Saiki R, Horn G, Erlich H (1986) Specific enzymatic amplification of DNA in vitro: the polymerase chain reaction. Cold Spring Harbor Symposia on Quantitative Biology 51: 263-273

Mullis K, Faloona F (1987) Specific synthesis of DNA in vitro via a polymerase-catalyzed chain reaction. Methods Enzymol 155: 335-350

Olson JM (1995) Multipoint linkage analysis using sib pairs: an interval mapping approach for dichotomous outcomes. Am J Hum Genet 56: 788-798

Paterson AH, Lander ES, Hewitt JD, Peterson S, Lincoln SE, Tanksley SD (1988) Resolution of quantitative traits into Mendelian factors by using a complete linkage map of restriction fragment length polymorphisms. Nature 335: 721-726

Rick CM (1991) Tomato paste: a concentrated review of genetic highlights from the beginnings to the advent of molecular genetics. Genetics 128: 1-5

Sanger F, Nicklen S, Coulson AR (1977) DNA sequencing with chain-terminating inhibitors. Proc Natl Acad Sci USA 74(12): 5463-5467

Sax K (1923) The association of size differences with seed coat pattern and pigmentation in *Phaseolus vulgaris*. Genetics 8: 552-560

Schmidt R, West J, Love K, Lenehan Z, Lister C, Thompson H, Bouchez D, Dean C (1995) Physical map and organization of *Arabidopsis thaliana* chromosome 4. Science 270: 480-483

Shizuya H, Birren B, Kim UJ, Mancino V, Slepak T, Tachiiri Y, Simon M (1992) Cloning and stable maintenance of 300-kilobase-pair fragments of human DNA in *Escherichia coli* using an F-factor-based vector. Proc Natl Acad Sci USA 89(18): 8794-8797

Smith HO, Wilcox KW (1970) A restriction enzyme from Haemophilus influenzae. I. Purification and general properties. J Mol Biol 51: 379-391

Southern EM (1975) Detection of specific sequence among DNA fragments separated by gel electrophoresis. J Mol Biol 98: 503-517

Stadler LJ (1928) Mutations in barley induced by X-rays and Radium. Science 58: 186-187

Sutton WS (1903) The chromosomes in heredity. Biol Bull 4: 231-251. Partial reproduction. In: JA Peters (ed) Classic Papers in Genetics (1959), Prentice-Hall, Englewood Cliffs, pp 27-41

Stam P (1993) Construction of integrated genetic linkage maps by means of a new computer package: JOINMAP. Plant J 3: 739-744

Sturtevant AH (1913) The linear arrangement of six sex-linked factors in *Drosophila*, as shown by their mode of association. J Exp Zool 14: 43-59

Suiter KA, Wendel JF, Case JS (1983) LINKAGE-1: a PASCAL computer program for the detection and analysis of genetic linkage. J Hered 74(3): 203-204

Tanksley SD, Orton TJ (1983) Isozymes in Plant Genetics and Breeding, Parts 1A and 1B. Elsevier, Amsterdam, The Netherlands

Tiselius A (1959) Introduction. In: M Bier (ed) Electrophoresis: Theory, Methods, and Applications. Academic Press, New York, USA

Watson JD, Crick FHC (1953a) A Structure for Deoxyribose Nucleic Acid. Nature 171: 737-738

Watson JD, Crick FHC (1953b) Genetical implications of the structure of the deoxyribonucleic acid. Nature 171: 964-967

Weaver R, Helms C, Mishra SK, Donis-Keller H (1992) Software for analysis and manipulation of genetic linkage data. Am J Hum Genet 50(6): 1267-1274

Welsh J, McClelland M (1990) Fingerprinting genomes using PCR with arbitrary primers. Nucl Acids Res 18: 7213-7218

Williams JGK, Kubelik AR, Livak KJ, Rafalski JA, Tingey SV (1990) DNA polymorphisms amplified by arbitrary primers are useful as genetic markers. Nucl Acids Res 18: 6531-6535

Wu C, Sun S, Lee M-K, Xu ZY, Ren C, Zhang H-B (2005) Whole genome physical mapping: An overview on methods for DNA fingerprinting. In: Meksem K and Kahl G (eds) The Handbook of Plant Genome Mapping: Genetic and Physical Mapping. Wiley-VCH Verlag GmbH, Weinheim, Germany, pp 257-284

2 Molecular Markers – History, Features and Applications

S. Doveri[1*], D. Lee[1], M. Maheswaran[2], and W. Powell[1]
[1]NIAB, Huntingdon Road, Cambridge, CB3 0LE, UK
[2]Department of Plant Breeding and Genetics, Agricultural College and Research Institute, Tamil Nadu Agricultural University, Coimbatore 641003
*Corresponding author: silvia.doveri@gmail.com

1   INTRODUCTION

Since the dawn of civilization, man has tried to improve livestock and crops by 'selective breeding': by choosing individuals with the best desirable characters to increase their yield. It was Mendel (1865) who first put forward the idea that traits were inherited as discrete characters but only in 1944, Avery et al. (1944) discovered that this information was encoded by deoxyribonucleic acid (DNA), and it took a further nine years for Watson and Crick (1953) to elucidate the structure of the hereditary material.

Advances in molecular biology over the last few years have provided researchers involved in the study of plant biology with a range of new tools for addressing issues from gene expression to genetic diversity. Initially, phenotypic markers were used to analyze variation in plants; these were replaced by proteins (or isozymes) which, in turn, have been augmented by DNA markers. Differences in DNA sequence can be directly observed and described to a degree of precision previously impossible to achieve. Many of the developed techniques have already been used to study the extent and distribution of variation in species and to investigate evolutionary and taxonomic questions. They have also shown their value in studies of accession identity and for the detection of novel useful variation.

Initially, genotyping of plant species and varieties using DNA analysis was confined to research purposes but it is now increasingly used to address a range of commercial needs. Thus, business critical issues can be addressed by DNA genotyping, but the suitability of the various methods deployed and the cost, speed and accuracy are of primary importance. Plant DNA analyses are conducted to address problems or questions with respect to:

- genetic mapping,
- product quality assurance,
- breach of intellectual property rights,
- support for plant breeders' rights and patents,
- guiding plant breeding programs,
- marker-assisted selection,
- forensic investigations,
- traceability of plant species and variety products in commercial food and trade chains,
- forward and reverse genetics for gene discovery.

1.1 Genetic Markers

Markers are entities that are heritable as simple Mendelian traits and are easy to score (Schulman et al. 2004). A genetic marker is 'a variant allele that is used to label a biological structure or process through the course of an experiment' (Griffiths et al. 1996). Essentially, it is a 'signature' in the DNA that allows diagnostic detection of DNA sequence variation existing between species and/or varieties. Today, genetic markers are used in both basic plant research and plant breeding to characterize plant germplasm, for gene isolation, marker-assisted introgression of favorable alleles, production of improved varieties (Henry 2001), and to obtain information about the genetic variation within populations. Genetic markers can be divided in to three classes: morphological (variation at phenotype level), biochemical (variation at gene product level) and molecular (variation at DNA level).

Morphological characters are the easiest of all markers to assay, and their use is as old as breeding and selection itself. Such characters range from organ shape and differences in the pigmentation of the plants, gross changes in development (such as vernalization requirement or dwarfness), to responses to individual races of a phytopathogen.

However, as the broad range of phenotypes is visible at different developmental stages, complete characterization can take up the whole growing season. Morphological characters are the predominant descriptors used for the assessment of varieties for statutory Distinctness, Uniformity and Stability (DUS) testing (Ardley and Hoptroff 1996). Many of the characters that are observed comprise continuous traits that are subject to the influence of the environment and require that all reference varieties must be grown together with the testing material for a full assessment, thus increasing the cost of DUS testing. There are strong arguments for replacing the testing procedure with a DNA-based system that would be less subjective, independent from environmental factors, season and cheaper to perform (Cooke and Reeves 2003; Donnenwirth et al. 2004). Much work has been performed to test the different marker types for the purposes of DUS testing (Lee et al. 1996a, b; Lombard et al. 2000; Tommasini et al. 2003) and all clearly demonstrate the utility of DNA markers for differentiating plant varieties. The biggest benefit is that a given variety needs to be tested only once and there would not be a need for repeated sowing of registered varieties for comparative purposes.

Despite their limitations, morphological markers are well-established tools for taxonomy, variety classification and breeding. However, the limitations to the application of morphological markers can be overcome by the application of biochemical and molecular markers.

Biochemical techniques detect variation at the gene product level: the variants are normally observed as electrophoretic protein polymorphisms (isozymes) and are generally breeder-neutral. The assays require the extraction of proteins from plant tissue, followed by their separation using gel electrophoresis or other means. The general utility of biochemical markers is hindered mainly by the relatively small number of genetic loci that can be assayed in any population, but also by the technical complexity of their use (since each protein system requires its own electrophoretic and staining protocol). Furthermore, as biochemical markers rely on gene expression, there can be problems associated with specificity to tissue type and, in some cases, the physiological age or developmental state of the tissue. However, in particular situations, they remain the marker of choice especially when their cost in routine testing is lower than using molecular markers.

Molecular markers are specifically developed to detect variation at the DNA level which can be diagnostics for a genotype, variety or plant species. Strictly defined, a molecular marker identifies changes in the DNA sequence. Molecular markers allow rapid identification of breeding lines, hybrids, cultivars and species, facilitate genetic diversity and relatedness estimations in germplasm, and they allow phylogenetic relationships to be established with more accuracy than was previously possible with morphological and biochemical techniques.

1.2   Molecular Markers

There are a large number of marker systems available and each one has advantages and disadvantages (Drabek 2001; Koebner et al. 2001; Table 1). The cost, sensitivity and suitability of each marker type vary substantially depending on the final application, plant species and instrumentation available. The technology related to the detection and analysis of variation at the DNA level is fast-changing. The driver for these developments is, in part, directed at making the technology more efficient, cheaper and robust.

Table I   Chronological evolution of molecular markers.

Acronym	Nomenclature	Reference
RFLP	Restriction Fragment Length Polymorphism	Grodzicker et al. 1974
VNTR	Variable Number Tandem Repeat	Jeffreys et al. 1985
ASO	Allele Specific Oligonucleotides	Saiki et al. 1986
OP	Oligonucleotide Polymorphism	Beckmann 1988
AS-PCR	Allele Specific Polymerase Chain Reaction	Landegren et al. 1988
SSCP	Single Stranded Conformational Polymorphism	Orita et al. 1989
STS	Sequence Tagged Site	Olsen et al. 1989
STMS	Sequence Tagged Microsatellite Sites	Beckmann and Soller 1990
OFLP	Oligo-Amplified Fragment Length Polymorphism	Lee et al. 1990
RAPD	Random Amplified Polymorphic DNA	Williams et al. 1990
AP-PCR	Arbitrarily Primed Polymerase Chain Reaction	Welsh and McClelland 1990
DAF	DNA Amplification Fingerprinting	Caetano-Anollés et al. 1991
RLGS	Restriction Landmark Genome Scanning	Hatada et al. 1991
SSR	Simple Sequence Repeat	Akkaya et al. 1992
CAPS	Cleaved Amplified Polymorphic Sequence	Akopyanz et al. 1992
DOP-PCR	Degenerate Oligonucleotide Primer - PCR	Telenius 1992
MAAP	Multiple Arbitrary Amplicon Profiling	Caetano-Anollés et al. 1993

Contd.

Contd.

SCAR	Sequence Characterized Amplified Region	Paran and Michelmore 1993
SNP	Single Nucleotide Polymorphism	Jordan and Humphries 1994
SAMPL	Selective Amplification of Microsatellite Polymorphic Loci	Morgante and Vogel 1994
ISSR	Inter-Simple Sequence Repeat	Zietkiewicz et al. 1994
ASAP	Allele Specific Associated Primers	Gu et al. 1995
AFLP	Amplified Fragment Length Polymorphism	Vos et al. 1995
CFLP	Cleavage Fragment Length Polymorphism	Brow et al. 1996
ISTR	Inverse Sequence-Tagged Repeat	Rohde 1996
DAMD-PCR	Directed Amplification of Minisatellite DNA-PCR	Bebeli et al. 1997
S-SAP	Sequence-Specific Amplified Polymorphism	Waugh et al. 1997
RBIP	Retrotransposon-Based Insertional Polymorphism	Flavell et al. 1998
IRAP	Inter-Retrotransposon Amplified Polymorphism	Kalendar et al. 1999
REMAP	Retrotransposon-Microsatellite Amplified Polymorphism	Kalendar et al. 1999
TE-AFLP	Three Endonuclease AFLP	van der Wurff et al. 2000
IMP	Inter-MITE Polymorphism	Chang et al. 2001
SRAP	Sequence-Related Amplified Polymorphism	Li and Quiros 2001
TRAP	Target Region Amplification Polymorphism	Hu and Vick 2003
TBP	Tubulin-Based Polymorphism	Bardini et al. 2004

Molecular markers can be classified into two major groups: those based on DNA-DNA hybridization between a DNA or RNA 'probe' and total genomic DNA (e.g., RFLP and dot-blot assay), and those based on the PCR amplification of genomic DNA fragments (e.g., RAPD, SCAR, SSR, AFLP, SNP, etc.). However, some assays combine features from both (e.g., RBIP).

More often, molecular markers are classified on a chronological basis (Figure 1). The first generation of the DNA markers included RFLP and RAPD and it has not lived up to initial expectations as universal genotyping assays. The technical limitations of RFLP (Karp et al. 1996) and lack of reproducibility of RAPD assay (Staub et al. 1996) have directed scientists towards the development of newer molecular markers that are more robust. In the latter half of the 1990s, three PCR-based marker systems gained popularity for harvesting the potential offered by variations at the DNA level in plants: these were AFLP, SSR and

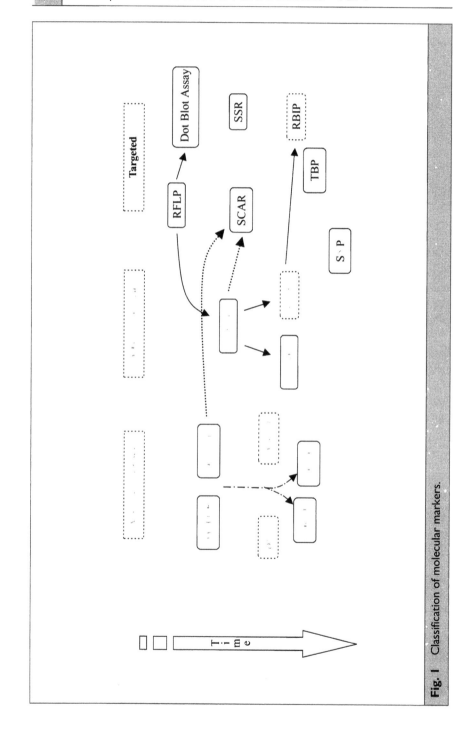

Fig. 1 Classification of molecular markers.

retrotransposon-based markers. Driven by the need to reduce the cost and increase the information content of molecular-based assays, the research community has begun to exploit the large amount of DNA sequence becoming freely available through the databases (www.ncbi.nlm.nih.gov) to generate a number of novel, so-called third generation marker assays. The primary focus has fallen on SNPs, each of which represents a defined position at a chromosomal site at which the DNA sequence of two individuals differs by a single base. SNPs that were first described by Jordan and Humphries (1994), have become the marker of choice by virtue of their genome coverage and the parallel testing procedures that enable thousands of loci to be assessed within a single experiment.

Here, we have described the most commonly used markers in a chronological order which best shows how markers have evolved.

2   THE MAJOR CATEGORIES OF MOLECULAR MARKERS

Restriction Fragment Length Polymorphism (RFLP)

RFLPs were the first widely used molecular markers for assessment of DNA variation in selected organisms (Grodzicker et al. 1974). In RFLP analysis, the DNA is first digested with restriction enzymes that recognize specific DNA sequence/motifs, generally 4-10 bp in length, and the resultant fragments are separated by gel electrophoresis (Southern 1975) (Figure 2). The separated DNA fragments are then transferred to a membrane filter by capillary blotting, a process termed as 'Southern blotting' (Southern 1979). To obtain some information on any differences that may be present within the smear that is generated from restriction digestion of genomic DNA, it is necessary to hybridize a 'probe' to the filter to highlight one or a set of bands. To perform the hybridization, a denatured labeled probe is mixed and allowed to anneal to the DNA on the filter, usually overnight. Salts such as dextran sulphate (Wahl et al. 1979) and SDS (Church and Gilbert 1984) can be added to accelerate the rate of hybridization. Temperature and salt conditions favor the formation of hybrids and the detection of specific fragments is determined by the wash conditions used to remove unbound probes and destabilize mismatched hybrids (Hames and Higgins 1985). Once excess probe has been removed, an X-ray film is placed over the surface of the filters (in the case of radio-labeled probes) or the filters are subjected to further processing to detect the labeled

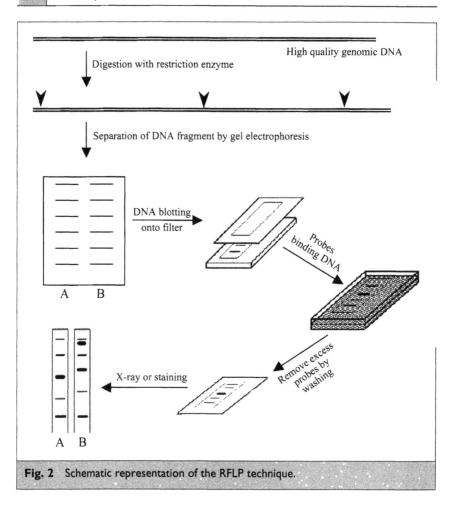

Fig. 2 Schematic representation of the RFLP technique.

probe (in the case of non-radioactive probes). Bands are visualized where the probe has hybridized to homologous DNA fragments: the pattern obtained is referred to as the restriction fragment pattern. Polymorphisms arise from sequence changes in the restriction site as well as from the detection of insertion/deletions in the restriction fragments detected by the probe.

As markers, RFLP fragments are distributed throughout the genomes of most species. They are co-dominant, so heterozygotes are distinguishable (Karp et al. 1996). However, technical limitations restrict the application of RFLPs. A large quantity (2-10 μg) of high quality DNA (i.e. of good integrity and quality to allow restriction digestion) is required for each restriction digestion. Therefore, RFLPs cannot be used when the quality or quantity of DNA that can be extracted is limited, as for example in foodstuffs, archaeological or forensic samples. The technique is also difficult to automate due to the blotting and hybridization steps (Karp et al. 1996). Furthermore, the development of probes that can be used for probing is time-consuming and labor-intensive. Random genomic DNA fragments have limited application due to the presence of highly repetitive elements in plant genomes. To reduce the frequency of repetitive sequences, probes made from expressed regions of the genome are targeted either by the use of cDNAs or by cloning hypomethylated DNA fragments using methylation-sensitive restriction enzymes such as *Pst*I (Sharpe et al. 1995).

In humans, one class of RFLPs, that exploited repeat lengths of a hypervariable minisatellite (Jeffreys et al. 1985), gained prominence in their contribution to the use of molecular markers in forensics, and can be viewed as the precursors/progenitors of the use of SSRs in plants. For most purposes, RFLPs have been superseded by marker systems that are more suitable for automation.

Dot Blot Assay

Where a probe sequence is present in one individual, but absent in another, the gel separation part of the RFLP assay can be omitted. Undigested genomic DNA is denatured and applied directly to the membrane, which is then hybridized with a probe in the normal way. This represents a drastic simplification of the RFLP technique, since neither restriction digestion, gel separation nor transfer is necessary. The technique finds application particularly where contrasts between species are relevant. In these cases, advantages of species-specific repetitive DNA can be exploited. Since the DNA is used directly without in vitro manipulation, purity is not a major issue. The use of multi-copied probes makes the assay very sensitive and exposure times can be reduced compared to the RFLP protocol.

PCR-based Markers

The advent of the Polymerase Chain Reaction (PCR) (Mullis et al. 1986) allowed the development of techniques that overcame some of the difficulties associated with RFLPs and originated a new generation of molecular markers. The PCR relies on a thermostable DNA polymerase enzyme (commonly *Taq* polymerase isolated from the thermophilic bacterium, *Thermophilus aquaticus*) and oligonucleotide primers to direct synthesis of DNA bound by the primers. PCR consists of three basic steps, which are summarized in Figure 3. The specificity of the reaction to a particular locus or loci depends on the choice of these oligonucleotides that serve as primers for the amplification. The amplification reaction is achieved in an automated thermocycling machine, in which the necessary parameters can be programed. During the three main steps: denaturation, annealing of primers and polymerization (25-50 cycles), DNA is synthesized de novo. After the initial product synthesis, the products generated from each round of amplification serve as templates for further amplification, making the whole process very efficient, and short regions of DNA can be amplified

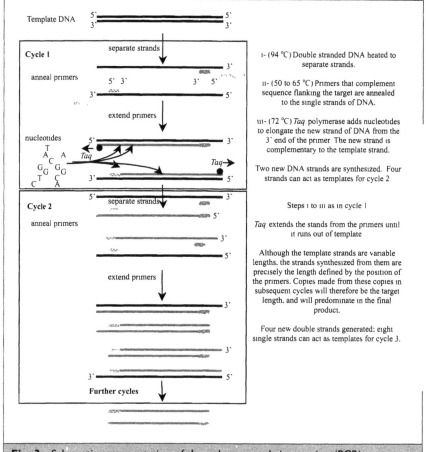

Fig. 3 Schematic representation of the polymerase chain reaction (PCR).

and purified without the need for cloning. The reaction products are then available for analyses: visualization after electrophoresis or further in vitro manipulations (e.g., restriction digestion). If the PCR procedure was 100% efficient, each cycle would double the number of copies of the template (Newton 1995) so theoretically, a single copy of a DNA fragment would be amplified exponentially. In reality, the reaction is limited by the activity of the *Taq* which declines during exposure to high temperature (50°C or more). However, enough DNA for electrophoretic analysis is produced in optimized reactions (Morell et al. 1995).

The major consideration for PCR is the design of primers: these interactions compromise PCR. Internal structures within the primers will reduce the efficiency at which they will bind to the template and interaction between the two primers may result in the formation of primer-dimers that are short and therefore efficiently amplified. However, there are many publicly available programs that have been developed to aid this aspect of PCR (e.g., Primer3, *http://frodo.wi.mit.edu/ cgi-bin/primer3/primer3_www.cgi*).

Random Amplified Polymorphic DNA (RAPD)

RAPD is a PCR technique that uses decamers, or other suitably short primers, of arbitrary sequence to amplify discrete fragments of the genome (Williams et al. 1990). The purpose of such short oligomers is to increase the likelihood that the primer will bind to two regions sufficiently close together in a manner to amplify the intervening region. Each product is derived from a stretch of DNA that is bordered by primer annealing sites. The amplified products, which typically range from 1-10 in number, are separated on an agarose gel and normally visualized using ethidium bromide-staining under UV light, although polyacrylammide gels are sometimes used, when higher resolution levels are required. Polymorphisms are detected as the presence or absence of bands. The RAPD technique is useful because it requires no sequence information for primer design and no probe development. The technique is simple, fast and relatively cheap (Williams et al. 1990). RAPD bands are dominant and the markers are suited to high throughput applications for plant breeding purposes, such as the construction of genetic maps and the tagging of desirable traits for marker-assisted selection, but are also useful in phylogenetic and population studies (Staub et al. 1996).

Problems arise with reproducibility of RAPDs because of the short primers and low stringency conditions used to increase the number of products. Results can differ according to the source of the tissue analyzed, DNA extraction techniques and PCR protocols. Band profiles have been demonstrated to show variations between laboratories if equipment or conditions are changed (Staub et al. 1996; Jones et al. 1997). Therefore, constant reaction conditions are crucial for reliable and reproducible results.

Other similar techniques to RAPD are represented by **Arbitrarily Primed PCR (AP-PCR**; Welsh and McClelland 1990) and **DNA amplification fingerprinting (DAF**; Caetano-Anolles et al. 1991). These techniques differ from RAPD principally in that the AP-PCR uses longer primers and after a few rounds of low annealing .temperature, the reaction is continued at higher, more specific annealing temperature. DAF differs by cutting the DNA first with a restriction enzyme, thereby, polymorphisms generated are due to restriction and primer binding site differences. The markers generated behave in a similar way to those of RAPDs and the techniques have the same reproducibility problems.

Amplified Fragment Length Polymorphism (AFLP)

The AFLP technique was developed by Zabeau and Vos (1993) and is essentially a combination of RFLP and PCR. Restriction fragments have adapters ligated to their ends allowing PCR amplification from primers derived from the adapters. As the potential numbers of bands is large, two methods are used to limit the numbers of observable fragments. Firstly the DNA is cut using a 'rare' cutting enzyme (*Pst*I and *Eco*RI, commonly used) and a frequent cutter (*Mse*I): labeling the primers that amplify from the rare cutter limits the visible products to the fragments generated by the rare cutter. Secondly, extra selective bases are added to the primers to restrict the products to those that contain the extra base(s). The addition of each selective base reduces the number of fragments amplified by a factor of 4 and the numbers required depends primarily on the size of the genome studied (Vos et al. 1995). The four steps of the AFLP technique are described in Figure 4 (Vos et al. 1995). Typically, 20-100 fragments are co-amplified (Maughan et al. 1996). Each band may be identified by its size and the selective bases used in the primers to generate them. Polymorphisms arise from sequence changes in the restriction sites or the selective bases used by the primers and insertions/ deletions (indels) in the fragments. Bands are generally detected as present or absent, making AFLPs a dominant marker system. AFLPs have a high marker index (MI) (Powell et al. 1996a) as many independent genetic loci are assayed in a single reaction. As the primers used to amplify fragments correspond to adapter sequences, the technique does not require a priori sequence data and may be easily applied to any

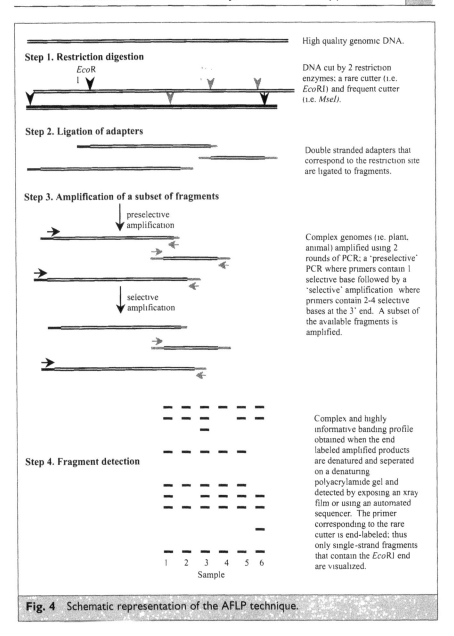

Step 1. Restriction digestion

High quality genomic DNA.

DNA cut by 2 restriction enzymes; a rare cutter (i.e. *Eco*RI) and frequent cutter (i.e. *MseI*).

Step 2. Ligation of adapters

Double stranded adapters that correspond to the restriction site are ligated to fragments.

Step 3. Amplification of a subset of fragments

preselective amplification

Complex genomes (ie. plant, animal) amplified using 2 rounds of PCR; a 'preselective' PCR where primers contain 1 selective base followed by a 'selective' amplification where primers contain 2-4 selective bases at the 3' end. A subset of the available fragments is amplified.

selective amplification

Complex and highly informative banding profile obtained when the end labeled amplified products are denatured and seperated on a denaturing polyacrylamide gel and detected by exposing an xray film or using an automated sequencer. The primer corresponding to the rare cutter is end-labeled; thus only single-strand fragments that contain the *Eco*RI end are visualized.

Step 4. Fragment detection

1   2   3   4   5   6
Sample

Fig. 4   Schematic representation of the AFLP technique.

species. This makes AFLPs particularly useful for minor crops and neglected species because highly informative genetic fingerprints may be generated with no previous sequence information knowledge. AFLPs have proved to be useful in assessing genetic relationships: its ease of use and high information level has resulted in an abundance of studies being conducted for assessing genetic diversity (Hill et al. 1996; Mackill et al. 1996). The AFLP technique requires 0.3-1.0 µg of good quality DNA and is technically demanding (Karp et al. 1997). Also, there is evidence that restriction fragments generated using *Eco*RI tend to cluster to centromeric regions (Gedil et al. 2001). This can be overcome by using a range of restriction enzymes to generate restriction fragments that would be more dispersed along the chromosomes. Different detection systems (for example, autoradiography and detection of fluorescently labeled products using automated genotypers, e.g. LI-COR) can be used to visualize the bands. The amplification of a large number of bands is one of the most useful features of AFLP but may give rise to problems of homoplasy: the co-migration of non-related bands. The impact of the number of selective bases on competition effects in the PCR was demonstrated by Vos et al. (1995).

One repeated mistake of users of AFLP is to assume that the commonly-used restriction enzyme *Eco*RI is methylation-insensitive. In fact the action of *Eco*RI can be impaired or blocked by 5-methyl C residue (see http://rebase.neb.com/rebase/rebase.html) which may serve to confuse the profiles generated. Methylation studies using AFLP-type assays are performed by comparing the profiles from two isoschizomers whose methylation status differs in a known manner (Xiong et al. 1999).

As a precautionary note, the adapters used for AFLP are usually made from commercially derived oligonucleotides. These are chemically synthesized and lack the 5' phosphate moiety. As such, ligation of adapters, generated by annealing two such oligonucleotides, to the cohesive ends generated by restriction enzymes-digested template will create incomplete joins: without the phosphate group, the 5' end of the adapter is not able to covalently bond to the 3'OH of the template. During the initial cycle of the AFLP reaction, the non-covalently attached oligonucleotide falls off during the initial denaturation step, allowing *Taq* polymerase to fill in the 'recess' before the two strands dissociate. This is critical as the synthesized DNA forms the binding sites for the primers during subsequent thermal cycling. The use of 'hot-start' protocols, either using antibody bound- or chemically-inactivated *Taq* polymerase, does not permit the synthesis of the binding sites in the preselective amplification step, and leads to experimental failure.

Sequence Characterized Amplified Region (SCAR)

Practicality of use requires that DNA fragments of interest generated by RAPD or AFLP (for example, linked to disease resistance or other traits

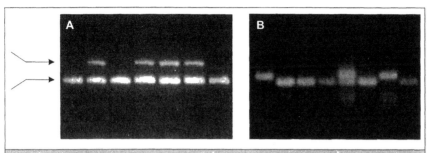

Fig. 5 A) Dominant and B) co-dominant SCAR markers amplified from olive DNA. In A), a positive control (lower band) is included to demonstrate that the reactions are working. Reproduced from Reale et al. (2006).

of interest) are converted to more robust PCR markers or SCARs (Figure 5). The fidelity of product amplification is improved by extending the primer length from 10 bp for RAPD markers to around 20 bp, which allows the PCR to be carried out under more stringent conditions (Paran and Michelmore 1993). In order to achieve this, it is necessary to sequence the fragments of interest: this process is often labor-intensive and may involve cloning the relevant fragments. The sequence information is then used to design a new pair of primers that, in general, will amplify only a single product. Though simple in theory, the conversion of DNA fragments, generated by RAPD and AFLP, to SCARs can be beset with problems. The sequence changes that determine the observed polymorphisms in the RAPD or AFLP assay may be lost when new primers are designed leading to monomorphic assays. Nevertheless, in combination with screening using RAPD or AFLP, SCAR development is an effective way to translate identified markers into robust PCR test for tagging purposes in a present/absent test.

Simple Sequence Repeat (SSR)

SSRs (also termed microsatellites or sequence-tagged microsatellite sites - STMS) are stretches of DNA, consisting of short tandemly repeated sequence motifs of units of 1-6 bp in length (Tautz and Schlötterer 1994). They seldom include more than 70 repeat units and are almost uniformly distributed over the entire genome, making them useful for genome mapping projects (Dib et al. 1996; Dietrich et al. 1996). For example, in mammals it has been estimated that the most common motif (GT/AC) occurs on average every 30 Kb (Wang et al. 1994). They are highly polymorphic DNA markers with discrete loci and co-dominant alleles.

Despite their advantages, isolation of useful SSR loci can be a time-consuming and expensive process. Many approaches have been used to isolate SSRs and their flanking sequences. Genomic clones containing SSRs can be isolated by screening with labeled oligonucleotides containing the desired repeat sequences. SSRs may also be obtained by screening sequences in databases or libraries of clones (Powell et al. 1996a). Polymorphism is revealed by PCR-amplification from total genomic DNA, using two unique primers that flank and hence define the microsatellite locus (Figure 6). Amplification products obtained from different individuals can be resolved on gels to reveal polymorphisms that are detected based on size differences and hence amplicon length, which predominantly result from differences in number of repeats. PCR primer sequences flanking microsatellite repeats are generally designed 17-22 nucleotide long, with a GC content of approximately 50% and a Tm about 60°C (McCouch et al. 1997). The ideal size for analysis of SSR amplification products is approximately 100-250 bp. PCR amplification of dinucleotide SSRs often leads to the production of numerous bands referred to as stutter. These bands, which differ in size by 2 bp increments, are most commonly caused by slippage of the DNA polymerase during the PCR cycles (Schlötterer and Tautz 1992). The

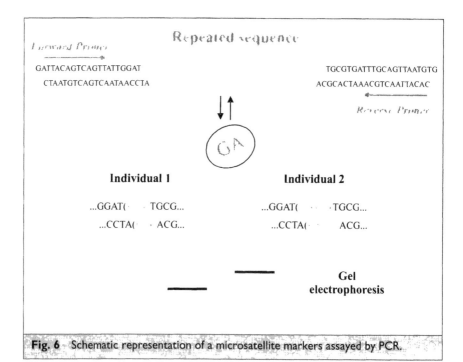

Fig. 6 Schematic representation of a microsatellite markers assayed by PCR.

presence of such artifacts can sometimes lead to problems in identifying the real allele size making it difficult to discriminate between two alleles differing by 2 bp. However the problem is usually solved by optimization of PCR conditions. When two alleles are sufficiently different in size, stutter peaks do not cause any problem. The use of SSRs containing trinucleotide or higher order repeats usually eliminates this problem; however, they are less abundant and less polymorphic than dinucleotide SSRs (Bryan et al. 1997). Another potential source of error in genotyping studies using SSRs, is due to the fact that DNA polymerase can catalyze non-templated addition of an extra 'A' to the 3' end of PCR products (Smith et al. 1995; Clark 1998). The extent of 'A-addition' is dependent on primer sequences and PCR conditions. Modification of PCR protocols can increase the proportion of products with an extra A to simplify the banding pattern (Smith et al. 1995).

Microsatellites have a wide range of applications; their high variability has made them a marker of choice in behavioral ecology, as they allow the determination of paternity and kinship (Bruford and Wayne 1993; Schlötterer and Pemberton 1994). They are also important in studying the structure of natural populations. The utility of microsatellites arises from two main factors: their high information content (which is a feature of the number and frequency of alleles detected) and ease of genotyping. The ability to distinguish between closely related individuals is particularly important for many crop species that tend to have a narrow genetic base. A further application of microsatellites is in the determination of 'hybridity'. For this purpose, the co-dominant nature of microsatellites is particularly important and allows the allelic contribution of each parent to be detected in sexual or somatic hybrids.

Expressed Sequence Tag-derived SSR (EST-SSR)

The use of microsatellites, derived from expressed sequence tag (EST) sequences, have gained popularity as markers for genetic studies. They represent functional molecular markers as 'putative' functions can be deduced for most sequences by protein database or translated sequences homology searches. The recent proteomics/metabolomics focus of biological research has generated a wealth of EST sequences from which EST-SSRs can be identified using 'in silico approaches' (Scott et al. 2000; Kantety et al. 2002). Furthermore, as the primers to amplify these sequences are often located with the coding regions, EST-SSR markers can be expected to be transportable across species.

Publicly-accessible software that identifies EST-SSRs from databases, are available, for example 'MISA' (MIcroSAtellite, *http://pgrc.ipk-gatersleben.de.misa*; Thiel et al. 2003) and 'Sputnik' (*http://cbi.labri.fr/outils/Pise/sputnik.html*). When these programs are coupled to other modules, then the resultant searches will not only yield potential polymorphic markers but also primers to amplify the repeats, the putative function of the loci and even their genetic locations (Love et al. 2004).

Inter-Simple Sequence Repeat (ISSR)

It is possible to exploit SSRs without the need to characterize and develop primers. Primers composed of SSRs can be used to amplify regions between two nearby inversely oriented SSRs (Zietkiewicz et al. 1994). Furthermore, the repeat length variability that makes SSR markers so polymorphic can be sampled by anchoring the SSR primers at the 5' end (Fisher et al. 1996). ISSR assays can be undertaken for any species that contains a sufficient number and distribution of SSR motifs and have the advantage that genomic sequence data are not required. This technique amplifies multiple DNA fragments per reaction, presumably representing loci from across the genome. However, in common with all single PCR primer assays, all the products generated have inverted repeats at their ends and, as such, are capable of forming hairpins structures. Competitive hybridization between the ends of the molecules with each other and with the primer, during PCR may explain DNA concentration effect in profiles obtained (Corbett et al. 2001).

Cleaved Amplified Polymorphic Sequence (CAPS)

CAPS markers are a form of genetic variation in the length of DNA fragments generated by restriction digestion of PCR products (Figure 7; Akopyanz et al. 1992; Koniecyzn and Ausubel 1993; Jarvis et al. 1994). Restriction enzyme digestion patterns that are polymorphic might be used to create CAPS markers, which are co-dominant molecular markers that amplify a short genomic sequence around the polymorphic endonuclease restriction site (Konieczny and Ausubel 1993). They are easily detected by agarose gel electrophoresis. The source of the sequence information for the primers can come from a genebank, genomic or cDNA clones, or cloned RAPD bands. The use of CAPS is thus affordable and practical for genotyping in positional or map-based cloning projects.

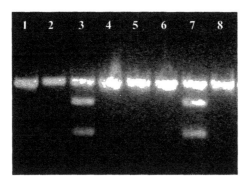

Fig. 7 CAP assay to genotype calcium-binding protein on olive cvs. The upper band represents the allele that does not contain the restriction enzyme recognition site: varieties 3 and 7 are heterozygotes. Reproduced from Reale et al. (2006).

Retrotransposon-based Markers

Retrotransposons are important vehicles for genome change and are frequently the causal agents of mutations, gene duplications and chromosome rearrangements. The insertions of retrotransposon into new sites are relatively stable because they replicate through reverse transcription of their RNA and integration of the resulting 'cDNA' into a new locus (Kumar and Bennetzen 1999). This mechanism of replication is shared with retroviruses, with the difference that retrotransposons do not form infectious particles that leave the cell to infect other cells (Havecker et al. 2004). This replicative mode of transposition can rapidly increase the copy number of elements and can thereby greatly increase plant genome size. Retrotransposons have been found in all eukaryotic genomes analyzed to date. They are distributed as interspersed repetitive sequences almost throughout the length of all host chromosomes and they can constitute a substantial fraction of the nuclear DNA. In maize, for example, over 50% of the genome are retrotransposon in origin (SanMiguel et al. 1996) and the 10,000-20,000 copies of BARE-1 constitute up to 5% of the barley genome (Manninen and Schulman 2000). Investigations of genome structure in flowering plants indicated that most variation in the size of nuclear genomes is caused by the differential amplification and/or retention of retrotransposons that contain long terminal repeats (LTRs) (SanMiguel et al. 1996; Bennetzen 2002).

The LTR retrotransposons (including retroviruses) are one class of sequences that replicate and transpose via an RNA intermediate (retroelements): non-LTR retrotransposons, LINEs (long dispersed repetitive elements) and SINEs (short interspersed repetitive elements) are also retroelements and together these sequences are found in high copy number in plant species (Bennetzen 1998).

LTR retrotransposons are located largely in intergenic regions and are the single largest components of most plant genomes (Kumar and Bennetzen 1999). They have been discovered in plants as sources of both spontaneous and induced mutations in maize and tobacco (Grandbastien et al. 1989; Marillonnet and Wessler 1998; Meyers et al. 2001). LTR retrotransposons are further subclassified into *copia*-like and the *gypsy*-like groups that differ from each other in both their degree of sequence similarity and their structure (i.e. the order of functional domains along the element-encoded polycistronic RNA) (Figure 8).

There are several advantages of using retrotransposon sequences as molecular markers. They are ubiquitous, present in moderate to high copy numbers as highly heterogenous populations, widely dispersed in chromosomes, and show insertional polymorphisms both within and between species in plant (Kumar et al. 1997). As they are stably inherited, retrotransposon positions in the genome act as temporal genetic markers that help in establishing pedigrees, phylogenies and for genetic diversity studies. Furthermore, the replicative mode that retrotransposons share with retroviruses means that the two LTRs that bound each element is derived from a single LTR (Varmus and Brown 1989): mutations begin to accumulate after transposition and thus divergence between these gives insight to the age of the elements.

The exploitation of retrotransposons as markers in plants was first described by Lee et al. (1990) where the members of a *copia*-like element were used as RFLP and PCR-based markers in pea. Subsequently, several molecular markers based on retrotransposons have been developed and exploited to study biodiversity in maize, pea and barley and to generate genetic linkage maps in barley (Waugh et al. 1997) and pea (Ellis et al. 1998).

Sequence-Specific Amplified Polymorphism (S-SAP)

S-SAP was first described by Waugh et al. (1997). The method exploits the unique junction created between retrotransposon and non-retrotransposon DNA by anchoring one end of a restriction DNA fragment to an adapter in a similar fashion to AFLP. Unlike AFLP only one restriction enzyme is required, a frequent cutter (*MseI* or *TaqI*), and adapters ligated to the ends of the fragments. PCR is performed using two primers, one derived from the sequence near the end of the

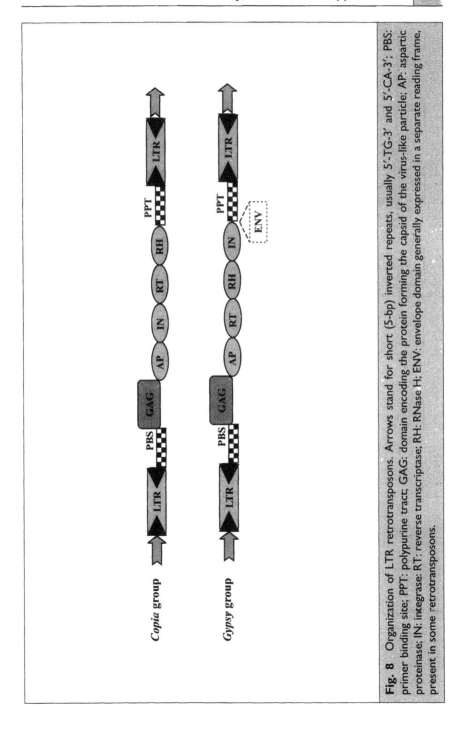

Fig. 8 Organization of LTR retrotransposons. Arrows stand for short (5-bp) inverted repeats, usually 5′-TG-3′ and 5′-CA-3′; PBS: primer binding site; PPT: polypurine tract; GAG: domain encoding the protein forming the capsid of the virus-like particle; AP: aspartic proteinase; IN: integrase; RT: reverse transcriptase; RH: RNase H; ENV: envelope domain generally expressed in a separate reading frame, present in some retrotransposons.

retroelement, usually within the LTR, and the other from the adapter. Depending on the number of copies in the genome, selective bases can be added to the anchored primer to reduce the number of bands when a retrotransposon family contains too many members, and thus can be optimized to individual retrotransposon families (Leigh et al. 2003). By labeling the LTR-derived primer, only LTR-related products are visualized. The method is so similar to AFLP that it is possible to use the same restriction-ligation reactions, even though it has been digested using two enzymes (Waugh et al. 1997; Taylor et al. 2004). In contrast, since PCR amplification from the retrotransposon primer is internal to the restriction-digested fragments, hot-start PCR is feasible for S-SAP and can reduced background as it eliminates all adapter-adapter amplified fragments.

S-SAP, like AFLP, is capable of efficiently generating a large amount of markers distributed over the whole genome. In fact the polymorphic information contents (PICs) of S-SAP markers, a measure of their genetic diversity, are higher than those for AFLP (Waugh et al. 1997). Moreover, the band intensities are fairly uniform unlike for AFLP, where organellar and repetitive DNAs give rise to intense, often monomorphic bands. The limiting step in the development of the marker system is the availability of terminal sequences from retrotransposons from which to design primers. The conservation of the amino acid sequence between very divergent retrotransposons and the uniform order of the functional domains allowed Pearce et al. (1999) to use a technique for the rapid isolation of *Ty1-copia* retrotransposon terminal sequences.

S-SAP has been applied in barley (Kumar et al. 1997; Waugh et al. 1997), peas (Ellis et al. 1998), wheat (Gribbon et al. 1999), Medicago (Porceddu et al. 2002), tobacco (Melayah et al. 2001) and grapevine (Pelcy and Merdinoglu 2002).

The high density of *BARE-1* retrotransposons in the cereal genomes has allowed the development of other retrotransposon PCR-based markers that do not require the restriction-ligation of DNA.

Inter-Retrotransposon Amplified Polymorphism (IRAP)

The IRAP method (Kalendar et al. 1999) amplifies DNA fragments between two LTR sequences sufficiently close to one another in the genome to permit PCR amplification of the region. In individuals where one or other of the two LTRs are missing, they will not amplify the corresponding fragment. Insertional polymorphisms are thus generally visualized as polymorphic, dominant bands by electrophoresis on either sequencing or high-resolution agarose gels. The amplified products range from under 100 bp to over several kilobase pairs, with the minimum size depending on the placement of the PCR priming sites with respect to the ends of the retrotransposons.

This method was first implemented in *Hordeum* by Kalendar et al. (1999) and since then has been used by Kalendar et al. (2000) and Vicient et al. (2001) for fingerprinting and biodiversity analysis, and for mapping of agronomically important traits (Manninen et al. 2000; Boyko et al. 2002). Retrotransposon markers have been used to investigate genomic stability in an allopolyploid (*Spartina anglica*) using this method (Baumel et al. 2002).

Retrotransposon-Microsatellite Amplified Polymorphism (REMAP)

REMAP is conceptually very similar to IRAP but instead of amplifying retrotransposon-retrotransposon fragments, it exploits the association between microsatellite sequences and retrotransposons (Kalendar et al. 1999). This method combines LTR primers with SSR primers to generate amplicons when the two sequences are sufficiently close in the genome which allows amplification.

Both IRAP and REMAP require sequence families of sufficient copy numbers to make it probable that they are close enough in the genome to produce amplicons in PCR. To date they have been predominantly used for cereals, though IRAP has been applied to pea (Smýkal 2006), oil palm (Price et al. 2004) and a variant using SINEs in *Brassica* has also been successfully developed (Tatout et al. 1999).

Retrotransposon-Based Insertional Polymorphism (RBIP)

RBIP is a method to survey whether a retrotransposon is present at a given site. Two primers flanking the insertion site will amplify when there is no element present. The presence of a retrotransposon displaces the two primers and the product is too large to be amplified. A third primer is included in the reaction that is derived from the element and will only be amplified with one of the external primers when the element is present (Flavell et al. 1998). Since both empty and filled sites will generate different products, the markers are co-dominant. Each PCR reaction surveys a single locus but as detection can be performed using hybridization assays rather than gel-based detection, RBIP can be automated to increased throughput using a TaqMan™ or DNA chip technology in order to increase sample throughput (Flavell et al. 2003). The major drawback to this technique is the requirement to isolate and sequence insertion sites.

Sequence Related Amplification Polymorphism (SRAP)

The SRAP assay involves use of primers designed to amplify the open reading frames (ORFs) of genes (Li and Quiros 2001). It uses two primers of 17 or 18 nucleotides in length comprising of the following elements: core sequences, which are 13 to 14 bases long, where the first 10 or 11 bases starting at the 5' end, are sequences of no specific constitution ('filler' sequences); followed by the sequence CCGG in the forward primer and AATT in the reverse primer. The core is followed by three selective nucleotides at the 3' end. The filler sequences of the forward and reverse primers must be different from each other and can be 10 or 11 bases long. For the first five cycles the annealing temperature is set at 35°C; the following 35 cycles are run at 50°C. The amplified DNA fragments are separated by denaturing acrylamide gels. The purpose for using the 'CCGG' sequence in the core of the first set of SRAP primers is to target exons or ORFs. This rationale is based on the fact that exons are normally in GC-rich regions. SRAP profiling was shown to target expressed regions of the *Brassica* genome (Li and Quiros 2001) though its use in *Olea europaea* (Reale et al. 2006) gave rise to an excess of retrotransposon sequences.

Target Region Amplification Polymorphism (TRAP)

TRAP uses 2 primers of 18 nucleotides to generate markers specific to a DNA sequence of interest (Hu and Vick 2003). One of the primers, the fixed primer, is designed from the targeted sequence in the database (e.g., an Expressed Sequence Tag, EST); the second primer, the arbitrary primer, is of arbitrary sequence with either an AT- or GC-rich core to anneal with an intron or exon, respectively. PCR amplification is run for the first 5 cycles with an annealing temperature of 35°C, followed by 35 cycles with an annealing temperature of 50°C. For different plant species, each PCR reaction can generate as many as 50 scorable fragments with sizes ranging from 50-900 bp when separated on a denaturing polyacrylamide gel. The TRAP technique can be used for genotyping germplasm collections and for tagging genes governing desirable agronomic traits in crop plants.

Single Nucleotide Polymorphism (SNP)

Single base changes and short insertion/deletion represent the most abundant source of DNA polymorphisms in organisms (Kwok et al. 1996; Collins et al. 1997; Kruglyak 1997) and they are both generally classified as single nucleotide polymorphisms or SNPs (Cho et al. 1999).

In principle, SNPs could be bi-, tri-, or tetra-allelic polymorphism. However, tri-allelic and tetra-allelic SNPs are rare almost to the point of non-existence, and so SNPs are sometimes simply referred to as bi-allelic markers (or di-allelic to be etymologically correct). In humans Kwok et al. (1996) found an average of one SNP every 1,000 bp. The frequency and nature of SNPs in plants is beginning to receive considerable attention. For instance, Tenaillon et al. (2001) found an average of one SNP every 104 bp in maize between two randomly sampled sequences; similar results have been obtained surveying sequence polymorphisms in eight lines of *Beta vulgaris* (Schneider et al. 2001). These polymorphisms could be used as simple genetic markers that may be identified within or near every gene. There is also great potential for these SNPs in the detection of associations between allelic forms of a gene and phenotypes (Rafalski 2002). SNP detection and discovery methods are many and various (Landegren et al. 1998).

SNP discovery relies on finding differences between two sequences. Direct sequencing of DNA fragments (amplified by PCR) from several individuals is the most direct way to identify SNP polymorphisms (Shattuck-Eidens et al. 1990; Gaut and Clegg 1993). PCR primers are designed to amplify 400-700 bp segments of DNA that are frequently derived from genes of interest or ESTs from the database. PCR is performed on a set of diverse individuals that represent the diversity in the species/population of interest. The PCR products are sequenced directly in both directions (Bhattramakki et al. 2002) and the resulting sequences are aligned and polymorphisms identified.

In addition, the dramatic increase in the number of DNA sequences submitted to the databases has made it possible to identify SNPs for several crops by electronic mining (e-mining:) without the need for sequencing (Taillon-Miller et al. 1998; Somers et al. 2003). This approach consists in the identification and alignment of sequences from the same locus from different sources (genotypes) allowing the detection of SNPs along these DNA sequences. The prerequisite is to have ample sequences for their alignment to identify polymorphisms and to distinguish real genetic changes from those generated by sequencing errors (Barker et al. 2003). The availability of EST databases makes it possible to target polymorphisms to functional regions of the genomes and even to specific genes (Useche et al. 2001).

Validation of SNPs can be performed by a number of different protocols. The choice of the method for a particular assay depends on many factors, including cost, throughput, equipment needed, difficulty of assay development, and potential for multiplexing. Typically, genotyping protocols start with target amplification and follow with allelic discrimination and product detection/identification: some or all of these steps could be combined and processed in parallel (Chen and Sullivan 2003).

Methods employed for SNP genotyping include:

- denaturing high-pressure liquid chromatography (DHPLC) (Underhill et al. 1997; Kuklin et al. 1998; O'Donovan et al. 1998),
- tetra-primers ARMS-PCR (Ye et al. 2001),
- single-strand conformational polymorphism (SSCP) (Orita et al. 1989a, b),
- enzymatic cleavage methods (CAPS) (Cotton 1993; Mashal et al. 1995; Youil et al. 1996) if the SNP alters a restriction enzyme recognition sequence,
- MALDI-TOF mass spectrometry-based systems,
- pyrosequencing (Tsuchihashi and Dracopoli 2002),
- single-based extension (SBE) or single nucleotide extension (SNuPE) assays,
- ligase chain reactions (LCR) (Landegren et al. 1988),
- Taqman™ assays,
- hybridization assays – allele-specific hybridization (Connor et al. 1983), microarray (Wang et al. 1998).
- molecular inversion probes (MIP; Hardenbol et al. 2003),
- RSCA (Argüello et al. 1998; Turner et al. 1999).

3   FEATURES OF MOLECULAR MARKERS

There are a number of methods that can be used to explore variation between genetic materials. Clearly, each marker system has advantages and disadvantages and none of the currently available fulfils all the requirements expected from a 'perfect' marker (Weising et al. 1995; Table 2).

Table 2 Comparative assessment of the molecular markers.

Marker	Type	Reproducibility	Development Time and Cost	Technical	Skills and Equipment Requirements	Genome	Utility And Applications
RFLP	Hybridization-based	High	Markers need to be screened for usefulness and therefore time-consuming and high cost	Labor-intensive; high quantity (2-5 µg) and quality DNA required per track	Size-fractionation of digested DNA fragments by gel electrophoresis, Southern blotting, hybridisation and detection of bound probes (usually radioactive)	Limited genome coverage per hybridisation: widely distributed but generally restricted to expressed regions	Co-dominant markers most useful for aligning genetic maps (intra- and inter-specific). Many steps make this procedure difficult to automate
Dot Blot Assay	Hybridization-based	High	Time-consuming and medium cost	1-2 µg DNA required per experiment, quality not an issue	Hybridization, washing and detection of bound probes	One locus per hybridization	Dominant markers most suitable for diagnostic assay
RAPD/ AP-PCR	PCR-based	Low	Fast and cheap	No sequence information needed for primer design	Thermocycler, fractionation of products (by agarose gel or polyacrylamide) electrophoresis	Multiple polymorphic loci (up to 10) per reaction with good distribution throughout genome	Dominant markers most useful providing markers to saturate genetic maps, the tagging of desirable traits for MAS, phylogenetic and population studies
AFLP	PCR-based	Medium	Fast and low	Does not require a priori sequence data, easily applied to any species, 0.3-1.0 µg of good quality DNA and	Thermocycler, polyacrylamide gel electrophoresis	Multiple polymorphic loci (up to 30) per reaction with good distribution throughout genome	Dominant markers most useful providing markers to saturate genetic maps, the tagging of desirable traits for MAS, phylogenetic and population studies

Contd.

Contd.

SCAR	PCR-based	High	Time-consuming and high cost	is technically demanding	Thermocycler, electrophoresis	One locus per reaction	Dominant markers most suitable for diagnostic assay
SSRs	PCR-based	High	Time-consuming and high cost	Labor-intensive, may involve cloning the relevant fragments	Thermocycler, electrophoresis	Good distribution throughout genome	High levels of polymorhism make SSRs highly suitable for varietal identification; high reproducibility make them suitable as anchorage points between different genetic maps
EST-SSRs	PCR-based	High	Medium and Low (by e-mining)	Low quantity DNA required	Thermocycler, electrophoresis	Good distribution throughout genome	High levels of polymorhism make SSRs highly suitable for varietal identification; high reproducibility make them suitable as anchorage points between different genetic maps. EST-SSRs are more transportable between species
ISSR	PCR-based	Medium	Fast and Low	No sequence information needed for primer design	Thermocycler, electrophoresis	Limited genome distribution	Dominant markers useful for mapping or inter-genomic comparison
CAPS	PCR-based	High	Time-consuming and high cost	PCR plus restriction digestion	Thermocycler, electrophoresis	One locus per reaction	Co-dominant marker

Contd.

Contd.

S-SAP	PCR-based	Medium	Medium	Sequence info required for the LTR primer, 0.3-1.0 µg of good quality DNA and is technically demanding	Thermocycler, electrophoresis	Multiple polymorphic loci (up to 30) per reaction with good distribution throughout genome	Dominant markers most useful providing markers to saturate genetic maps, the tagging of desirable traits for MAS, phylogenetic and population studies
IRAP	PCR-based	Medium	Fast and low cost where high density retroelements are found	Sequence info required for the LTR primers	Thermocycler, electrophoresis	Limited genome distribution	Dominant markers most useful providing markers to saturate genetic maps, the tagging of desirable traits for MAS, phylogenetic and population studies. Limited to species where retroelements are found in sufficiently high densities
REMAP	PCR-based	Medium	Fast and low cost where high density retroelements are found	Sequence info required for the LTR and SSR primers	Thermocycler, gel electrophoresis on polyacrylamide or high-resolution agarose	Limited genome distribution	Dominant markers most useful providing markers to saturate genetic maps, the tagging of desirable traits for MAS, phylogenetic and population studies. Limited to species where retroelements are found in sufficiently high densities
RBIP	PCR-Hybridization	High	Time-consuming and high cost	Sequence info required for empty sites	Thermocycler and hybridization assay	Good genome coverage	Development may be costly to find single copy locus to find insertion sites but high throughput potential using microarray justifies input if

Contd.

Contd.

							sufficient genotyping is performed.
SRAP	PCR-based	Low	Low	Does not require *a priori* sequence data, easily applied to any species	Thermocycler, gel electrophoresis on polyacrylamide	Primers target intron/exon regions	Dominant markers targets expressed regions
TRAP	PCR-based	Medium	Low	Sequence info required	Thermocycler, gel electrophoresis on polyacrylamide	Gene targeted	Dominant markers, particularly useful for walking from a sequence where linkage has already been established
SNP	PCR-based/ hybridization/ electrophoresis	High	Medium-high	Sequence info required	Most versatile of markers can be detected using PCR, hybridization or by conformational polymorphism	Genome coverage second to none	Wide genome coverage and assaying platforms/methods make this the marker of choice

An ideal molecular marker system would possess a number of desirable properties:

- high level of polymorphism and reproducibility,
- co-dominant inheritance (allowing discrimination between homo- and heterozygote in diploid organisms),
- clear designation of alleles,
- frequently occurrence in the genome,
- even distribution throughout the genome,
- selectively neutrality,
- straightforward and cheap development of assay,
- easy/rapid procedure (amenable to automation if required),
- possibility of exchange between laboratories,
- contained costs in routine analyses following marker development.

The choice of an appropriate technique for a particular investigation depends on a number of factors such as on the material being studied and the question being addressed (Karp and Edwards 1977). Technical limitations include throughput and speed, equipment and skills required, the need for automation, cost effectiveness, ease of access to the technique; relevant training and cost also need to be considered (Avise 1994; Weising et al. 1995). Other considerations relate to the technique itself, the informativeness and sensitivity of the marker system, and its overall reliability. Finally, the individual application and its demands on accuracy and data-analysis must also be considered.

It is possible to identify important criteria by which to judge the value of any particular marker system for a chosen application. Numerous types of DNA markers based on the indirect detection of sequence-level polymorphism have been developed (Henry 2001; Phillips and Vasil 2001). Among them, RFLPs are the markers of choice for comparative genomic studies. Although RFLPs are one of the earlier markers still used in many laboratories, examining numerous marker-genotype combinations is laborious and time-consuming. RAPD markers do not require a priori sequence information and are advantageous in terms of simplicity and cost and are, therefore, frequently used for species in which no other markers have been developed. On the other hand, AFLP markers can be used to quickly create well-saturated genetic maps.

SSR are the markers that are currently most widely used. Frequently, highly informative SSR markers are preferred (Powell et al. 1996a) even if they are less suitable for association studies because of the homoplasy:

the occurrence of SSR alleles of identical size but different evolutionary origin (Estoup et al. 1995; Viard et al. 1998). SSRs are easier to use than RFLPs owing to the smaller amount of DNA required, the technology is amenable to automation and they can easily be exchanged between researchers because each locus is defined by the primer sequences. They are particularly attractive for distinguishing between cultivars because the level of polymorphism detected at SSR loci is higher than that detected with any other molecular marker assay (Saghai Maroof et al. 1994; Powell et al. 1996b).

In creating genetic maps, SSRs have replaced RFLP markers because of their ease of use. If their map positions have been determined in other crosses, they are often used as genetic landmarks forming the backbone or skeleton. In-filling of the map can be performed relatively quickly using anonymous multi-loci markers, such as RAPDs, AFLPs and S-SAPs to provide excellent coverage: the number of reactions depends on required marker density required. These maps allow the identification of the genetic location of genes of interest for simple characters or by quantitative trait loci (QTL) mapping for multi-determinant phenotypes (Maheswaran et al. 2000).

For forensic-type analyses, for example provenance testing, microsatellites are often the markers of choice for their high information content, since the quantity and quality of DNA recoverable from processed materials may be limiting (Doveri et al. 2006).

Powell et al. (1996b) examined the utility of RFLP, RAPD, AFLP and SSR markers for soybean germplasm analysis by evaluating information content (expected heterozygosity), number of loci simultaneously analyzed per experiment (multiplex ratio) and effectiveness in assessing relationships between accessions. SSR markers revealed the highest expected heterozygosity while AFLP markers showed the highest effective multiplex ratio.

In contrast, SNPs have a low heterozygosity value because of bi-allelism but are rapidly becoming the markers of choice for several reasons. SNP assays do not require DNA separation by size, can be automated in an assay-plate format or on microchips, and are easier to locate in most single-copy regions of the genome than SSRs. The information provided by SNPs is most useful when several, closely spaced SNPs define haplotypes in the region being examined. Not many systematic efforts to discover SNPs have yet been undertaken in plants although there has been a big effort in the pharmaceutical sector to study small populations in the hope of discovering SNPs markers linked to diseases. The potential financial rewards offered, makes such a study feasible in humans but not in plant species. There is great potential for SNPs to be tested in high throughput screening platforms, such as those relying on DNA chips, which will ultimately reduce the cost of each assay while dramatically increasing the testing capacity.

When markers are used to estimate genetic diversity and to build molecular phylogenies, the taxonomical level of analysis becomes of great importance. Amplification-based scanning techniques are useful here for distinguishing individuals and closely related organisms, usually below the species level (cultivars, accessions, lines, clones, etc.), whereas the presumed neutrality and high allele number of SSR markers makes them superior to isoenzymes and ideally suited for the study of populations. In contrast, RFLP, direct sequencing and amplification-based profiling techniques are useful for cross-species analysis and phylogenetic reconstruction. Moreover, the use of a marker system in one species does not necessary indicate its applicability in another species. SSRs can be informative but polymorphic primers identified in one species are generally not useful in another (Staub and Serquen 1996). Marker systems also differ in their utility across populations, species, and genera and their efficacy in the detection of polymorphisms. Molecular markers have been used infrequently in assisting breeding of ornamental species and especially woody plants mostly because minor crops/species have seen less investment in this direction. However, cultivars of ornamental species are typically generated by selection and propagation of sports and somaclonal variants, and are thus usually poorly characterized phenotypically. However, in these cases, because of the very close genetic relationship between the sports, even DNA markers have limited practical applications.

4   FUTURE PERSPECTIVES

Many of the methods that have been applied to research on plants were first developed in the human research arena. It is clear that there are many methods that can assay genomes in a very efficient manner using highly parallel genotyping systems (Fan et al. 2006). The systems developed allow thousands of genetic loci (SNPs) to be simultaneously interrogated, providing data to perform studies that were not previously feasible, such as whole genome linkage disequilibrium (LD) association studies (Hirschhorn and Daly 2005). In the plant research world, whole genome assays would facilitate and speed up research into association mapping, which exploits historic recombination events to identify genes of interest, and the search for domestication genes, those responsible for traits that our early ancestors selected. Such genes would exhibit genetic bottlenecks that are significantly narrower than for domestication alone (Wright et al. 2005). However, the availability of useful SNPs limits the application of these highly parallel genotyping systems, and currently there are only a handful of crops for which sufficient data are available to exploit the multiplexing capabilities anywhere near their potential. Microarray-based methods, such as diversity array technology (DArT), described as 'AFLP on a chip' (Jaccoud et al. 2001), offer a cost-effective way of genome scanning.

In parallel with the emphasis on technology leveraged from mammalian genomics, emphasis needs to be given to population development and phenotyping to ensure that rapid gene-phenotype relationships can be determined and validated (Mackay and Powell 2007). However with the continued development of highly parallel sequencing such as massively parallel signature sequencing (MPSS) (Brenner et al. 2000), DNA sequencing by hybridization with oligonucleotide matrix (SHOM) (Khrapko et al. 1991) and sequencing by synthesis (SBS) approaches, such as pyrosequencing (Ronaghi et al. 1998), sequencing of single DNA molecules (Braslavsky et al. 2003) and polymerase colonies (Mitra et al. 2003), sequencing costs will continue to decline. If costs were reduced to the commercial target of US$1000 per genome (Bennett et al. 2005), it would become cheaper to sequence than genotype whole genomes. Genotyping or genome scanning will no longer be surrogates for sequence data. Then the next stage in plant genetics will not be marker development but the need to extract useful data from the billions and billions of bases that will be generated.

References

Akkaya MS, Bhagwat AA, Cregan PB (1992) Length polymorphisms of simple sequence repeat DNA in soybean. Genetics 132: 1131-1139

Akopyanz N, Bukanov NO, Westblom TU, Berg DE (1992) PCR-based RFLP analysis of DNA sequence diversity in the gastric pathogen *Helicobacter pylori*. Nucl Acids Res 20: 6221-6225

Ardley J, Hoptroff CGM (1996) Protecting plant 'invention': the role of plant-variety rights and patents. Trends Biotechnol 14: 67-69

Argüello JR, Little A-M, Pay AL, Gallardo D, Rojas I, Marsh SGE, Goldman JM, Madrigal JA (1998) Mutation detection and typing of polymorphic loci through double-strand conformation analysis. Nat Genet 18: 192-194

Avery OT, MacLeod CM, McCarthy M (1944) Studies on the chemical nature of the substance inducing transformation of pneumococcal types. I. Induction of transformation by a deoxyribonucleic acid fraction isolated from pneumococcus type III. J Exp Med 79: 137-158

Avise JC (1994) Molecular Markers, Natural History and Evolution. Chapman and Hall, London, UK

Bardini M, Lee D, Donini P, Mariani A, Gianì S, Toschi M, Lowe C, Breviario D (2004) TBP: a new tool for testing genetic diversity in plant species based on functionally relevant sequences. Genome 47: 281-291

Barker G, Batley J, O'Sullivan H, Edwards KJ, Edwards D (2003) Redundancy based detection of sequence polymorphisms in expressed sequence tag data using autoSNP. Bioinformatics 19: 421-422

Baumel A, Ainouche M, Kalendar R, Schulman AH (2002) Retrotransposons and genomic stability in populations of the young allopolyploid species *Spartina anglica* C.E. Hubbard (*Poaceae*). Mol Biol Evol 19: 1218-1227

Bebeli PJ, Zhou Z, Somers DJ, Gustafson JP (1997) PCR primed with minisatellite core sequences yields DNA fingerprinting probes in wheat. Theor Appl Genet 95: 276-283

Beckmann JS (1988) Oligonucleotide polymorphisms: A new tool for genomic genetics. Bio/Technology 6: 161-164

Beckmann JS, Soller M (1990) Toward a unified approach to genetic mapping of eukaryotes based on sequence tagged microsatellite sites. Bio/Technology 8: 930-932

Bennett ST, Barnes C, Cox A, Davies L, Brown C (2005) Toward the $1000 human genome. Pharmacogenomics 6: 373-382

Bennetzen JL (1998) The structure and evolution of angiosperm nuclear genomes. Curr Opin Plant Biol 1: 103-108

Bennetzen JL (2002) Mechanisms and rates of genome expansion and contraction in flowering plants. Genetica 115: 29-36

Bhattramakki D, Dolan M, Hanafey M, Wineland R, Vaske D, Register JC, Tingey SV, Rafalski A (2002) Insertion-deletion polymorphisms in 3' regions of maize genome occur frequently and can be used as highly informative genetic markers. Plant Mol Biol 48: 539-547

Boyko E, Kalendar R, Korzun V, Fellers J, Korol A, Sculman AH, Gill BS (2002) A high-density cytogenetic map of the *Aegilops tauschii* genome incorporating retrotransposons and defence-related genes: insights into cereal chromosome structure and function. Plant Mol Biol 48: 767-790

Braslavsky I, Hebert B, Kartalov E, Quake SR (2003) Sequence information can be obtained from single DNA molecules. Proc Natl Acad Sci USA 100: 3960-3964

Brenner S, Johnson M, Bridgham J, Golda G, Lloyd DH, Johnson D, Luo S, McCurdy S, Foy M, Ewan M, Roth R, George D, Eletr S, Albrecht G, Vermaas F, Williams SR, Moon K, Burcham T, Pallas M, DuBridge RB, Kirchner J, Fearon K, Mao J, Corcoran K (2000) Gene expression analysis by massively parallel signature sequencing (MPSS) on microbead arrays. Nat Biotechnol 18: 630-634

Brow MA, Oldenburg MC, Lyamichev V, Heisler LM, Lyamicheva N, Hall JG, Eagan NJ, Olive DM, Smith LM, Fors L, Dahlberg JE (1996) Differentiation of bacterial 16S rRNA genes and intergenic regions and *Mycobacterium tuberculosis* katG genes by structure-specific endonuclease cleavage. J Clin Microbiol 34: 3129-3137

Bruford MW, Wayne RK (1993) Microsatellites and their application to population genetic studies. Curr Opin Genet Dev 3: 939-943

Bryan GJ, Collins AJ, Stephenson P, Orry A, Smith JB, Gale MD (1997) Isolation and characterisation of microsatellites from hexaploid bread wheat. Theor Appl Genet 94: 557-563

Caetano-Anollés G, Bassam BJ, Gresshoff PM (1991) DNA amplification fingerprinting using very short arbitrary oligonucleotide primers. Bio/Technology 9: 553-557

Caetano-Anollés G, Bassam BJ, Gresshoff PM (1993) Enhanced detection of polymorphic DNA by multiple arbitrary amplicon profiling of endonuclease-digested DNA: identification of markers tightly linked to the supernodulation locus in soybean. Mol Gen Genet 241: 57-64

Chang RY, O'Donoughue LS, Bureau TE (2001) Inter-MITE polymorphisms (IMP): a high throughput transposon-based genome mapping and fingerprinting approach. Theor Appl Genet 102: 773-781

Chen X, Sullivan PF (2003) Single nucleotide polymorphism genotyping: biochemistry, protocol, cost and throughput. Pharmacogenomics J 3: 77-96

Cho RJ, Mindrinos M, Richards DR, Sapolsky RJ, Anderson M, Drenkard E, Dewdney J, Lynne Reuber T, Stammers M, Federspiel N, Theologis A, Yang W-H, Hubbell E, Au M, Chung EY, Lashkari D, Lemieux B, Dean C, Lipshutz RJ, Ausubel FM, Davis RW, Oefner PJ (1999) Genome-wide mapping with biallelic markers in *Arabidopsis thaliana*. Nat Genet 23: 203-207

Church GM, Gilbert W (1984) Genomic Sequencing. Proc Natl Acad Sci USA 81: 1991-1995

Clark JM (1998) Novel non-templated nucleotide addition reactions catalyzed by procaryotic and eucaryotic DNA polymerases. Nucl Acids Res 16: 9677-9686

Collins FS, Guyer MS, Chakravarti A (1997) Variations on a theme: cataloguing human DNA sequence variation. Science 278: 1580-1581

Connor BJ, Reyes AA, Morin C, Itakura K, Teplitz RL, Wallace RB (1983) Detection of sickle cell BS-globin allele by hybridization with synthetic oligonucleotides. Proc Natl Acad Sci USA 80: 278-282

Cooke RJ, Reeves JC (2003) Plant genetic resources and molecular markers: variety registration in a new era. Plant Genet Resour 1: 81-87

Corbett G, Lee D, Donini P, Cooke RJ (2001) Identification of potato varieties by DNA profiling. Acta Hort 546: 387-390

Cotton RG (1993) Current methods of mutation detection. Mutat Res 285:125-144

Dib C, Faure S, Fizames C, Samson D, Drouot N Vignal A, Millasseau P, Marc S, Hazan J, Seboun E, Lathrop M, Gyapay G, Morisette J, Weissenbach J (1996) A comprehensive genetic map of the human genome based on 5,264 microsatellite. Nature 380: 152-154

Dietrich WF, Miller J, Steen R, Merchant MA, Damron-Boles D, Husain Z, Dredge R, Daly MJ, Ingalls KA, O'Connor TJ (1996) A comprehensive genetic map of the mouse genome. Nature 380: 149-152

Donnenwirth J, Grace J and Smith S (2004) Intellectual Property Rights, patents, plant variety protection and contracts: a perspective from the private sector. IP Strategy Today 9: 19-34

Doveri S, O'Sullivan DM, Lee D (2006) Non-concordance between genetic profiles of olive oil and fruit: a cautionary note to the use of DNA markers for provenance testing. J Agri Food Chem: doi: 10.1021/jf061564a (in press)

Drábek J (2001) A commented dictionary of techniques for genotyping. Electrophoresis 22: 1024-1045

Ellis THN, Poyser SJ, Knox MR, Vershinin AV, Ambrose MJ (1998) Polymorphism of insertion sites of *Ty1-copia* class retrotransposons and its use for linkage and diversity in pea. Mol Gen Genet 260: 9-19

Estoup A, Tailliez C, Cornuet JM, Solignac M (1995) Size homoplasy and mutational processes of interrupted microsatellites in two bee species, *Apis mellifera* and *Bombus terrestris* (Apidae). Mol Biol Evol 12: 1074-1084

Fan J-B, Chee MS, Gunderson KL (2006) Highly parallel genomic assays. Nat Genet 7: 632-644

Fisher PJ, Gardner RC, Richardson TE (1996) Single locus microsatellites isolated using 5' anchored PCR. Nucl Acids Res 24: 4369-4371

Flavell AJ, Knox MR, Pearce SR, Ellis THN (1998) Retrotransposon-based insertion polymorphisms (RBIP) for high throughput marker analysis. Plant J 16: 643-650

Flavell AJ, Bolshakov VN, Booth A, Jing R, Russell J, Ellis THN, Isaac P (2003) A microarray-based high throughput molecular marker genotyping method: the tagged microarray marker approach. Nucl Acids Res 31: e115

Gaut BS, Clegg MT (1993) Nucleotide polymorphism in the *Adh1* locus of pearl millet (*Pennisetum glaucum*) (Poaceae). Genetics 135: 1091-1097

Gedil MA, Wye C, Berry S, Segers B, Peleman J, Jones R, Leon A, Slabaugh MB, Knapp SJ (2001) An integrated restriction fragment length polymorphism - amplified fragment length polymorphism linkage map for cultivated sunflower. Genome 44: 213-221

Grandbastien MA, Spielmann A, Caboche M (1989) *Tnt1*, a mobile retroviral-like transposable element of tobacco isolated by plant cell genetics. Nature 337: 376-380

Gribbon BM, Pearce SR, Kalendar R, Schulman AH, Paulin L, Jack P, Kumar A, Flavell AJ (1999) Phylogeny and transpositional activity of *Ty1-copia* group retrotransposons in cereal genomes. Mol Genet Genom 261: 883–891

Griffiths AJF, Miller JH, Suzuki DT, Lewontin RC, Gelbart WM (1996) An Introduction to Genetic Analysis. 6th edn. WH Freeman, New York, USA

Grodzicker T, Williams J, Sharp P, Sambrook J (1974) Physical mapping of temperature sensitive mutations. Cold Spring Harbor Symp Quart Biol 39: 439-446

Gu WK, Weeden NF, Yu J, Wallace DH (1995) Large-scale, cost-effective screening of PCR products in marker-assited selection applications. Theor Appl Genet 91: 465-470

Hames BD, Higgins SJ (1985) Nucleic Acid Hybridisation: A Practical Approach. IRL Press, Oxford, UK

Hardenbol P, Banér J, Jain M, Nilsson M, Namsaraev EA, Karlin-Neumann GA, Fakhrai-Rad H, Ronaghi M, Willis TD, Landegren U, David RW (2003) Multiplexed genotyping with sequence-tagged molecular inversion probes. Nat Biotechnol 21: 673-678

Hatada I, Hayashizaki Y, Hirotsune S, Komatsubara H, Mukai T (1991) A genome scanning method for higher organism using restriction sites as landmarks. Proc Natl Acad Sci USA 88: 397-400

Havecker ER, Gao X, Voytas DF (2004) The diversity of LTR retrotransposons. Genome Biol 5: 225

Henry RJ (ed) (2001) Plant Genotyping. The DNA Fingerprinting of Plants. CABI Publ, Wallingford, UK

Hill M, Witsenboer H, Zabeau M, Vos P, Kesseli R, Michelmore R (1996) PCR based fingerprinting using AFLPs as a tool for studying genetic relationships in *Lactuca* spp. Theor Appl Genet 93: 1202-1210

Hirschhorn JN, Daly MJ (2005) Genome-wide association studies for common diseases and complex traits. Nat Genet 6: 95-108

Hu J, Vick BA (2003) Target region amplification polymorphism: A novel marker technique for plant genotyping. Plant Mol Biol Rep 21: 289–294

Jaccoud D, Peng K, Feinstein D, Kilian A (2001) Diversity Arrays: a solid state technology for sequence information independent genotyping. Nucl Acids Res 29: e25

Jarvis P, Lister C, Szabo V, Dean C (1994) Integration of CAPs markers into the RFLP map generated using recombinant inbred lines of *Arabidopsis thaliana*. Plant Mol Biol 24: 685-687

Jeffreys AJ, Wilson V, Thein SL (1985) Hypervariable minisatellite regions in human DNA. Nature 314: 67-73

Jones CJ, Edwards KJ, Castaglione S, Winfield MO, Sala F, Van De Wiel C, Bredemeijer G, Vosman B, Matthes M, Daly A, Brettschneider R, Bettini P, Buiatti M, Maestri E, Malcevschi A, Marmiroli N, Aert R, Volckaert G, Rueda J, Linacero R, Vazquez A, Karp A (1997) Reproducibility Testing of RAPD, AFCP and SSR markers in Plants by a Network of European Laboratories. Mol Breed 3: 381-390.

Jordan SA, Humphries P (1994) Single nucleotide polymorphism in Exon 2 of the BCP Gene on 7 931-935. Hum Mol Genet 3: 1915.

Kalendar R, Grob T, Suoniemi A, Schulman AH (1999) IRAP and REMAP: Two new retrotransposon-based DNA fingerprinting techniques. Theor Appl Genet 98: 704-711

Kalendar R, Tanskanen J, Immonen S, Nevo E, Schulman AH (2000) Genome evolution of wild barley (*Hordeum spontaneum*) by *Bare-1* retrotransposon dynamics in response to sharp microclimatic divergence. Proc Natl Acad Sci USA 97: 6603-6607

Kantety RV, Rota ML, Matthews DE, Sorrells ME (2002) Data mining for simple sequence repeats in expressed sequence tags from barley, maize, rice, sorghum and wheat. Plant Mol Biol 48: 501-510

Karp A, Edwards KJ (1977) DNA markers: a global overview. In: G Caetano-Anollés and PM Gresshoff (eds) DNA Markers: Protocols, Applications and Overviews. John Wiley, New York, USA, pp 1-13

Karp A, Seberg O, Buiatti M (1996) Molecular techniques in the assessment of botanical diversity. Ann Bot (Lond) 78: 143-149

Karp A, Kresovich S, Bhat KV, Ayad WG, Hodgkin T (1997) Molecular tools in plant genetic resources conservation, a guide to the technologies. Tech Bull no 2, IPGRI

Khrapko KR, Lysov YuP, Khorlin AA, Ivanov IB, Yershov GM, Vasilenko SK, Florentiev VL, Mirzabekov AD (1991) A method for DNA sequencing by hybridization with oligonucleotide matrix. DNA Seq 1: 375-388

Koebner RMD, Powell W, Donini P (2001) Contributions of DNA molecular marker technologies to the genetics and breeding of wheat and barley. Plant Breed Rev 21: 181-219

Koniecyzn A, Ausubel FM (1993) A procedure for mapping Arabidopsis mutation using co-dominant ecotype-specific PCR-based markers. Plant J 4: 403-410

Kruglyak L (1997) The use of a genetic map of biallelic markers in linkage studies. Nat Genet 17: 21-24

Kuklin A, Munson K, Gjerde D, Haefele R, Taylor P (1998) Detection of single nucleotide polymorphism with WAVE (tm) DNA fragment analysis system. Genet Test 1: 201-206

Kumar A, Pearce SR, McLean K, Harrison G, Heslop-Harrison JS (1997) The *Ty1-copia* group of retrotransposons in plants: genomic organisation, evolution, and use as molecular markers. Genetica 100: 205-217

Kumar A, Bennetzen JL (1999) Plant retrotransposons. Annu Rev Genet 33: 479-532

Kwok PY, Deng Q, Zakeri H, Nickerson DA (1996) Increasing the information content of STS-based genome maps: identifying polymorphisms in mapped STSs. Genomics 31: 123-126

Landegren U, Kaiser R, Sanders J, Hood L (1988) A ligase-mediated gene detection technique. Science 241: 1077-1080

Lee D, Ellis THN, Turner L, Hellens RP, Cleary WG (1990) A *copia*-like element in *Pisum* demonstrates the uses of dispersed repeated sequences in genetic analysis. Plant Mol Biol 15: 707-722

Lee D, Reeves JC, Cooke RJ (1996a) DNA profiling and plant variety registration: 1. The use of random amplified polymorphisms to discriminate between varieties of oilseed rape. Electrophoresis 17: 261-265

Lee D, Reeves JC, Cooke RJ (1996b) DNA profiling and plant variety registration: 2. Restriction fragment length polymorphisms in varieties of oilseed rape. Plant Var Seeds 9: 181-190

Leigh F, Kalendar R, Lea V, Lee D, Donini P, Schulman AH (2003) Comparison of the utility of barley retrotransposon families for genetic analysis by molecular marker techniques. Mol Genet Genom 269: 464-474

Li G, Quiros CF (2001) Sequence-related amplified polymorphism (SRAP), a new marker system based on a simple PCR reaction: its application to mapping and gene tagging in *Brassica*. Theor Appl Genet 103: 455-461

Lombard V, Baril CP, Dubreuil P, Blouet F, Zhang D (2000) Genetic relationships and fingerprinting of rapeseed cultivars by AFLP consequences for varietal registration. Crop Sci 40: 1417-1425

Love CG, Batley J, Lim G, Robinson AJ, Savage D, Singh D, Spangenberg GC, Edwards D (2004) New computational tools for Brassica genome research. Comp Funct Genom 5: 276-280

Mackay I, Powell W (2007) Methods for Linkage Disequilibrium Mapping in Crops. Trends Plant Sci 12: 57-63

Mackill DJ, Zhang Z, Redona ED, Colowit PM (1996) Level of polymorphism and genetic mapping of AFLP markers in rice. Genome 39: 969-977

Maheswaran M, Huang N, Sreerangasamy SR, McCouch SR (2000) Mapping quantitative trait loci associated with days to flowering and photoperiod sensitivity in rice (*Oryza sativa* L.). Mol Breed 6: 145-155

Manninen I, Schulman AH (2000) *BARE-1*, a *copia*-like retroelement in barley (*Hordeum vulgare* L.). Proc Natl Acad Sci USA 97: 6603-6607

Manninen O, Kalendar R, Robinson J, Schulman AH (2000) Application of *BARE-1* retrotransposon markers to the mapping of major resistance gene for net blotch in barley. Mol Genet Genom 264: 325–334

Marillonnet S, Wessler SR (1998) Extreme structural heterogeneity among the members of a maize retrotransposon family. Genetics 150: 1245-1256

Mashal RD, Koontz J, Sklar J (1995) Detection of mutations by cleavage of DNA heteroduplexes with bacteriophage resolvases. Nat Genet 9: 177-183

Maughan PJ, Saghai Maroof MA, Buss GR, Huestis GM (1996) Amplified fragment length polymorphism (AFLP) in soybean: Species diversity, inheritance, and near-isogenic line analysis. Theor Appl Genet 93: 393-401

McCouch SR, Chen X, Panaud O, Temnykh S, Xu Y, Cho YG, Huang N, Ushii T, Blair M (1997) Microsatellite marker development, mapping and applications in rice genetics and breeding. Plant Mol Biol 35: 89-99

Melayah D, Bonnivard E, Chalhoub B, Audeon C, Grandbastien MA (2001) The mobility of the Tobacco *Tnt1* retrotransposon correlates with its transcriptional activation by fungal factors. Plant J 28: 159-168

Mendel G (1865) Versuche über pflanzen-hybriden. Verhandlungen des naturforshenden Vereins in Brünn 4: 3-47

Meyers BC, Tingey SV, Morgante M (2001) Abundance, distribution, and transcriptional activity of repetitive elements in the maize genome. Genome Res 11: 1660-1676

Mitra RD, Shendure J, Olejnik J, Olejnik EK, Church GM (2003) Fluorescent *in situ* sequencing on polymerase colonies. Anal Biochem 320: 55-65

Morell MK, Peakall R, Appels R, Preston LR, Lloyd HL (1995) DNA profiling techniques for plant variety identification. Aust J Exp Agri 35: 807-819

Morgante M, Vogel J (1994) Compound microsatellite primers for the detection of genetic polymorphisms. US Patent Appl 08/326456

Mullis K, Faloona F, Scharf S, Saiki R, Horn G, Erlich H (1986) Specific enzymatic amplification of DNA *in vitro*: the Polymerase Chain Reaction. Cold Spring Harb Symp Quant Biol 51: 263-273

Newton CR (ed) (1995) PCR Essential Data. John Wiley, Chichester, UK

O'Donovan MC, Oefner PJ, Roberts SC, Austin J, Hoogendoorn B, Guy C, Speight G, Upadhyaya M, Sommer SS, McGuffin P (1998) Blind analysis of denaturing high-performance liquid chromatography as a tool for mutation detection. Genomics 52: 44-49

Olsen M, Hood L, Cantor C, Botstein D (1989) A common language for physical mapping of the human genome. Science 245: 1434-1435

Orita M, Iwahana H, Kanazawa H, Hayashi K, Sekiya T (1989) Detection of polymorphism of human DNA by gel electrophoresis as single-strand conformation polymorphisms. Proc Natl Acad Sci USA 86: 2766-2770

Paran I, Michelmore RW (1993) Development of reliable PCR-based markers linked to downy mildew resistance genes in lettuce. Theor Appl Genet 85: 985-993

Pearce SR, Stuart-Rogers C, Knox MR, Kumar A, Ellis THN, Flavell AJ (1999) Rapid isolation of plant Ty1-copia group retrotransposon LTR sequences for molecular marker studies. Plant J 19: 711-717

Pelcy F, Merdinoglu D (2002) Development of grape molecular markers based on retrotransposons. In: Plant, Animal and Microbe Genome X Conf, San Diego, CA, USA

Phillips RL, Vasil IK (eds) (2001) DNA-based Markers in Plants. Kluwer Acad Publ, Dordrecht, The Netherlands

Porceddu A, Albertini E, Barcaccia G, Marconi G, Bertoli FB, Veronesi F (2002) Development of S-SAP markers based on an LTR-like sequence from *Medicago sativa* L. Mol Genet Genom 267: 107-114

Powell W, Machray GC, Provan J (1996a) Polymorphism revealed by simple sequence repeats. Trends Plant Sci 1: 215-222

Powell W, Morgante M, Andre C, Hanafey M, Vogel J, Tingey S, Rafalski A (1996b) The comparison of RFLP, RAPD, AFLP and SSR (microsatellite) markers for germplasm analysis. Mol Breed 2: 225-238

Price Z, Schulman AH, Mays S (2004) Development of new marker methods – an example from oil palm. Plant Genet Res 1: 103-113

Rafalski A (2002) Applications of single nucleotide polymorphisms in crop genetics. Curr Opin Plant Biol 5: 94-100

Reale S, Doveri S, Diaz A, Angiolillo A, Lucentini L, Pilla F, Martin A, Donini P, Lee D (2006) SNP-based markers for discriminating olive (*Olea europaea* L.) cultivars. Genome 49: 1193-1205

Rohde W (1996) Inverse sequence-tagged repeat (ISTR) analysis, a novel and universal PCR-based technique for genome analysis in the plant and animal kingdom. J Genet Breed 50: 249-261

Ronaghi M, Uhlen M, Nyren P (1998) A sequencing method based on real-time pyrophosphate. Science 281: 363-365

Saghai Maroof MA, Biyashev RM, Yang GP, Zhang Q, Allard RW (1994) Extraordinarily polymorphic microsatellite DNA in barley: Species diversity, chromosomal locations, and population dynamics. Proc Natl Acad Sci USA 91: 5466-5470

Saiki RK, Bugawan TL, Horn GT, Mullis KB, Erlich HA (1986) Analysis of enzymatically amplified beta-globin and HLA-DQ alpha DNA with allele-specific oligonucleotide probes. Nature 324: 163-166

SanMiguel P, Tikhonov A, Jin YK, Motchoulskaia N, Zakharov D, Melake-Berhan A, Springer PS, Edwards KJ, Lee M, Avramova Z, Bennetzen JL (1996) Nested retrotransposons in the intergenic regions of the maize genome. Science 274: 765-768

Schlötterer C, Pemberton J (1994) The use of microsatellites for genetic analysis of natural populations. In: B Schierwater, B Streit, GP Wagner and R DeSalle Birkhäuser (eds) Molecular Ecology and Evolution: Approaches and Applications. Verlag, Basel, pp 203-214

Schneider K, Weisshaar B, Borchardt DC, Salamini F (2001) SNP frequency and allelic haplotype structure of Beta vulgaris expressed genes. Mol Breed 8: 63-74

Schlötterer C, Tautz D (1992) Slippage synthesis of simple sequence DNA. Nucl Acids Res 22: 285-288

Schulman AH, Flavell AJ, Ellis THN (2004) The application of LTR retrotransposons as molecular markers in plants. Methods Mol Biol 260: 145-173

Scott KD, Eggler P, Seaton G, Rossetto M, Ablett EM, Lee LS, Henry RJ (2000) Analysis of SSRs derived from grape ESTs. Theor Appl Genet 100: 723-726

Sharpe AG, Parkin IAP, Keith DJ, Lydiate DJ (1995) Frequent nonreciprocal translocations in the amphidiploid genome of oilseed rape (*Brassica Napus*). Genome 38: 1112-1121

Shattuck-Eidens DM, Bell RN, Neuhausen SL, Helentjaris T (1990) DNA sequence variation within maize and melon: observations from polymerase chain reaction amplification and direct sequencing. Genetics 126: 207-217

Smith JR, Carpten JD, Brownstein MJ, Ghosh S, Magnuson VL, Gilbert DA, Trent JM, Collins FS (1995) Approach to genotyping errors caused by non-template nucleotide addition by *Taq* DNA polymerase. PCR Methods Appl 5: 312-317

Smýkal P (2006) Development of an efficient retrotransposon-based fingerprinting method for rapid pea variety identification. J Appl Genet 47: 221-230

Somers DJ, Kirkpatrick R, Moniwa M, Walsh A (2003) Mining single-nucleotide polymorphism from hexaploid wheat ESTs. Genome 49: 431-437

Southern EM (1975) Detection of specific sequences among DNA fragments separated by gel electrophoresis. J Mol Biol 98: 503-517

Southern E (1979) Gel electrophoresis of restriction fragments. Meth Enzymol 68: 152-176

Staub JE, Serquen FC (1996) Genetic markers, map construction, and their application in plant breeding. HortScience 31: 729-741

Staub J, Bacher J, Poetter K (1996) Sources of potential errors in the application of Random Amplified Polymorphic DNAs in cucumber. HortScience 31: 262-266

Taillon-Miller P, Gu Z, Li Q, Hillier L, Kwok PY (1998) Overlapping genomic sequences: a treasure trove of single-nucleotide polymorphisms. Genome Res 8: 748-754

Tatout C, Warwick S, Lenoir A, Deragon JM (1999) SINE insertions as clade markers for wild crucifer species. Mol Biol Evol 16: 1614-1621

Tautz D, Schlötterer C (1994) Simple sequences. Curr Opin Genet Dev 4:832-837

Taylor EJA, Konstantinova P, Leigh F, Bates JA, Lee D (2004) Gypsy-like retrotransposons in *Pyrenophora* species: an abundant and informative class of molecular markers. Genome 47: 519-525

Telenius H, Carter NP, Bebb CE, Nordenskjold M, Ponder BJ, Tunnacliffe A (1992) Degenerate oligonucleotide-primed PCR: general amplification of target DNA by a single degenerate primer. Genomics 13: 718-725

Tenaillon MI, Sawkins MC, Long AD, Gaut RL Doebley JF, Gaut BS (2001) Patterns of DNA sequence polymorphism along chromosome 1 of maize (*Zea mays* ssp *mays* L). Proc Natl Acad Sci USA 98: 9161-9166

Thiel T, Michalek W, Varshney RK, Graner A (2003) Exploiting EST databases for the development and characterization of gene-derived SSR-markers in barley (*Hordeum vulgare* L.). Theor Appl Genet 106: 411-422

Tommasini L, Batley J, Arnold G, Cooke R, Donini P, Lee D, Law J, Lowe C, Moule C, Trick M, Edwards K (2003) The development of multiplex simple sequence repeat (SSR) markers to complement distinctness, uniformity and stability testing of rape (*Brassica napus* L.) varieties. Theor Appl Genet 106: 1091-1101

Tsuchihashi Z, Dracopoli NC (2002) Progress in high-throughput SNP genotyping methods. Pharmacogenomics J 2: 103-110

Turner DM, Poles A, Brown J, Arguello JR, Madrigal JA, Navarrete CV (1999) HLA-A typing by reference strand-mediated conformation analysis (RSCA) using a capillary-based semi-automated genetic analyser. Tiss Antigens 54: 400-404

Underhill PA, Jin L, Lin AA, Mehdi SQ, Jenkins T, Vollrath D, Davis RW, Cavalli-Sforza LL, Oefner PJ (1997) Detection of numerous Y chromosome biallelic polymorphisms by denaturing high-performance liquid chromatography. Genome Res 7: 996-1005

Useche FJ, Gao G, Harafey M, Rafalski A (2001) High-throughput identification, database storage and analysis of SNPs in EST sequences. Genome Inform Ser Workshop 12: 194-203 van der Wurff AWG, Chan YL, van Straalen NM, Schouten J (2000) TE-AFLP: combining rapidity and robustness in DNA fingerprinting. Nucl Acids Res 28: 105-109

Varmus H, Brown P0 (1989) Retroviruses. In: DE Berg and MM Howe (eds) Mobile DNA. Am Soc Microbiol, Washington DC, USA, pp 53-108

Viard F, Franck P, Dubois MP, Estoup A, Jarne P (1998) Variation of microsatellite size homoplasy across electromorphs, loci, and populations in three invertebrate species. Mol Evol 47: 42-51

Vicient CM, Jääskeläinen M, Kalendar R, Schulman AH (2001) Active retrotransposons are a common feature of grass genomes. Plant Physiol 125: 1283-1292

Vos P, Hogers R, Bleeker M, Reijans M, van de Lee T, Hornes M, Friters A, Pot J, Peleman J, Kuiper M, Zabeau M (1995) AFLP: a new technique for DNA fingerprinting. Nucl Acids Res 23: 4407-4414

Wahl GM, Stern M, Stark GR (1979) Efficient transfer of large DNA fragments from agarose gels to diazobenzyloxymethyl-paper and rapid hybridization by using dextran sulfate. Proc Natl Acad Sci USA 76: 3683-3687

Wang DG, Fan JB, Siao CJ, Berno A, Young P, Sapolsky R, Ghandour G, Perkins N, Winchester E, Spencer J, Kruglyak L, Stein L, Hsie L, Topaloglou T, Hubbell E, Robinson E, Mittmann M, Morris MS, Shen N, Kilburn D, Rioux J, Nusbaum C, Rozen S, Hudson TJ, Lipshutz R, Chee M, Lander ES (1998) Large-scale identification, mapping, and genotyping of single-nucleotide polymorphisms in the human genome. Science 280: 1077-1082

Wang Z, Weber JL, Zhong G, Tanksley SD (1994) Survey of plant short tandem DNA repeats. Theor Appl Genet 88: 1-6

Watson JD, Crick FHC (1953) Genetical implications of the structure of deoxyribonucleic acid. Nature 171: 964-967

Waugh R, McLean K, Flavell AJ, Pearce SR, Kumar A, Thomas BT, Powell W (1997) Genetic distribution of *BARE-1*-like retrotransposable elements in the barley genome revealed by sequence-specific amplification polymorphisms (S-SAP). Mol Gen Genet 253: 687-694

Weising K, Nybom H, Wolff K, Meyer W (1995) DNA Fingerprinting in Plants and Fungi. CRC Press, Boca Raton, USA

Welsh J, McClelland M (1990) Fingerprinting genomes using PCR with arbitrary primers. Nucl Acids Res 18: 7213-7218

Williams JGK, Kubelik AR, Livak KJ, Rafalsk JA, Tingey SV (1990) DNA polymorphisms amplified by arbitrary primers are useful as genetic markers. Nucl Acids Res 18: 6531-6535

Wright SI, Bi IV, Schroeder SG, Yamasaki M, Doebley JF, McMullen MD, Gaut BS (2005) The effects of artificial selection on the maize genome. Science 308: 1310-1314

Xiong LZ, Xu CG, Maroof MAS, Zhang QF (1999) Patterns of cytosine methylation in an elite rice hybrid and its parental lines, detected by a methylation-sensitive amplification polymorphism technique. Mol Gen Genet 261: 439-446

Ye S, Dhillon S, Ke X, Collins A, Da I (2001) An efficient procedure for genotyping single nucleotide polymorphisms. Nucl Acids Res 29:17:e88

Youil R, Kemper B, Cotton RG (1996) Detection of 81 known mouse beta-globin promoter mutations with T4 endonuclease VII – the EMC method. Genomics 32: 431-435

Zabeau M, Vos P (1993) Selective restriction fragment amplification: a general method for DNA fingerprinting. Intl Publ No WO93/06239. International application published under the patent cooperation treaty (PCT) accessible through esp@cenet database: www.espacenet.com

Zietkiewicz E, Rafalski A, Labuda D (1994) Genome fingerprinting by simple sequence repeat (SSR)-anchored polymerase chain reaction amplification. Genomics 20: 176-183

3 | Mapping Populations: Development, Descriptions and Deployment

Shailaja Hittalmani[1], T. N. Girish[1], Hanamareddy Biradar[1] and Peter J. Maughan[2]*

[1]Department of Genetics and Plant Breeding, University of Agricultural Sciences, GKVK, Bangalore-560 065, India

[2]Department of Plant and Wildlife Sciences, 285 WIDB, Brigham Young University, Provo, Utah 84602, USA

*Corresponding author: Jeff_Maughan@byu.edu

1 INTRODUCTION

Most agronomically important traits (e.g., yield, abiotic and biotic stress tolerance, lodging, etc.) are quantitatively controlled (controlled by the action of many genes). A first step towards cloning these quantitative trait loci (QTL) is to locate them on specific chromosomal regions (locus) via genetic linkage mapping. Once mapped and their phenotypic effect is known, the QTLs can be cloned based on their map position via map-based cloning and candidate gene approaches. Virtually all of the genes that have been cloned and reported in the literature are single genes with qualitative or major effects on a single trait. It has only been in the last few years that researchers have reported the cloning of QTLs, which include QTLs for flowering time (El-Din El Assal et al. 2001; Werner et al. 2005) and root morphology (Mouchel et al. 2004) in Arabidopsis; plant architecture in maize (Doebley et al. 1997); heading time (Yano et al. 2000; Takahashi et al. 2001; Doi et al. 2004) in rice; and fruit dimension in tomato (Cong et al. 2002; Frary et al. 2002; Liu et al. 2002; Fridman et al. 2004). QTL mapping projects often involve screening of large populations consisting of thousands of progenies using polymorphic

DNA markers to generate genotypic data (genotyping) and trait evaluation to generate phenotypic data (phenotyping). Several computer software packages have been developed to aid in the analysis of genetic linkage data and phenotypic data for the identification of major and minor genes (for details see Chapters 6 and 11 in this volume).

Critical to mapping of QTLs is the development of appropriate mapping populations; generation of segregation data across the population(s), including the phenotypic trait and molecular marker data; calculation of multipoint recombination frequencies; and the establishment of gene order and linkage groups. In this chapter, we will describe the most common mapping populations and their advantages and disadvantages for linkage map development and trait analysis.

2   MAPPING POPULATIONS

A population used for genetic linkage mapping and QTL analysis is commonly referred to as a mapping population. Mapping populations are usually obtained from controlled crosses between selected parents. Development of a population for genetic mapping requires the researcher to choose mapping parents and determine a mating scheme for population development. The decision regarding the parents and the mapping population structure should be made based on the objectives of the experiment. Parents of mapping populations must differ genetically for not only the trait(s) of interest, but also for the type of genetic markers to be employed in the map construction. If the mapping parents differ phenotypically for the trait of interest, it is a reasonable expectation that the parents differ genetically. However, the lack of phenotypic variation between parents does not always indicate a lack of genetic variation, as different sets of genes could result in the same phenotype. Genetic variation is essential to accurately map QTLs and to produce dense genetic maps – especially when genetic markers are limited. The more genetic variation that exists, the easier it is to find informative (polymorphic) markers and thus produce saturated genetic maps.

3   SELECTION OF PARENTS

Selection of parents for developing a mapping population is critical to successful map construction. Maximizing genetic diversity between the parents of the mapping populations facilitates the placement of markers on the map and thus the generation of dense genetic maps. However, care must be taken while attempting to maximize this genetic diversity. Interspecific hybrids may exhibit an aberrant chromosome pairing and severely suppressed recombination rates, resulting in reduced linkage distances and distorted linkage disequilibrium (LD). The use of populations developed through hybridization of exotics with adapted material (intraspecific wide crosses) will usually produce segregating populations with high levels of genetic diversity and are often the cross of choice when adapted genetic materials lack genetic polymorphism. Finally, the researcher needs to understand the implications of selecting parents that are true breeding (complete homozygosity at all or most loci) as opposed to segregating (heterozygous at many loci). Inbred parents, often derived from inbreeding species, (e.g., soybeans – *Glycine max*) have a maximum of two alleles per diploid locus, and each line is homozygous at all loci. First filial (F_1) progeny from these crosses will be heterozygous for all polymorphic loci and all markers are equally informative, with no among-family variance being present in the segregating progeny population. Inbred parents will breed true and therefore lend themselves to multi-year, multi-environment replicated experiments. In contrast, heterozygous diploid parents, which is usually the situation in allogamous (outcrossing) autotetraploid parents that do not tolerate inbreeding (e.g. alfalfa – *Medicago sativa*), can have a maximum of four alleles per locus, each line may be highly heterozygous for alternating alleles, F_1 progeny will be a mix of segregating alleles (showing genetic ratios of 1:2:1, 1:1 or fixed at different loci) and markers in the segregating population will vary for information content based on family structure. Thus, not only is genetic mapping in populations derived from heterozygous parents computationally more difficult, but also the parents and the populations may not readily lend themselves to multi-year or multi-environment experimentation.

4   POPULATIONS DERIVED FROM HOMOZYGOUS PARENTS

Common mapping populations derived from homozygous parents and used in linkage mapping experiments are: (i) F_2 populations; (ii) F_2-derived F_3 (F_2:F_3) populations; (iii) backcrosses; (iv) doubled haploids (DHs); (v) recombinant inbred lines (RILs); and (vi) near-isogenic lines (NILs). The characteristic features, advantages and disadvantages of each of these populations are briefly presented below.

4.1 F_2 Population

$P_1 \times P_2 \rightarrow F_1$ (self pollinate or intercross with identical F_1) $\rightarrow F_2$ population

True-breeding parents (P_1 and P_2) with contrasting traits are crossed giving rise to F_1 progeny, which are then selfed to produce a segregating F_2 population. F_2 individuals are the products of a single meiotic cycle and are expected to follow normal Mendelian segregation patterns for the marker alleles. Dominant markers should segregate in a 3:1 ratio (dominant:recessive alleles), while co-dominant markers should segregate in a 1:2:1 ratio. A completely described F_2 population is the most informative, with two gametes providing information in each individual (i.e., both the male and female gametes are subject to recombination). If, for instance, two co-dominant loci are considered, nine genotypes are obtained in the F_2 (Figure 1) and the best estimate of recombination (r) is calculated by the method of maximum likelihood. Other likelihood equations are used if there is dominance for one of the two loci (six possible genotypic classes in the F_2 population), or dominance for both loci (four possible genotypic classes).

Advantages

- Easy to develop in plants which don't suffer large effects of inbreeding depression;
- Short development time. An F_2 population is developed in two generations;
- Segregation of all possible genotypic combination (including both homozygotes and the heterozygote) permits the inference of the mode of gene action at each locus (dominant, recessive, additive, or incomplete dominance).

Disadvantages

- Linkage established using F_2 population is based on a single meiotic event;
- F_2 populations are segregating and have limited life spans unless they are asexually propagated;
- Interspecific F_2 populations are often limited in size and are subject to selection (distorted segregation) due to seedling vigor and fertility (pre-zygotic/post-zygotic selection);
- F_2 populations do not lend themselves to replicated experiments, since each F_2 individual is genetically unique and cannot be evaluated in replicated experiments. As a consequence, G × E interactions on the expression of quantitative traits cannot be precisely estimated;

- F_2 populations have limited use for high density mapping experiments, since limited amount of tissue (DNA) may be available from each F_2 individual.

A traditional argument against the use of F_2 populations in basic genetic studies is the difficulty in distinguishing whether heterozygotes at consecutive marker loci represent double-parentals or double-recombinants. The use of maximum likelihood algorithms, however, alleviates this problem and is incorporated into most genetic mapping programs (e.g., Mapmaker, JoinMap).

4.2 F_2-derived F_3 ($F_{2:3}$) Population

$P_1 \times P_2 \rightarrow F_1$ (self) $\rightarrow F_2$ (self pollinate each F_2 plant) $\rightarrow F_{2:3}$ population

$F_{2:3}$ populations are obtained by selfing F_2 individuals. The F_3 families are termed F_2-derived F_3 ($F_{2:3}$) families, since each family pedigree traces back to a single F_2 individual. Since each F_2 individual can produce a large $F_{2:3}$ family (depending on the seed yield from the F_2 individual) replicated experiments are possible. Thus, $F_{2:3}$ populations are suitable for mapping of quantitative traits and qualitative traits where replication is needed for accurate phenotypic scoring. The most common use of the $F_{2:3}$ mapping population is, therefore, to provide replication and a more precise estimation of the trait phenotype. In practice, $F_{2:3}$ lines consisting of 20-40 plants are planted in each replication of the experiment and an average value across the replicate (averaged across all plants in the family) is used to estimate the phenotypic value. It is important to note that plants within an $F_{2:3}$ family are not homogeneous since polymorphic loci in the original F_2 plant segregate independently to each $F_{2:3}$ family member, but when averaged (pooled) they should theoretically reflect the genotype of the original F_2 plant. Similarly, genetic mapping can be performed using DNA derived from the F_2 plants themselves or from pooled DNA of several (usually between 8-12) individuals in the $F_{2:3}$ family. Based on the genotype of each F_2 plant a linkage map can be constructed. Co-dominant markers should segregate in a 1:2:1 ratio, while dominant markers should segregate in a 3:1 ratio.

Advantages

- $F_{2:3}$ populations can be used in replicated experimental designs that facilitate QTL mapping experiments. Thus, G × E interactions can be estimated;
- $F_{2:3}$ populations can be developed in only three generations;
- DNA bulks from $F_{2:3}$ lines can be used to reconstitute the F_2 DNA.

Disadvantages

- $F_{2:3}$ populations, like F_2 populations, are segregating and have limited life spans. Hence, the source of tissue from which to isolate DNA will become exhausted at some point;
- Care must be taken to accurately pool equal amounts of DNA to reconstitute the original F_2 plant genotypes;
- The number of plants needed to reconstitute the true segregation of a quantitative trait in an $F_{2:3}$ family will vary according to the number of genes affecting the trait and their relative contribution to the overall phenotype.

4.3 Backcross (BC) Population

$P_1 \times P_2 \rightarrow F_1 \times RP$ (Recurrent Parent = P_1 or P_2) $\rightarrow BC_1$

Backcross populations are generated by (back) crossing F_1 individual(s) with one of the two inbred parents used in the initial cross (termed the recurrent parent or RP). The choice of which parent (P_1 or P_2) to use as the RP is usually determined by the gene action (dominant or recessive) of the trait being studied, or for QTL studies by which parent represents the most-adapted genotype. For recessive traits, the backcross is usually done with the recessive parent (also called a testcross). With respect to dominant and co-dominant molecular markers, a backcross with the 'dominant' parent will segregate in a ratio 1:0 and 1:1 (dominant:recessive), respectively. In contrast, a backcross with the recessive parent would segregate in a ratio of 1:1, irrespective of the nature of the marker (Figure 1). The expectations of this population are exactly the same as for a doubled haploid population (section 4.4).

If the recurrent parent is a homozygous inbred line, then mapping is done on a gametic basis, since the only recombination events come from the gametes provided by the F_1 and, therefore, the linkage data is based on only one parent (the F_1). The F_1 will produce four types of gametes, which will lead to the segregation observed in the BC progeny. The proportion of recombinant individuals can be counted to estimate r and to determine gene order. A unique problem arises if dominant markers are used in the linkage analysis of backcross populations. The parental and recombinant progeny classes can only be distinguished if the dominant alleles are derived from the F_1 hybrid. If the dominant alleles are derived from the recurrent parent they will mask any recombinant gametes from the hybrid parent.

Backcross populations have been especially useful for the introgression of exotic germplasm from wild relatives (e.g., landraces,

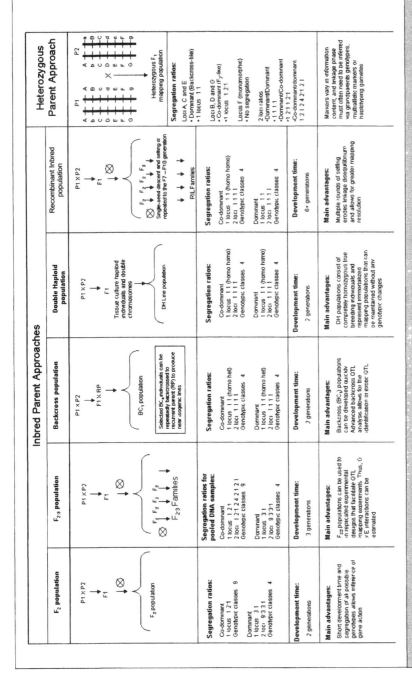

Fig. I Comparison of different mating types for different pedigree designs in control crosses with inbred parents.

weedy species, often referred to as the donor parent) into adapted material. Advanced backcross QTL (AB-QTL) analysis, first described by Tanksley and Nelson (1995), utilizes negative selection during the development of advanced backcross populations (populations that are advanced to BC_2, BC_3, etc.) to reduce the frequency of deleterious alleles and linkage drag around targeted QTLs. QTL analysis in these populations has identified numerous cryptic alleles derived from exotic germplasm that have beneficial effects in the adapted material. In brief, the advantages of AB-QTL analysis are i) genotypes (and phenotypes) of the BC individuals are more similar to the elite recurrent parent, making measurements of quantitative traits more meaningful; ii) by delaying the QTL analysis, the researcher can apply negative selection to the population to reduce the frequency of deleterious alleles from the donor parent; and iii) epistatic interactions are less likely to be detected in advanced BC populations due to the overall lower frequency of donor alleles in individual BC families – thus the QTLs that are detected are likely to have similar effects once transferred to the cultivated material (Tanksley et al. 1996; Xiao et al. 1996; Ming et al. 2002).

Advantages

- Backcross (BC_1) populations can be developed in two generations. Advanced BC populations require only one or two additional generation(s) of breeding;
- BC_1 populations have more individuals in the recessive trait class (50%) than do F_2 population (25%). Increased class size has a direct effect on the power of the statistical test (e.g., regression analysis) between phenotypic class means and marker genotypic classes;
- Scoring of marker data is simpler as only two genotypic classes exist (heterozygotes and one homozygote class);
- Once QTLs have been identified in a backcross population, additional backcrossing assisted with marker data can quickly generate near-isogenic lines (NIL) with the putative QTLs facilitating the verification of the QTL in replicated field studies.

Disadvantages

- The backcross populations are not 'immortal'. Like F_2 and $F_{2:3}$ populations BC populations have limited life spans. Hence, the source of tissue from which DNA is to be isolated will become exhausted at some point in species that cannot be propagated asexually;

- The mode of gene action that accounts for a particular QTL cannot be determined, since only two of the three possible genotypes are present;
- The effects of recessive donor parent alleles cannot be determined.

4.4 Doubled Haploid Lines (DHL)

$P_1 \times P_2 \to F_1$: Tissue culture haploid individuals and treat with colchicine $\to$ DHL

A doubled haploid (DH) population is generated via chromosome doubling of F_1 anther- or microspore-derived haploid plants. Doubled haploids are used to make completely homozygous recombinant lines in a single generation, eliminating several generations in the breeding process. Indeed, lines generated by the DH process may be placed immediately into replicated field trials. The necessary culture conditions, and therefore the utility of the approach, are often taxon-specific – nonetheless, DH populations do exist for a number of crop species, including maize (*Zea mays*, Murigneux et al. 2004), barley (*Hordeum vulgare*, Devaux et al. 1995), cotton (*Gossypium hirsutum*, Zhang et al. 2002), pepper (*Capsicum annuum*, Lefebvre et al. 1995), and *Brassica oleracea* (Sebastian et al. 2004) and have been used for genetic linkage map construction. Genetically, individual plants within a DH population contain two identical gametes, and therefore each contains only one uniquely informative gamete. Thus, DH lines are equivalent to first-generation backcross individuals, in terms of their recombinant information content. The expected ratio for the segregating genetic marker is 1:1, irrespective of genetic nature of the marker (whether dominant or co-dominant; Figure 1). The genetic nature of the markers makes no difference since no heterozygous individuals exist in the population.

Advantages

- A fixed homozygous state is reached in only a few generations;
- Since DH populations consist of completely homozygous, true-breeding individuals, they represent an immortalized mapping population that can be maintained without any genotypic changes;
- DH seeds can easily be multiplied to facilitate replicated field trials and the dissemination of populations to other researchers;
- Scoring of genetic marker data is simpler, since only the two homozygous genotypic classes exist.

Disadvantages

- Recombination is only estimated from the parent from which the haploids are developed;
- Crop specific tissue culture expertise is required for developing DH lines;
- Tissue culturing and haploid production methods are often species- or even genotype-specific. Only a limited number of proven DH methods have been reported in the literature;
- Over-dominance cannot be measured in DH populations as heterozygotes do not exist;
- Gene action at specific QTLs cannot be measured, since all loci are homozygous;
- Development of DH mapping populations is costly as compared to other mapping populations. The production of DH populations is more costly than that of an F_2, $F_{2:3}$ or BC population, since they require specific laboratory facilities (tissue culture);
- Potential for tissue-culture induced (gametoclonal) variation or pollen selection during the tissue culturing process.

4.5 Recombinant Inbred Lines (RIL)

P1 × P2 → F1 → F2 → F3 … F7 → RIL
  (self-pollinate from F_1 to RIL)

Recombinant inbred (RI) lines are developed by single-seed descent from selfed F_2 plants. Single-seed descent and selfing is repeated through the F_7 – F_{10} generations, at which point the RILs are true-breeding and fixed (homozygous mosaics of the parental genomes). Thus an RI population consists of nearly homozygous lines that can be multiplied to produce large quantities of seed for each individual RIL and are ideal for QTL studies, where replication across different environments and/or years is needed for accurate phenotyping. Apart from single-seed descent method, bulk and pedigree breeding methods without selection can also be used to generate RI populations.

Since RI populations are produced by multiple rounds of selfing, each RIL represents multiple meiotic events. Thus recombination breakpoints in RILs are denser, which leads to greater mapping resolutions than those that occur in any one meiosis event (e.g., F_2). Segregation ratios of genetic markers in RI populations for both dominant and co-dominant markers should be 1:1 (Figure 1). When RI populations are used for mapping, the probability of fixing a recombinant genotype (R) is higher than that experienced in other crosses where only single meiosis is represented (r), as there are several chances for recombination with each successive generation of inbreeding. Indeed, R is not an estimate of r, although they are related mathematically as $R = 2r/(1 + 2r)$ (Taylor 1978). Ultimately, the use of an RI population results in an expansion of map distance for linked markers, which may present some difficulties in the initial stages of mapping since more markers will be needed to establish linkage groups. However, this expansion of map distances also gives an advantage over a conventional segregating mapping population since recombination between closely linked markers is more readily detected (Burr et al. 1988).

RILs are the recombinant output from which superior stabilized segregants can be directly used as breeding lines. They serve as a means for obtaining recombinants with desirable traits from both the parents involved in the cross and also transgressive segregants of superior quality. Therefore, they offer several unique advantages over other types of mapping populations and find wide usage in both breeding and molecular work (Girish et al. 2006).

RI populations can be used to create immortalized F_2 (IF_2) populations. IF_2 populations are developed by paired crossing of randomly chosen RILs in all possible combinations excluding reciprocals. The set of RILs used for crossing, along with the F_1s produced, provide a true representation of all possible genotype combinations (including the heterozygotes) expected in the F_2 of the cross from which the RILs were derived. The RILs can be maintained by selfing and the required quantity of F_1 seed can be produced at will by new hybridizations. The IF_2 population, therefore, provides an opportunity to map heterotic QTLs as well as determine dominance interaction among alleles (Gardiner et al. 1993).

Advantages

- RI populations are inexpensive to develop for self-pollinated crops;
- Genotyping needs to occur only once for each RIL and can be done in any generation once homozygosity is reached;
- RILs can be propagated indefinitely without genetic change;
- RILs can reproduce themselves for replicated experiments;
- The degree of dominance for genes linked to markers can be estimated;

- RILs are easily disseminated to other researchers (as is the genotypic information for each RIL);
- Multiple rounds of selfing erode linkage disequilibrium and allows for greater mapping resolution;
- IF$_2$ populations derived from RI populations provide a unique opportunity to study heterosis in replicated field trials (multi-environments and -year tests).

Disadvantages

- Developing RILs from outcrossing species with high inbreeding depression is difficult, if not impossible;
- Interspecific or wide cross RI populations may experience inadvertent selection due the segregation of deleterious alleles from the unadapted germplasm, creating distorted segregation ratios at some marker/gene loci.

4.6 Near-Isogenic Lines (NIL)

RP × DP → F$_1$ × RP → BC$_1$ (If dominant trait, apply selection in the BC$_1$)
$\downarrow$ (If recessive trait, self and apply selection in the BC$_1$F$_2$)

Selected BC$_N$ individuals are backcrossed to recurrent parent (RP).
$\downarrow$ (Cycle repeated for 7 to 10 generations)

NILs

The objective of NIL development is to recover a line that is genetically nearly identical to the RP, except for the one or few gene(s) being transferred from the donor parent (DP). Most NILs are produced by repeated backcrossing of hybrid progeny to one of the original parents (the RP). The parent contributing the trait [i.e., gene(s)]-of-interest is the DP and is used only in the initial cross. All subsequent backcrosses are between the hybrid BC progeny and the RP. For a single polymorphic locus, only heterozygotes and homozygotes-RP genotypes (in a 1:1 ratio; Figure 1) are present in the hybrid BC$_N$ (where N represents backcross generation) individuals. Hence, phenotypic screening of the hybrids is needed to ensure the selection of a heterozygous BC hybrid and the subsequent maintenance of the gene-of-interest through the backcrossing process. If the trait is controlled by a dominant allele and the phenotype can be evaluated prior to flowering (e.g., disease resistance), selected heterozygous BC$_N$ hybrids are directly crossed to the RP. If, however, the trait is controlled by a recessive allele, a progeny test must be applied to identify the heterozygous hybrids. In brief, each BC$_N$ hybrid is crossed to the RP to produce a new BC generation (BC$_{N+1}$) and is also allowed to self-pollinate to produce a segregating F_2 population ($BC_N F_2$). The $BC_N F_2$ population is used for the progeny test to determine which BC_N individuals were heterozygous and thus which BC_{N+1} population to proceed forward with in the NIL development.

In the absence of selection, the average percentage of genes from the DP decreases with each backcross by half and can be determined by the equation $(\frac{1}{2})^{n+1}$, where n is backcross generation number. Thus, after 6 rounds of backcrossing the $BC_6 F_1$ population should have on an average 0.8% DP and 99.2% RP genome content and can be accurately termed a near-isogenic line with the RP. Once the desired level of isogenicity between the RP and the NIL is achieved, the NIL is selfed and true-breeding homozygous lines are selected. At this point, the RP and its NIL are essentially identical at all loci except for the selected gene(s) and the immediate region surrounding the selected gene(s) – which are also maintained as the DP genotype due to linkage drag. From a genetic mapping perspective any polymorphisms detected between the NIL and its recurrent parent are likely to be near the selected locus.

A second method for creating NILs is through the use of single seed descent (SSD) and selection to maintain heterozygosity at the gene-of-interest. After 7-10 generations of SSD, heterozygous lines for the gene-of-interest are selfed and contrasting true breeding, homozygous sister-lines are selected. In this method, the contrasting homozygous lines are isogenic sisters (genetically identical, except for the gene-of-interest), and genomically are a mosaic of the DP and RP genomes. Similar to a traditional BC-generated NIL, any polymorphisms detected between the sister-line NILs are likely to be linked to the selected gene.

Advantages

- NILs are 'immortal';
- Many NILs (and their corresponding RP and DP) already exist in germplasm collections of cultivated species;
- Increased backcross generations reduce linkage drag and facilitate tight mapping of the gene-of-interest.

Disadvantages

- Transfer of multiple genes via the NIL process requires large number of backcross progenies;
- Detectable genetic diversity must exist between the RP and DP to detect linkage between the marker and allele of interest;
- NIL development can be time consuming, especially if recessive or multiple loci are to be transferred via the backcross method;

- NILs are directly useful only for molecular tagging of the gene concerned. Additional information is needed to produce linkage distance information;
- Complete linkage maps cannot be produced from NILs.

5 CONSIDERATIONS FOR QTL MAPPING FROM HETEROZYGOUS PARENTS

The aforementioned approaches to generate mapping populations are generally applicable only to species that are self-compatible and are capable of producing homozygous pure lines. In species that are self-incompatible or suffer large effects due to inbreeding depression, the development of mapping populations is complicated by the required use of heterozygous parental lines. These complications include addressing the genotype of each parent at each locus, measuring disequilibrium across single families (*versus* whole populations) and the complications associated with the segregation of up to four alleles at a single genetic locus.

5.1 Genetic Load, Heterozygosity and Mating Type

Genetic load (and therefore inbreeding depression) tends to be higher for outcrossing species and makes the development of inbred lines nearly impossible. As a consequence of the enforced outcrossing, higher levels of heterozygosity are common in many outcrossing species. In homozygous inbred crosses, mating configuration refers to all loci (all polymorphic loci are heterozygous and identical in the F_1 generation), whereas mating configuration in an outcross pedigree refers to each individual locus. Indeed, individual loci in the F_1 progeny genotypes may show intercross (1:2:1), backcross (1:1) or be fixed depending on the genotype of the parents (Figure 1). For a marker to be informative, one or both parents must be heterozygous (i.e., $A_1A_2 \times A_2A_2$, $A_1A_1 \times A_1A_2$, or $A_1A_2 \times A_1A_2$) at the marker locus. If two loci are considered, there are 81 possible mating configurations for a two-locus co-dominant model; however, only 17 of these provide information about linkage and only 7 of 9 mating configurations provide linkage information for dominant makers (Liu 1998). Using maximum likelihood methods, recombination estimates (r) and linkage order can be estimated for all mating configurations (Ritter et al. 1990).

Linkage phase must also be determined for each pair of loci to obtain accurate estimates of recombination. This is not as easy as with controlled inbred crosses where linkage phase can be inferred simply by the genotype of the parents. When co-dominant markers are tightly linked, the linkage phase is easily determined based on the predominant progeny type. In loose linkages (> 0.30 recombination frequency), linkage phase is much more difficult to determine, since no predominant progeny class exists. Liu (1998), however, stated that "it is not critical to determine linkage phase when linkage is loose, because the loci will be located far away on the genome", and that loosely linked loci have little to no effect on the order or multipoint map distance measures.

However, in mixtures of dominant and co-dominant markers, consideration of linkage phase is needed. Consider the case where both parents are double heterozygotes for two dominant loci linked in coupling (AB/ab × AB/ab) or repulsion (Ab/aB × Ab/aB) phase. In these situations, linkage phase cannot be distinguished from the distribution of phenotypes of the population. Thus, a distribution in which [AB] = 0.51, [Ab] = 0.24, [aB] = 0.24 and [ab] = 0.01 gives r = 0.2 if there is repulsion in the two parents, or r = 0.04 if there is coupling for one parent and repulsion for the other. In these situations, haplotype data from the parental genotypes can be used to accurately determine linkage phase. Parental haplotypes can be determined by i) use of grandparent genotype data; ii) tracking multiple alleles over generations (from parent to segregating progeny); and iii) direct marker haplotyping of the parental gametes. A full review of the linkage phase models in mixtures of selfs and random mating populations is given by Liu (1998).

This higher level of heterogeneity of mating configurations makes QTL mapping less efficient and more complex. The increase in heterozygosity (more informative loci), however, can compensate for some of the loss of power due to the lack of inbred parental lines.

6 ESTIMATION OF ADDITIVITY AND DOMINANCE

The degree of dominance for genes linked to markers can be estimated only from the F_2 populations with co-dominant markers. This is due to the fact that only the F_2 population contains all three possible genotypes at a single locus (both homozygotes and the heterozygote classes). In DH and RI lines, additivity (a) can be estimated, but not dominance. In RILs, recombination between the marker and the QTL will reduce the estimated additive effect more rapidly than with other types of populations; however, several individuals can be analyzed for each genotype, as is the case in DHs. In an F_2 population, half the individuals are expected to be heterozygous at a given locus and therefore are not informative for the estimation of additivity (a). Hence, for estimates of additivity, the power of statistical tests (from populations of equal size) should be clearly superior in DH or RI populations, since more individuals are present in each of only two genotypic classes. This gain, however, seems to be partly negated by the fact that the within-class variation is higher for these populations because of the absence of heterozygotes at other QTLs. When the phenotypic value of the F_2 individuals is estimated by the average of their F_3 offspring, as is often done, the degree of dominance estimated will be half the real degree of dominance, because the progeny of a heterozygous F_2 genotype comprises only 50% heterozygotes.

7 MAPPING ACCURACY

For a given population size, the accuracy of estimation of r differs according to the type of population used. The variance of r can be calculated from the likelihood equation and is equal to the opposite of the inverse of the second derivative of the likelihood equation (Fisher 1937):

$$Var(r) = -dr^2/d^2(\ln L)$$

and is classically written in the form:

$$Var(r) = 1/Ni_r$$

Where, N is the number of individuals in the population, and i_r is the 'individual information' (i.e., the gain in precision contributed by each individual). The value i_r depends on the type of population as well as the inheritance (dominance vs. co-dominance). Allard (1956) calculated the variances of r for all of the common mapping populations. Summarized below are Allard's conclusions:

- The F_2 population with co-dominant loci provides the most accurate estimations of r.
- The r values estimated from DH and backcross populations have a variance about twice those from F_2 populations for r values < 0.15. This is likely because only one effective meiosis is at the origin of these populations and not two as with the F_2 population. Thus, to obtain a given precision, twice as many individuals are needed in a BC or DH population when compared to an F_2 population (especially for dense linkage maps).
- When r is very small, the proportion of F_2 individuals having received a recombined gamete from one parent is about $2r$, which is about twice the chance of observing a recombination between nearby loci in an F_2 population as in DH or BC populations.

However, for larger intervals, the increased possibility of multiple crossovers progressively blurs this difference, thus, bringing the values of variances closer.

- RI populations approach the precision of F_2 populations only for small values of r (lower than ~0.05). Interestingly, however, as r increases, its variance increases quickly, and at $r = 0.45$ is three times higher than that of the F_2 population. This rapid increase in variance at larger r values occurs because of the successive generations of self-pollination and additional crossovers.

- In an F_2 population, if there is dominance for one of two loci, the variance of r is nearly twice that observed when both loci are co-dominant. If there is dominance for the two loci, the situation differs greatly according to the linkage phase. In coupling, the variance is practically the same as in the preceding case for low values of r and then diverges for the high values. In repulsion, the variance is high when r is small. When $r = 0$, the variance is as high as the variance observed when $r = 0.5$ with the RI population, since the distribution of genotypic frequencies expected for independent assortment or total linkage are very similar (Table 1). Thus, mapping in an F_2 population with dominant markers is clearly less precise than with co-dominant markers (with all other things being equal).

- In the case of populations developed from heterozygous parents, mating configurations of the types AB/ab × Ab/ab and Ab/aB × Ab/ab are similar to that described for dominant markers in repulsion in an F_2 population and mating configurations of the type $A_1B_1/A_2B_2 \times A_1B_1/A_3B_3$ (multiple alleles) are exactly twice as precise as those described for a BC population, at all values of r.

Table I Segregation in an F_2 population when both loci are dominant. For each locus, allele A and B are dominant over the alleles a and b, respectively. The genotypes are indicated in brackets.

Population Type	Linkage Phase	
	Coupling	Repulsion[a]
F_1	AB/ab	Ab/aB
F_2 (independence)	9[A_B_]:3[A_bb]:3[aaB_]: I[aabb]	9[A_B_]:3[A_bb]:3[aaB_]: I[aabb]
F_2 (total linkage)	3[A_B_]:I[aabb]	2[AaBb]:I[AAbb]:I[aaBB]

[a]In repulsion phase, the estimation of the linkage is not precise, since the ratios expected for independence and complete linkage are very similar.

7.1 Segregation Distortion in Linkage Mapping

Significant deviation from expected segregation ratios for a given marker is referred to as segregation distortion. Segregation distortion of markers may be a result of random chance or the result of linkage disequilibrium with genes that ultimately reduce or enhance the viability of the gamete and/or zygote (meiotic drive). The presence of clusters of markers skewed to one parent or the other is suggestive of chromosomal regions containing such gametic or zygotic factors (Lu et al. 2002). Alternatively, distorted clusters of markers may be associated with genes conferring a selective advantage for specific genotypes under the particular growing conditions utilized to produce the mapping populations (post-zygotic viability). Similarly, differential responses of specific genotypes to tissue culture may also cause segregation distortion during DH population development.

While segregation distortion is generally believed to be greater in interspecific crosses, reaching levels as high as 68.5% (Paterson et al. 1988), levels can also be high in intraspecific crosses. For example, Hall and Willis (2005) observed similar levels of distortion (nearly 50%) in both interspecific and intraspecific crosses, an observation attributed to the high level of genomic divergence between the parents of the intraspecific cross. Thus, the extent of segregation distortion appears to be only indirectly related to the type of cross, and more directly related to the extent of genome divergence between the lines being crossed. Chi-square analyses are used to test for segregation distortion and are standard statistical subroutines in many mapping programs (e.g., JoinMap; Ooijen and Voorrips 2001).

8 COMBINING MARKERS AND POPULATIONS

Genetic segregation ratios for specific types of markers are determined by the nature of the marker (dominant/co-dominant) and the structure of the mapping population. A thorough understanding of the nature of genetic markers and their expected segregation ratio within specific mapping populations is crucial for correctly analyzing linkage data. Markers such as restriction fragment length polymorphism (RFLP), and microsatellites (also termed simple sequence repeat or SSR) show principally co-dominant inheritance, while amplified fragment length polymorphism (AFLP) and random amplified polymorphic DNA (RAPD) are usually inherited in a dominant fashion. Recombinant inbred (RI) and doubled haploid (DH) mapping populations equalize marker

Table 2 General characteristics for different genetic markers types in different mapping populations.

| Marker type[a] | Nature | Polymorphism | Expertise/ Equipment needed | Cost | Segregation Ratio | | | | |
					F_2 & $F_{2:3}$	RIL	DH	NIL	BC_1
RFLP	Co-dominant	Medium	Medium-high	Medium	1:2:1	1:1	1:1	1:1	1:1
RAPD	Dominant[b]	Medium-high	Low	Low	3:1	1:1	1:1	1:1	1:0
AFLP	Dominant[b]	Medium-high	Medium-high	Medium	3:1	1:1	1:1	1:1	1:0
SSR	Co-dominant	High	Medium	Medium[d]	1:2:1	1:1	1:1	1:1	1:1
SNP	Co-dominant	Medium	Medium[c]	Low-Medium[cd]	1:2:1	1:1	1:1	1:1	1:1

[a]Marker type abbreviations: RFLP = restriction fragment length polymorphism; RAPD = Random amplified polymorphic DNA; AFLP = Amplified fragment length polymorphism; SSR = Simple sequence repeat (also termed microsatellite); SNP = Single nucleotide polymorphism
[b]Infrequently co-dominant due to indel polymorphisms
[c]Several new automated detection systems are available that simplify SNP analysis and can dramatically reduce cost
[d]SSR and SNP markers are based on sequence information and initially can be expensive to develop

type because these populations consist of fixed (homozygous) lines, resulting in 1:1 segregation ratios at marker loci irrespective of the genetic nature of markers. F_2 mapping populations segregate in 1:2:1 ratio for a co-dominant marker and in 3:1 ratio for a dominant marker. Depending upon the segregation pattern, statistical analysis of marker data will vary. A summary of specific genetic marker characteristics in different mapping populations is given in Table 2.

Conclusions

We have described the most common mapping populations and their advantages and disadvantages for linkage map development and trait analysis. The cost and time commitment associated with population development underscores the need for careful evaluation of several aspects of the population development, including parent selection, breeding strategy, and the need for replicated phenotypic trait measurement. The lack of polymorphism between the parents and the mortal nature of the mapping population, which can effectively limit replicated analysis for phenotypic trait measurement, are two important limitations that each investigator should explore prior to investing time and resources into the development of a mapping population. While molecular genotypes are independent of the environment, trait phenotypes can be heavily influenced by the environment (G × E interactions), particularly in cases of quantitatively controlled traits. Therefore, it becomes imperative to accurately estimate the trait value by evaluating the phenotypes in replicated field trials. Immortalized mapping populations (RI, IF_2, DH) are best suited for the genetic mapping of quantitative traits. The investigator also needs to understand the types and availability of molecular markers for his/her particular species. Almost all sequence-independent markers (e.g., RAPD and AFLP) are dominantly inherited and have specific limitations. Sequence-based markers (e.g., microsatellites and SNPs), while co-dominant and more informative, are costly and time-consuming to develop. It should be obvious that molecular geneticists and breeders may differ in their selection of the most appropriate mapping population since their end objectives may differ. For example, molecular geneticists often wish to proceed to map-based cloning, whereas plant breeders may only need linked genetic markers to proceed to marker-assisted selection. Regardless, a careful consideration of all desired outcomes and the limitations of specific population types is needed to ensure the full utility of a population and successful completion of a linkage mapping project.

References

Allard RW (1956) Formulas and tables to facilitate the calculation of recombination values in heredity. Hilgardia 24: 235-278

Burr B, Burr FA, Thompson KH, Albertson MC, Stuber CW (1988) Gene mapping with recombinant inbreds in maize. Genetics 118: 519-526

Cong B, Liu J, Tanksley SD (2002) Natural alleles at a tomato fruit size quantitative trait locus differ by heterochronic regulatory mutations. Proc Natl Acad Sci USA 99: 13606-13611

Devaux P, Kilian A, Kleinhofs A (1995) Comparative mapping of the barley genome with male and female recombination-derived, doubled haploid populations. Mol Gen Genet 249: 600-608

Doebley J, Stec A, Hubbard L (1997) The evolution of apical dominance in maize. Nature 386: 485-488

Doi K, Izawa T, Fuse T, Yamanouchi U, Kubo T, Shimatani M, Yoshimura A (2004) *Ehd1*, a B-type response regulator in rice, confers short-day promotion of flowering and controls *FT-like* gene expression independently of *Hd1*. Genes Dev 18: 926-936

El-Assal SED, Alonso-Blanco C, Peeters AJM, Raz V, Koornneef M (2001) A QTL for flowering time in *Arabidopsis* reveals a novel allele of *CRY2*. Nat Genet 29: 435-440

Fisher RA (1937) The Design of Experiments, 2nd edn. Oliver and Boyd, Edinburgh, London, UK, 260 p

Frary A, Nesbitt TC, Frary A, Grandillo S, van der Knaap E, Cong B, Liu J, Meller J, Elbere R, Alpert KB, Tanksley SD (2000) Fw2.2: A quantitative trait locus key to the evolution of tomato fruit size. Science 289: 85-88

Fridman E, Carrari F, Liu Y-S, Fernie AR, Zamir D (2004) Zooming in on a quantitative trait for tomato yield using interspecific introgressions. Science 305: 1786-1789

Gardiner JM, Coe EH, Melia-Hancock S, Hosington DA, Chao S (1993) Development of a core RFLP map in maize using an immortalized F_2-population. Genetics 134: 917-930

Girish TN, Gireesha TM, Vaishali MG, Hanamareddy BG, Hittalmani S (2006) Response of a new IR50/Moroberekan recombinant inbred population of rice (*Oryza sativa* L.) from an indica x japonica cross for growth and yield traits under aerobic conditions. Euphytica 152: 149-161

Hall MC, Willis JH (2005) Transmission ratio distortion in intraspecific hybrids of *Mimulus guttatus*: implications for genomic divergence. Genetics 170: 375-386

Lefebvre V, Palloix A, Caranta C, Pochard E (1995) Construction of an intraspecific integrated linkage map of pepper using molecular markers and doubled-haploid progenies. Genome 38: 112-121

Liu J, Eck JV, Cong B, Tanksley SD (2002) A new class of regulatory genes underlying the cause of pear-shaped tomato fruit. Proc Natl Acad Sci USA 99: 13302-13306

Liu BH (1998) Statistical Genomics: Linkage, Mapping, and QTL analysis. CRC Press, LLC, Boca Raton, Florida, USA

Lu H, Romero-Severson J, Bernardo R (2002) Chromosomal regions associated with segregation distortion in maize. Theor Appl Genet 105: 622-628

Ming R, Del Monte TA, Hernandez E, Moore PH, Irvine JE, Paterson AH (2002) Comparative analysis of QTLs affecting plant height and flowering among closely-related diploid and polyploid genomes. Genome 45: 794-803

Mouchel CF, Briggs GC, Hardtke S (2004) Natural genetic variation in *Arabidopsis* identifies BREVIS RADIX, a novel regulator of cell proliferation and elongation in the root. Genes Dev 18: 700-714

Murigneux A, Baud S, Beckert M (2004) Molecular and morphological evaluation of doubled-haploid lines in maize. 2. Comparison with single-seed-descent lines. Theor Appl Genet 87: 278-287

Ooijen JW van, Voorrips RE (2001) JoinMap 3.0, software for the calculation of genetic linkage maps. Plant Res Intl, Wageningen, The Netherlands

Paterson AH, Lander ES, Hewitt JD, Peterson S, Lincoln SE, Tanksley SD (1988) Resolution of quantitative trait into Mendelian factors by using a complete linkage map of restriction fragment length polymorphisms. Nature 335: 721-726

Ritter E, Gebhardt C, Salamini F (1990) Estimation of recombination frequencies and construction of RFLP linkage maps in plant from crosses between heterozygous parents. Genetics 125: 645-654

Sebastian RL, Howell EC, King GJ, Marshall DF, Kearsey MJ (2004) An integrated AFLP and RFLP *Brassica oleracea* linkage map from two morphologically distinct doubled-haploid mapping populations. Theor Appl Genet 100: 75-81

Takahashi Y, Shomura A, Sasaki T, Yano M (2001) *Hd6*, a rice quantitative trait locus involved in photoperiod sensitivity, encodes the α subunit of protein kinase CK2. Proc Natl Acad Sci USA 98: 7922-7927

Tanksley SD, Nelson JC (1995) Advanced backcross QTL analysis: a method for the simultaneous discovery and transfer of valuable QTLs from unadapted germplasm into elite breeding lines. Theor Appl Genet 92: 191-203

Tanksley SD, Grandillo S, Fulton TM, Zamir D, Eshed Y, Petiard V, Lopez J, Beck-Bunn T (1996) Advanced backcross QTL analysis in a cross between an elite processing line of tomato and its wild relative *L. pimpinellifolium*. Theor Appl Genet 92: 213-224

Taylor BA (1978) Recombinant inbred strains: use in gene mapping. In: HC Morese (ed) Origins of Inbred Mice. Academic Press, New York, USA, pp 423-438

Werner JD, Borevitz JO, Warthmann N, Trainer GT, Ecker JR, Chory J, Weigel D (2005) Quantitative trait locus mapping and DNA array hybridization identify an *FLM* deletion as a cause for natural flowering-time variation. Proc Natl Acad Sci USA 102: 2460-2465

Xiao J, Li J, Yuan L, Tanksley SD (1996) Identification of QTLs affecting traits of agronomic importance in a recombinant inbred population derived from a subspecific rice cross. Theor Appl Genet 92: 230-244

Yano M, Katayose Y, Ashikari M, Yamanouchi U, Monna L, Fuse T, Baba T, Yamamoto K, Umehara Y, Nagamura Y, Sasaki T (2000) *Hd1*, a major photoperiod sensitivity quantitative trait locus in rice, is closely related to the *Arabidopsis* flowering time gene *CONSTANS*. Plant Cell 12: 2473-2484

Zhang J, Guo W, Zhang T (2002) Molecular linkage map of allotetraploid cotton (*Gossypium hirsutum* L. × *Gossypium barbadense* L.) with a haploid population. Theor Appl Genet 105: 116-1174

4 Construction of Genetic Linkage Maps

Elisa Mihovilovich, Reinhard Simon and Merideth Bonierbale*

International Potato Center (CIP) Apartado 1558 Lima 12, Peru

*Corresponding author: m.bonierbale@cgiar.org

Overview

Rapid developments in modern molecular genetics and computing science have allowed researches to generate and test theoretical models and algorithms that have facilitated the construction of a series of genetic maps that are elucidating patterns of the evolution, architecture and function of genetic diversity at an unprecedented rate.

In this sense, we have attempted to capture and highlight key contributions, concepts and strategies of a broad community of scientists addressing the range of genetic complexities presented by diverse higher organisms in the interest of advancing the field and the utility of genetic mapping.

This chapter attempts to introduce both the basic concepts required for constructing linkage maps, and the range of key strategies and theoretical models that have been elaborated for mapping to date in a simple and intuitive manner. This treatment seeks to contribute to awareness of mathematical procedures required for mapping, without delving deeply into the complex formulae that underlie them.

The content of this chapter is oriented to readers with basic but not extensive statistical knowledge, who wish to design mapping projects or merge available and new genotypic data into maps using available software tools. Reference will be made to plants, with examples drawn from potato on which the authors have gained most of their practical experience.

1 INTRODUCTION

In the last 20 years, the rapid development of molecular markers, and with them of genetic linkage maps, have enabled a refreshment of classical plant science, since new light has been cast on the black box of plant genetics (Watanabe 1994).

Consequently, plant breeding has benefited in different ways, namely, facilitating the targeting of individual genes responsible for quantitative traits, accurate identification of introgressed foreign genes, efficiency in following selection-recombination cycles, and selection of genotypes with specific combinations of favorable traits without need for progeny testing. In other words, genetic linkage maps have become an indispensable tool in basic genetic studies and modern applied breeding programs. Currently, they are also the vertebral column for the isolation of genes by map-based cloning.

Linkage mapping in self-pollinating species already offers several advantages and practical tools over other systems and therefore retains theoretical attraction. For outcrossing and autopolyploid systems, statistics models are more complex and software tools are only recently becoming available.

2 FROM MARKERS TO LINKAGE MAPPING

A genetic linkage map represents the relative order of genetic markers along a chromosome. Recombination frequencies are used to determine the relative distance between the markers.

2.1 Basic Concepts

A genetic marker is a measurable character that can detect variation in an observable genetic trait, a protein or DNA sequence. This difference, whether phenotypic or genotypic, may act as a genetic marker if its inheritance can be followed through different generations. In molecular biology, a genetic marker is a DNA polymorphism that serves experimentally as a probe descriptor to classify or describe an individual or set of individuals.

Genetic markers located on the same chromosome may exhibit distorted Mendelian ratios, i.e., they do not sort independently into gametes as predicted by Mendel's second law. This is because genes on the same chromosome undergo recombination. Recombination is the result of exchange of genetic information between homologous chromosomes due to physical exchange of chromosomal material or crossing over during meiosis. The effect of this physical exchange is the rearrangement of heterozygous homologous chromosomes into new combinations.

Recombination can occur between alleles at any two loci or markers on a chromosome, the amount of crossing over is a function of how close these loci are to each other. Loci that are closer undergo fewer crossing over events and the number of non-recombinant gametes will exceed those of recombinant gametes (Figure 1).

For linkage analysis, identification of recombinant and non-recombinant genotypes depends on the allelic configuration of parents involved in the original cross of a segregating population. Two allelic configurations of original parents at two linked loci are possible, coupling or repulsion. The figure below (Figure 2) depicts gametes of an F_1 genotype of a cross between two diploid lines, in which in the upper part, alleles at two linked loci A and B are in coupling (AB/ab) and in the lower part are in repulsion (Ab/aB) (the slash here separates the two homologous chromosomes).

If crossing over does not occur, the products are **parental gametes**. If crossing over occurs, the products are **recombinant gametes**. If two loci are far apart on the chromosome, a recombination is likely to occur every time that pairing occurs and an equal number of parental and recombinant chromosomes will be produced. But as two loci are closer and closer on the chromosome, it is less likely that recombination events will occur between them and thus fewer recombinant chromosomes will

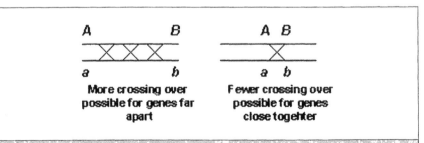

Fig. 1 Crossing over between alleles at different loci. 'A', 'a' and 'B', 'b' denote alternative alleles at two linked loci.

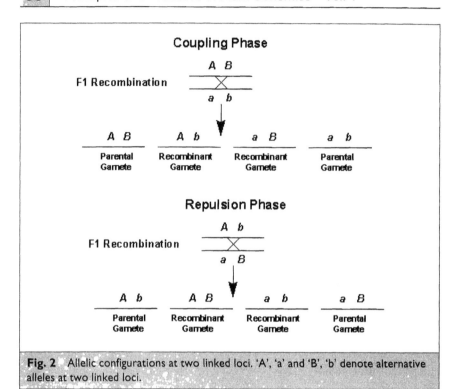

Fig. 2 Allelic configurations at two linked loci. 'A', 'a' and 'B', 'b' denote alternative alleles at two linked loci.

be derived. Hence, the gametes found in the lowest frequency are recombinants as they result from the variably limited recombination between two loci. Likewise, by looking at the gametes that are most abundant, it is possible to determine if the original cross was a coupling or repulsion phase cross.

2.2 Two- and Three-point Analysis

The frequency of recombination events that occur during meiosis is the base for calculating genetic distances between loci. As fewer recombination events occur between two loci physically close together on a chromosome, the lower the frequency of recombinant genotypes will be seen. If a testcross is performed, a deviation from the expected 1:1:1:1 ratio - seen when an equal number of parental and recombinant chromosomes are produced- will be obtained.

For example, in Table 1 it is seen that the large chi-square value indicates that the ratio does not fit the 1:1:1:1. To calculate the observed

Table 1   Chi square value calculated from the number of observed and expected parental and recombinant genotypes of a testcross (AC/ac × ac/ac) assuming independent segregation between loci 'A' and 'C' (Slash separates the two homologous chromosomes)

F_2 Genotypes	Observed	Expected	$(O-E)^2/E$
AC/ac	583	362	134.9
Ac/ac	183	362	88.5
aC/ac	85	362	212.0
ac/ac	597	362	152.6
Total	**1,448**	**1,448**	$x^2 = 5828.0$

recombination frequency between loci A and C, one can simply divide the number of recombinant genotypes into the total genotypes analyzed. In this case of a total of 1,448 genotypes analyzed, of which 268 (183 Ac/ac and 85 aC/ac) were the result of recombination events, the estimate of the observed recombination frequency between the two loci is 18.5% [(268/1,448)*100].

However, the recombination frequency calculated in this example only considered recombinant products that resulted in an odd number of crossovers between the two loci, i.e., one, three, depending on how far apart these loci are. When an additional locus is added between them, additional recombinant products from double or any even number of crossovers can be detected, and these may change the previously calculated recombination frequency. Figure 3 shows the different recombinant products that are possible when a third locus, B, is added between A and C.

Now if a testcross with an F_1, is performed to an individual that is *aabbcc*, deviation from the expected 1:1:1:1:1:1:1:1 ratio, indicating linkage, will be found as occurred in the two point-analysis described above. Table 2 shows the type and number of parental and recombinant genotypes observed in the same 1,448 segregating individuals when a third, intermediate locus is considered, permitting the detection of products from double-crossovers between A and C. The genotypes found most frequently are the parental genotypes. The double-crossover genotypes are always in the lowest frequency.

For the calculation of the observed recombination frequencies of both intervals A → B and B → C, we proceed as with the two-point analyses, but this time including the double-crossovers. The observed recombination frequency between locus A and B is 13.2% [100*((45 + 40 +

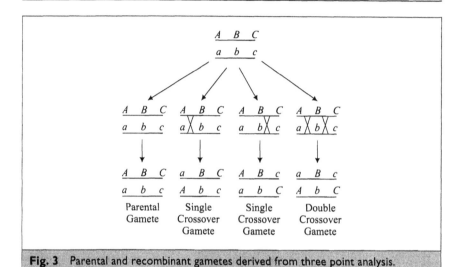

Fig. 3 Parental and recombinant gametes derived from three point analysis.

Table 2 Type and number of observed parental and recombinant genotypes at three consecutive linked loci in 1,448 individuals from a hypothetical testcross

Genotype	Observed	Type of F_1 gamete
ABC/abc	580	Parental
abc /abc	592	Parental
ABc/abc	45	Single-crossover between loci B and C
AbC/abc	40	Single-crossover between loci B and C
AbC /abc	3	Double-crossover
aBc /abc	5	Double-crossover
Abc/abc	89	Single-crossover between loci A and B
aBC/abc	94	Single-crossover between loci A and B
Total	**1,448**	

3 + 5)/1448)], and that between B and C is 6.4% [100*((89 + 94 + 3 + 5)/ 1448)]. Now, the observed recombination frequency between A and C in the three-point analysis is 19.2% (13.2% + 6.4%), which is higher than the 18.5% calculated with the two-point analysis. This is because double-crossover products between A and C cannot be detected in the two-point analysis since they appear as parental types (see allelic configuration for loci A and C in double-crossover products), and consequently the recombination frequency was underestimated.

2.3 Genetic Distance and Mapping Functions

Genetic distances are estimated from the frequency of recombination events (r) between loci, which is the probability of each crossover occurring between them in a single meiosis. The observed recombination frequencies (θ) are the result of this frequency of recombination events. For the estimation of genetic distances, these recombination frequencies are commonly converted into map units or centiMorgans by applying a mapping function which carries certain assumptions. Without considering these assumptions, one centiMorgan (cM) is the genetic distance between two loci with a recombination frequency of 1%. Hence, the genetic distance in centiMorgans is numerically equal to the recombination frequency expressed as a percentage. In the examples depicted above, distance between loci A and C by two-point analysis was 18.5 cM, or 19.6 cM by three-point analysis. This difference shows that recombination frequencies are not additive, that is, when new loci are added, previously obtained distances need to be adjusted. An additional consideration is that because of the way in which the calculations of recombination frequencies are performed, it is never possible to obtain more than 50% of recombinant genotypes. Therefore the maximum distance between two genes that can be measured is just under 50 cM. If two genes are greater than 50 cM apart, then we can not determine if they reside on the same or different chromosomes unless other types of evidence are available.

However, because this direct linear relationship between recombination frequency and genetic distance is limited to cases of close linkage (< ~7 cM), the previous concept that one centiMorgan equals to 1% of recombination frequency needs a statistical correction. This occurs because as distances become greater between two loci, for which there are not additional loci segregating in between, the probability of double- (or other multiple-) crossovers that cannot be detected, will increase. The minimum distance at which double-crossover may occur is variable and depends on the species, chromosomal regions, and genetic control. Thus, researchers have developed mapping functions that can correct this difficulty. These mathematical models impose, as mentioned above, one of two assumptions; the presence, or absence, of 'interference'. Interference is the probability that the presence of a crossover in one region prevents another from taking place nearby. If no interference is assumed, then all possible double-crossovers would be observed. In the example described above, the recombination frequency between loci A and B was 13.2% or 0.132 and 6.4% or 0.064 between loci B and C. Therefore, we would expect 0.84% [100*(0.132 × 0.064)] double recombinants. With a sample size of 1,448, this would amount to 12 double recombinants, but actually only 8 were detected. So this partial interference can be measured as '1 – coefficient of coincidence'. The coefficient of coincidence is the ratio of observed to expected double recombinants. The interference value in our example is 33% [1 – (100*8/12)].

There are two mapping functions commonly used to correct this departure from the linear relationship between recombination fraction and genetic distance. The Haldane (1919) function assumes that crossovers occur randomly and independently over the entire chromosome i.e., crossovers occur at random without interference, and the Kosambi (1944) function allows for partial interference. Haldane's mapping function was obtained from the Poisson distribution of the number of crossing overs, so that, the function that relates the genetic distance (m) and the observed recombination fraction (θ) containing an odd number of crossovers is $m = -\ln(1 - 2\theta)/2$. On the other hand, Kosambi's function which allows for partial interference is given by $m = 1/4 \ln[(1 + 2\theta)/(1 - 2\theta)]$. The latter is used more frequently than the former.

If the calculations described above are written in an algebraic function of the recombination fractions (θ), the effect of interference represented by the parameter ϕ, can be more clearly understood. So that $\theta_{AC} = \theta_{AB} + \theta_{BC} - (1 - \phi) \theta_{AB} \times \theta_{BC}$; where ϕ ranges from zero if crossovers are independent (Haldane's model), to one, if the presence of a crossover in one region partially or completely suppresses crossovers in adjacent regions (Kosambi's model). Thus, in the absence of very strong interference, the recombination frequencies can only be considered to be additive if they are small enough for the product $2(1 - \phi) \theta_{AB} \times \theta_{BC}$ to be neglected. This is not surprising given that the recombination frequencies measures only a fraction of all recombinant events, those that resulted in an odd number of crossovers. A map or genetic distance (m), on the other hand, attempts to measure the total number of crossovers (both odd and even) between two loci. Mapping functions attempt to predict the number of crossovers between loci that are *m* map units apart from observed recombination frequencies (θ). As observed in Figure 4, when the genetic distance is small (less than 10 cM), both Kosambi and Haldane mapping functions provide essentially the same value.

There is no universal relationship between genetic and physical distances between loci. Depending on the species and varying between and within chromosomes, a centiMorgan corresponds to a span of about 10 thousand (10 kb) to a million (1,000 kb) nucleotide base pairs. Crossovers are often suppressed (which increases the number of base

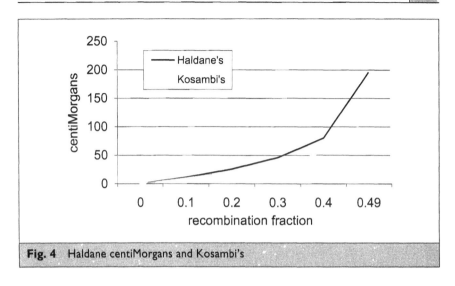

Fig. 4 Haldane centiMorgans and Kosambi's

pairs per cM) near centromeres and telomeres (Tanksley et al. 1992). Further, as it seems that the rate of recombination is under genetic control, there may be considerable variation among individuals and among populations in the strength of linkage (Dvorák et al. 1998).

2.4 Estimating Genetic Distance from Experimental Data

Mapping populations consist of individuals of one species, or come from crosses between parents from related species that differ in a trait or traits to be studied. The type of mapping populations suited to mapping depends on the reproductive mode of the target species, i.e. whether it is self-fertilizing or cross-pollinating. Mapping populations suitable for self-fertilizing species are generated by crossing two available pure lines and consist of F_2 plants, recombinant inbred lines (RIL), backcrosses (BC), or doubled haploid (DH) lines, among others. If pure lines cannot be generated due to self-incompatibility or inbreeding depression, heterozygous parental plants are used to generate F_1 offspring or backcross lines. What is important is that the selection of parents used to generate a mapping population should maximize the probability of detecting DNA polymorphism for the type of molecular markers to be used.

Mapping populations should comprise several tens of individuals that may represent a large enough sample of recombination events, and be genotyped with some hundreds of markers previously selected to be polymorphic in the parents. Hence experimental data comprise large amounts of segregation data that are routinely processed by computer programs to construct a genetic map.

Figure 5 shows an example of an experimental data set (data taken from BCT family) comprising 158 segregating progenies and 80 RFLP markers (Bonierbale et al. 1994). The data set was used to construct a framework map with the MAPMAKER V3.0 software. The 80 RFLP markers represent a subset of a larger number of RFLP markers used on the expanded molecular map of potato (Tanksley et al. 1992).

Estimation of genetic distances from large data sets is tedious and rather complex if attempted manually. Moreover, statistical problems associated with features underlying estimates of recombination frequency (θ), not considered in the examples described above, can reduce their accuracy. These include the number of segregating individuals tested, obvious recombinations not observed, and missing data. Fortunately, several available mapping packages with suitable algorithms take these problems into account and provide more accurate estimations.

These programs proceed by first tabulating the distinct genotype combinations observed between loci. Each genotype combination will have a frequency which depends on the probability of each crossover or recombination event (r) occurring between each pair of loci, and on the crossing design of the experiment. Then by applying an appropriate maximum likelihood (ML) method, a value for 'r' is found that maximizes the probability of the observed frequencies (θ) of these combinations.

The likelihood normally used in linkage analysis is the binomial as applied to the recombination fraction. To apply ML, an adequate ML-equation for calculating recombination frequencies should be built based on the different types of allelic configurations. Ritter et al. (1990) describe formulas and ML-equations for classical allelic configurations from backcross and F_2 populations derived from inbred lines as well as for other several coupling and repulsion configurations common to F_1 populations from crosses between heterozygous parents, as used in outcrossing.

What the ML method does, is determine a series of likelihood (L) values that the genotypic data for the two loci analyzed (observed frequencies) resulted from a given probability of recombination event (r) value. This necessitates the application of a statistical test to define the best r value (the maximun likelihood value). The LOD Score (likelihood of odds) is a widely used technique that was first applied in human research and then in plant and animal linkage analysis. The method

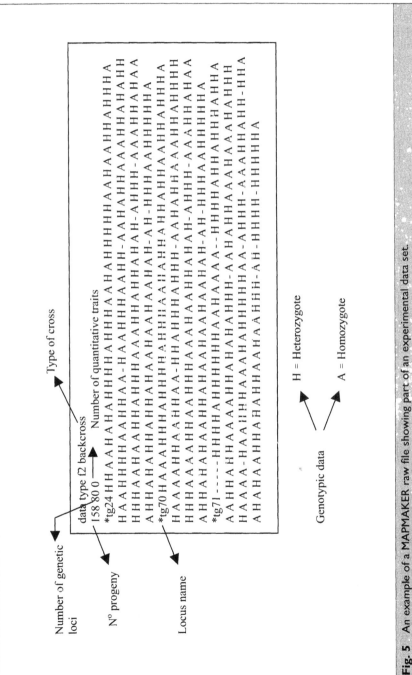

Fig. 5 An example of a MAPMAKER raw file showing part of an experimental data set.

developed by Newton E. Morton is an iterative approach, whereby a series of these LOD scores are calculated from a number of proposed r values. The LOD score is the log base 10 of the likelihood ratio under the hypotheses of linkage and non-linkage, that is, it compares the likelihood for each value of r to the likelihood of r for unlinked loci (r = 0.5).

For instance, we have 25 sibs, four of which are recombinants (R) with a probability equal to r, and 21 non recombinant (NR) with a probability (1 − r), we would write the likelihood of this data given an unknown r value, as,

$$L(r = x) = r^4(1 - r)^{21},$$

The likelihood under no linkage i.e., $r = 0.5$ is

$$L(r = 0.5) = L(0.5) = 0.5^{25}$$

According to LOD definition

$$\text{LOD} = \log_{10}[L(r = x)/L(r = 0.5)]$$

So for θ = 10%

$$\text{LOD} = \log_{10}[(0.1^4 \times (1 - 0.1)^{21})/0.5^{25}]$$
$$\text{LOD} = 2.56$$

For several r values

r	LOD
1	−0.56
5	1.85
10	2.56
15	2.74
20	2.69

The highest LOD score in this example is 2.74, indicating that the most likelihood value of r is 15 cM. A LOD score of 3.0 means that the likelihood of linkage occurring at that distance or probability of a recombinant event (r) is 1,000 times greater than no linkage. As more individuals are added, the LOD score increases and becomes a better indicator of linkage and map distance. For many likelihood ratio tests, twice the natural log of the likelihood ratio is asymptotically distributed as a chi-squared statistic. So, a LOD score can be converted into an equivalent chi-squared statistic with the following simple formula: LR = 2ln(10)LOD. An important software package, MAPMAKER, that is widely used in plant mapping research is based in part on the LOD score method (Lander et al. 1987). Finally, once a value of r is defined for each pair of loci, a mapping function can be applied, i.e., cM (Haldane's) or cM (Kosambi's) to obtain a map distance (m).

2.5 Formation of Linkage Groups

Once r for each pair of loci or markers have been estimated and converted into distances, the next step is to 'group' the loci into candidate chromosome segments or linkage groups. Ideally the number of linkage groups in a genetic map equals the haploid chromosome number of the experimental organism. Grouping must fullfil some usual admission rules based on upper linkage thresholds, i.e., distance less than or equal to a specified maximum; and lower limit of detection thresholds, i.e., LOD score exceeding a specified minimum, for considering linkage between pairs of loci. Hence, if two markers are significantly linked they belong to the same linkage group. A computerized search through all pairs of markers, using previously defined threshold values, will then produce a grouping of the markers, within which linkage is considered fully transitive (e.g., if A is linked to B, and B is linked to C, then A, B, and C will be included in the same linkage group).

A conservative threshold of the LOD score may lead to more linkage groups than the haploid chromosome number. On the other hand, a relaxed threshold may cause group of markers from different chromosomes to be incorrectly assigned to a single linkage group.

The researcher should set these admission thresholds to give a balance between number and size of groups based on his experience and knowledge of the genetic material.

Figure 6 shows four linkage groups of BCT framework map (Bonierbale et al. 1994) obtained using the Group command of MAPMAKER v3.0 software (Lander et al. 1987). Markers within linkage groups had a minimum LOD 4.0 with a maximally likely map distance of no more than 30 cM to at least one other locus in the group.

2.6 Ordering of Markers within a Linkage Group

Ordering of markers is a problem analogous to the classical traveling salesman's problem (TSP) of trying to find the shortest (or a nearly shortest) trip route connecting a number of locations (perhaps hundreds of cities), on his sales route.

Finding the correct order in a TSP is a classical hard-to-solve problem. Software packages assist in finding a 'best' correct order and

```
1> centimorgan function Kosambi
centimorgan function: Kosambi

2> informative criteria 4 140
Informativeness Criteria: min Distance 4.0, min #Individuals 144

3> default linkage criteria 4 30
default LOD score threshold is 4.00
default centimorgan distance threshold is 30.00

4> sequence all
sequence #1= all

10> group
Linkage Groups at min LOD 4.00, max Distance 30.0

group1= tg24 tg70 tg71 tg326 tg116 tg237 tg430a tg259
-------
group2= tg276 tg306 tg462 tg20a ct75 tg449A tg34 tg141
-------
group3= tg135 tg449b tg130 tg74 tg42 tg411 tg244
-------
group4= tg123 tg208 tg65 tg155 tg443 tg450
```

Fig. 6 An output of MAPMAKER showing some sentences and four potato linkage groups after performing the 'group' command.

calculating distances. The number of possible orders for a number n of markers equals $(\frac{1}{2})n!$, hence it is not hard to see how rapidly this number increases with the number of markers. Researchers on diploid species have developed several different criteria (objective functions) to define a 'best' order and several algorithms to find the order with the optimal criterion value. Based on the criterion adopted, a number is calculated for any given order which indicates the 'goodness-of-fit' of that order.

The criteria that have been proposed include the maximum likelihood (Lander et al. 1987; Jansen et al. 2001), the minimum sum of adjacent recombination fractions (SARF), the maximum sum of adjacent LOD scores (SALOD) (Liu and Knapp 1990), the minimum number of crossovers (Thompson 1987) and the 'least square locus order' (Stam 1993). As a mean to provide some insights into these objective functions, for example, the SARF uses the estimates of the recombination frequencies between adjacent markers. The order with the smallest value of SARF is regarded as 'best'. On the other hand, SALOD calculates the sum of the LOD scores for each adjacent marker pair in the sequence, and the order with the highest value of SALOD is regarded as 'best'. As the LOD score is in some way the 'amount of linkage information', the SALOD criterion considers all pieces of information in a sequence and adds these. From a number of alternative sequences, the most informative (with respect to linkage) is considered 'best'.

Various software packages for linkage mapping have implemented these criteria, combined with an algorithm for ordering markers. Most of these algorithms allow these software to apply shortcuts and 'tricks' to avoid the computational challenge of testing all possible orders for a high number of markers - e.g., for 100 markers = 4.6×10^{157} orders - and thus arrive at the best order or an order that is 'almost best'. One of the simplest algorithms to achieve this, known as seriation (Doerge 1996; Crane and Crane 2005), is a greedy algorithm that works by adding one marker at a time starting with the most tightly linked locus pair. Each successive addition is made to optimize the current order without consideration of the loci not yet added, or removal of any previously added. For example, starting with two tightly-linked markers A and B, the different orders generated by a third marker C to be added, i.e., CAB, ACB and, ABC are compared and the best fitting one is chosen, say ACB. When adding a next marker, say D, the orders $DACB$, $ADCB$, $ACDB$ and $ACBD$ are compared, and, again, the best one is chosen. This continues until all markers have been placed in the sequence. A more elaborate greedy algorithm is that of MAPMAKER (Lander et al. 1987) of finding all three-marker orders, then excluding the most unlikely and proceeding by evaluating permissible multilocus orders built from the remaining ones. To find the multilocus order with the highest likelihood, this software performs an exhaustive search for an initial order (seed order) using a subset of highly informative markers, then applies the branch-and-bound algorithm with likelihood as the criterion for bounding, with the most informative loci tried first.

The branch and bound algorithm (Thompson 1987) consists of branch points corresponding to the insertion of information on additional marker, the alternative branches corresponding to alternative positions for insertion. At a certain level when a i-th marker is to be inserted, every node has i-branch points and all subtrees from a given level must correspond to insertion of the same set of markers. The search for the best order involves the evaluation of the score increase on each branch of the tree, and comparison with a previously determined value which is the current minimum found for a full order (at some tree tip). The search of any subtree can be terminated as soon as the value at the root of that subtree is not strictly smaller than the current comparison value. So this required that some order (seed order) with low total score were found early in the search, to give a small value for comparison, and that high score increases were acquired at low levels in the tree to reduce the chance of having to search many subtrees to high levels. MAPMAKER's final optimizing is done by rippling.

Other software, such as JoinMap (Stam 1993; Stam and van Ooijen 1995) uses a stepwise search which is a combination of seriation and branch and bound, to minimize the least squares order criteria. This method chooses the markers to be added to the sequence based on the 'amount of information' they contain - joint data scored pairwise - and accepts them only if a goodness-of-fit test shows an improvement. After a marker is added, a 'local reshuffling' is applied in order to prevent that the previous order will not be changed any more and the algorithm is trapped in a local optimum from which it cannot escape.

A different algorithm, known as simulated annealing, is used by the GMENDEL software (Liu and Knapp 1990) to minimize SARF. The simulated annealing algorithm employs a 'temperature' parameter that governs the amount of change in a configuration that may be applied at each step, as well as the probability of acceptance of a configuration with a lower (more unfavorable) score than the current one. As the configuration stabilizes at some temperature, the system is 'cooled', changes become less extreme, and unfavorable changes are less readily accepted.

CarthaGene (Schiex and Gaspin 1997) offers simulation annealing and greedy algorithms, and a genetic algorithm. This latter algorithm aims to emulate nature's system for combinational optimization of species fitness using sexual recombination with some mutation rate, followed by selection based on a fitness function.

Once a locus order has been obtained, interlocus distances have to be updated. MAPMAKER and CarthaGene do this directly, using the expectation-maximization (EM) algorithm, a class of convergent algorithm for estimating mutually dependent unknown parameters by updating each in turn from the others (Dempster et al. 1977). JoinMap calculates the distances having least square errors from the two point distances, while giving more weight to distance estimates based on more information. The likelihood of the final map is increased under both methods.

Figure 7 shows in brief how the 'Order' command of MAPMAKER finds a linear order of a subset of markers from chromosome 3 of BCT framework map (Bonierbale et al. 1994). The order command first searches for a starting map order of highly-informative markers that has

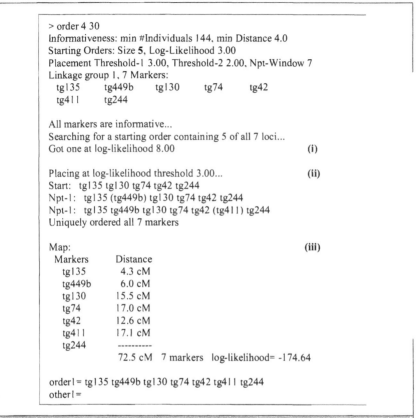

```
> order 4 30
Informativeness: min #Individuals 144, min Distance 4.0
Starting Orders: Size 5, Log-Likelihood 3.00
Placement Threshold-1 3.00, Threshold-2 2.00, Npt-Window 7
Linkage group 1, 7 Markers:
   tg135      tg449b      tg130      tg74      tg42
   tg411      tg244


All markers are informative...
Searching for a starting order containing 5 of all 7 loci...
Got one at log-likelihood 8.00                              (i)

Placing at log-likelihood threshold 3.00...                 (ii)
Start:  tg135 tg130 tg74 tg42 tg244
Npt-1:  tg135 (tg449b) tg130 tg74 tg42 tg244
Npt-1:  tg135 tg449b tg130 tg74 tg42 (tg411) tg244
Uniquely ordered all 7 markers

Map:                                                        (iii)
   Markers      Distance
    tg135         4.3 cM
    tg449b        6.0 cM
    tg130        15.5 cM
    tg74         17.0 cM
    tg42         12.6 cM
    tg411        17.1 cM
    tg244        ----------
                 72.5 cM   7 markers   log-likelihood= -174.64

order1 = tg135 tg449b tg130 tg74 tg42 tg411 tg244
other1 =
```

Fig. 7 An output of MAPMAKER showing the order of markers of potato linkage group 3 after performing the 'order' command'.

only one plausible map order supported with a high log likelihood (i). Having found this seed order, other markers are added to the order one at a time (ii). To accept a placement in the order as unique, a specified threshold is defined by the analyst <strict threshold used = 3.0 >, <ordinary threshold used = 2.0 >. The analysis stops when it can no longer add any markers to the order, and reports the final unique map order as well as the possible relative positions of any remaining markers (iii). The 'order command' uses the multipoint analysis feature of MAPMAKER which takes into account the primary genotype data (raw data) for all loci considered in the linkage group simultaneously.

The number of markers to use for the starting order is defined by the analyst (we chose 5 for 7 markers). Finding a starting (seed) order depends on the ability to find a good highly informative subset. So the analyst should define the informative criteria parameters according to the nature of the data set, namely, type of cross, mapping population size, type of markers (dominant or co-dominant). The parameters involve specification of the minimum distance between any pair of markers and the minimum number of individuals that are informative. In the example above, a minimum distance of 4.0 cM and a minimum number of individuals of 144 of the 158 comprising the whole population were chosen.

2.7   Special Considerations

In the course of this chapter, we have seen how linkage mapping uses the frequency of recombination events that occur during meiosis as a basis for calculating genetic distances between loci. By following the inheritance of genetic markers in a meiotic population, recombination events are then linearly ordered along a linkage group. This linear order defines segment of a set of chromosomes which varies in both physical and genetic size. These sizes are defined by the number of analyzed descendants in the mapping population and by the average number of recombinant events that occur during meiosis. So, as the number of markers scored in the population exceeds the number of recombination events that can be detected in a population, some segments will have multiple cosegregating markers. In other words, when our mapping population is not large enough, small distances (below 5 cM) may not be resolved. The size of the mapping population sets the limits to our ability to detect recombination, and thus to resolve small genetic distances.

Another issue related to success is errors in genotype scoring, or genotype classification. Misclassification of genotypes leads to an overestimation of recombination frequencies. This overestimation (= bias) relative to the true value of r increases as r decreases. When substantially different results are obtained each time we run a software package for a map order, this is an indication of data set errors. Several mapping software packages have modules that enable the user to identify 'suspect' genotype scores. Although these error detection routines are helpful, they are not the ultimate solution, because a 'suspect' genotype may be a correct one even when it is very unlikely ('suspectness' in error detection algorithms are based on likelihood). So double checking of gel readings or automation of gel reading may greatly increase the overall correctness of genotype scores.

Hence the outcome of a mapping experiment depends on the composition of the sample population. The larger the mapping population, the more confidence we have in the estimates of recombination frequencies and map distances. Allard (1956) proposed that for a given sample size of n, the standard error for a recombination fraction θ can be expressed as $[\theta (1 - \theta)/n]^{1/2}$ for backcross type data. Our sample size of 158 gave a standard error estimate at 0.04 for a maximum recombination fraction of 0.50. This is quite acceptable for a framework map. Populations of size in the range 80-400 are frequently used for this purpose. Population type also influences the standard errors of the estimates. For instance, a replicate experiment with 100 RILs will result in a slightly different map, while replicate sampling of an F_2 population of 100 descendants may result in a slightly more different map and in a quite variable total map length and inter-marker distances.

All these aspects, lead us to conclude first that there is not an 'ultimate true map'and second, that a calculated map is nothing more than the best statistical approximation, given the sample population.

3 MAPPING OUTBREEDING SPECIES

The main feature of outbreeders that is of importance for linkage mapping is their high level of heterozygocity, which is also reflected in the relatively large amount of DNA polymorphism found in studied populations. Due to this high level of heterozygocity, markers can segregate in either coupling or repulsion phase. Allelic phase or configuration has to be determined to avoid contradictions or ambiguities in order precision (Ritter et al. 1990). The strategy commonly used for mapping outbreeding species has been to construct a linkage map of each parent of a cross, relying on markers that are present in one parent and absent in the other, and segregate in the progeny testcross fashion, i.e. in a 1:1 pattern.

Markers originating from Parent 1 and absent in Parent 2 are scored for presence and absence in the progeny. Simultaneously but of course in a different data set, markers segregating from Parent 2 and absent in Parent 1 are treated in the same way.

The utilization of genetically distant species, or interspecific crosses has further increased the chance of a greater marker polymorphism. Thus, in potato, the author developed the first genetic linkage map for an outbred species of an interspecific cross between a South American native cultivated species *Solanum phureja* and a hybrid between the commercial potato *S. tuberosum* and the wild species *S. chacoense* using tomato-derived RFLP probes (Bonierbale et al. 1988). This approach was further expanded using an interspecific cross between *S. tuberosum* and a hybrid between this species and a wild relative of the potato (*S. berthaultii*) (Tanksley et al. 1992). Both backcross maps were based on one parent only.

Linkage maps for each parent of a *S. tuberosum* backcross population were developed by Gebhardt et al. (1991). The two maps were aligned using the allelic bridge approach (Ritter et al. 1990). This approach uses genetic loci common to both maps, with alleles segregating in both parents. These common alleles or allelic bridges segregate in a 3:1 pattern from dominant $<a_1o \times a_1o>$, co-dominant $<a_1a_2 \times a_1a_4>$ or mixed $<a_1a_2 \times a_1o>$ heterozygous markers in both parents.

This model for mapping in outbreeders has been described as 'pseudo-testcross' (Grattapaglia and Sederoff 1994). These authors explained that the F_1 generations of outbreeding species is genetically analogous to the BC or F_2 of inbreeders. Thus, if two different alleles segregate at a single locus, and the two parents were heterozygous $<a_1a_2 \times a_1a_2>$, the F_1 will segregate into 1:2:1 or 3:1, respectively for a co-dominant and dominant marker. On the other hand if one of them was heterozygote and the other homozygote $<a_1a_2 \times a_2a_2>$, the segregation would be 1:1 and both dominant and co-dominant markers would be equally informative. In this scenario, up to three or four different alleles can segregate at a single locus of a heterozygous parent, $<a_1a_2 \times a_1a_4>$ or $<a_1a_2 \times a_3a_4>$ for a co-dominant marker. Parental configurations, $<a_1a_2 \times a_2a_2>$ and $<a_1a_2 \times a_3a_4>$ provide alleles segregating 1:1 for mapping, but the latter is more informative as it provides alleles segregating 1:1 from each parent. Meanwhile, the parental configurations that provide alleles segregating 3:1 are used as allelic bridges to align both parental maps. This procedure has also been applied in potato (Ghislain et al. 2001) and several other outbred species such as eucalyptus (Grattapaglia and Sederoff 1994), grape (Doucleff et al. 2004), sugarcane (Sobral and Honeycutt 1993), and apple (Hemmat et al. 1994). Separate maps of both parents can be constructed with any mapping software while a suitable one that accepts multiple segregation patterns i.e., 3:1 and 1:1, such as JoinMap or CarthaGene should be used to build the composite, or third, map.

Two important issues to be aware of at this point are: first, that the genetic distance between the parents used can affect the recombination frequencies in the cross. Though the utilization of genetically distant parents has the advantage of a higher polymorphism, this may also result in a decrease of chromosome pairing due to a lack of homology. The second issue is the general phenomenon of inbreeding depression of highly heterozygous crops like potato. Distorted segregation of marker alleles in one of the parents may be found as a consequence of the presence of associated uncharacterized (sub) lethal alleles. As a consequence, clusters of markers with distorted segregation are expected to arise. Both, recombination suppression and linkage of markers to deleterious alleles affect distance estimation and the order of markers in linkage groups (van Os 2005). These genetic factors responsible for local aberrations have been reported and clearly discussed in potato and tomato (Bonierbale et al. 1988; Gebhardt et al. 1991).

4 MAPPING WITH A REDUCED POPULATION SIZE

Vision et al. (2000) developed a strategy that reduces genotyping costs, for adding new markers to a previously developed framework map. This saving is in direct proportion to the reduction in the size of the original mapping population, without sacrificing map resolution. The authors' main goal was to provide a method that would facilitate the development of high quality, denser linkage maps given the available resources of laboratories. The strategy, called 'selective mapping', is based on the selection of an optimized sample of genotypes bearing complementary recombinational crossover sites from the large random population used to construct an original map. As the strategy requires a framework map on which to place the new markers, the first phase is the development of this framework map, which is based on a subset of markers known to be evenly spaced in the genome, using the information and data from an original larger map. Then, using the information of the framework markers, a subset of individuals with optimal map resolution for a given size are selected from the entire population and used in the second phase, permitting the placement of new markers with almost the same precision obtained with the entire population in the framework map.

Clarification of the following concepts should contribute to better understanding of this strategy. The main one is map resolution, which is defined as the set of the lengths (distances) between adjacent recombination events or breakpoints in the population or sample. The interval between two consecutive recombination events or breakpoints is called a 'bin'. Within a bin no other breakpoints occur among any members of the given set of individuals (see Figure 8) contributing to a map. Therefore, bins are the smallest unit of resolution in a genetic map, within which two or more markers cannot be ordered relative to one another without supplementary information, because all members have the same genotype for the markers in the available data set. As offspring differ in the number and distribution of recombination events, the most

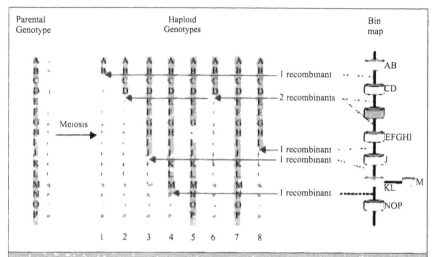

Fig. 8 The recombination bin mapping concept. A heterozygous parental chromosome pair and eight haploid genotypes are represented. Allelic marker loci, Aa to Pp are represented as upper and lowercase letters on a white or shaded background. The diagram shows the position of recombination events visualized as a linearly ordered set of bins, each separated by a single recombination event. Courtesy from Isidore et al. (2003).

informative sample is that in which the individuals of the selected sample complement one another in the order and position of bins (information), to provide the maximum mapping resolution. The mapping resolution of the sample will depend on the distribution of the bins in it, which in turn determines its precision for mapping new markers.

Hence, picking an optimal sample requires minimizing a function of the bin length distribution. Two objective functions found suitable for this purpose were the maximum bin length (MBL) and the sum of the squares of the bin length (SSBL). The former statistic computes the longest distance between consecutive breakpoints (including the ends of linkage groups) and the latter is equivalent to computing the expected length of the bin containing a marker chosen uniformly at random from the whole genome. The aim is to identify samples of minimum size whose objective function values are close to that of the full population. Researchers also found it useful to assume independence of recombination sites for breakpoint distribution in the sample selection process (Haldane mapping function). In other words, it is assumed that the sites of breakpoints are unique and independent of one another and that map distances are additive. This would not significantly affect map

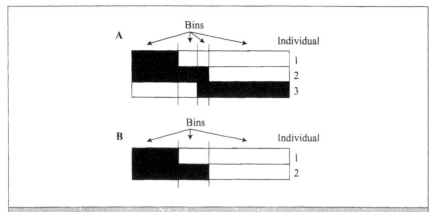

Fig. 9 Schematic representation of sample selection for selective mapping. The approach is based on complementation of sampled genotypes in the order and position of bins (see text). Boundaries between shaded and unshaded areas represent breakpoints. Bin lengths were drawn to scale. Individuals are assumed to be haploid. Courtesy from Vision et al. 2000).

distances because Haldane's function performs similarly to other mapping functions at small map distances, which are of main concern for the selective mapping purpose. Figure 9 taken from Vision et al. (2000) shows very simply how sample individuals are chosen from the entire population. In A, the inclusion of all three individuals breaks the interval into four unequal-length bins (boundaries between shaded and unshaded areas represent breakpoints or recombination events). In B, it is shown that removing the third individual causes the loss of the smallest bin, but the maximum bin length remains unchanged.

Now, in modeling the sample selection problem, the authors first assume a non-realistic situation using a deterministic approach in which the exact sites of all breakpoints (sites of recombination and chromosome boundaries) are known in all population members. The authors first analyzed this idealized case, and then modified the model to make it compatible with biological data. In this latter case, in which the exact location of breakpoints is not available, a stochastic model was developed. Here, we present a brief description of both approaches to provide the reader a good understanding of this mapping strategy.

4.1 The Deterministic Approach

Under this approach, simulated data with exactly specified breakpoints for each individual of the entire population, 'n' , are first generated. The aim is to seek the best sample subset of the population for a given size 'k'. Since it is not possible to evaluate all possible samples of size k from a population of size n, a computational algorithm is needed to construct a subset of size k which minimizes the MBL and/or the SSBL between consecutive breakpoints for maximum map resolution.

Two computational algorithms were found successfully in identifying samples with small maximum bin length (Vision et al. 2000). The first is a mixed greedy algorithm that employs the two objective functions sequentially, first minimizing the SSBL and then minimizing the MBL. It also exploits randomness to search a large space of possible good samples. The second involves linear programming with randomized rounding to convert fractional assignments into integral ones. Sample sets from both algorithms are improved by the inclusion of a clean-up routine that disposes of members that do not contribute to the quality of the sample and 'greedily' replaces them with other selections that do. Both algorithms are nearly identical in performance, but the greedy algorithm has the advantage that it does not require specialized linear programming software to implement. The greedy algorithm tends to be fast and give satisfactory solutions. To get a better taste of this algorithm a summary explanation is given below.

Greedy Algorithm

To construct a sample of size k, the algorithm starts with an empty sample. Since a mixed greedy algorithm is desired, a sample S_i of size k/2 is first built by seeking and adding those population member that improves the objective function until the size is reached. Then a switch is made to the greedy algorithm to minimize objective function MBL, augmenting S_1 until a sample S of size k is obtained. Minimizing the SSBL objective function first, forces all bins to be small rather than myopically focusing on the largest bin. A limited element of randomness into the choice of each member of the first half of the sample is performed by a randomized greedy procedure in which each next member is chosen from the 'r' choices that would most improve the objective function for some small values of r (e.g., 3 or 5). This is repeated a large number of times until the best sample set is found. The remaining half is produced by a non-randomized greedy algorithm, as described above.

4.2 The Stochastic Approach

Unlike the deterministic approach, in which the exact site of each breakpoint is known, the stochastic approach of real biological data counts only on the information on the sites of the markers flanking visible breakpoints. The marker genotypes in the data set allow the identification of those intervals bearing odd numbers of breakpoints. The approximate positions of these breakpoints are found by scanning each individual for differences in the genotypes at adjacent framework markers. Two consecutive markers showing different genotypes in the same individual is indicative of the occurrence of an odd number of breakpoints between them. Hence biological data consists of two sets of information: the length of the intervals between each pair of consecutive framework markers in the genome, and, for each inter-marker interval, the set of population members with a visible breakpoint in that interval. Given the uncertainty in precise breakpoint location, the sample sought is that, which minimizes the expectation of the MBL (E(MBL)) or SSBL (E(SSBL)), under the assumption that known breakpoints are uniformly and independent distributed within framework intervals. For E(SSBL) a closed–form solution was derived given known marker sites and a known number of breakpoints between consecutive markers, while for the E(MBL), it was not possible (Vision et al. 2000). In this latter case, a series of 100 replicates of the population was generated in which all breakpoints were randomly resolved to exact sites, and the mean quality of this sample computed, using the MBL as when breakpoints are known with precision. These two stochastic objective functions are sequentially minimized in the mixed greedy algorithm to derive the sample as with the deterministic model. However, as opposed to the deterministic model, the first half of the sample which minimizes the E(SSLB), is selected without the need to randomly resolve the breakpoint locations many times over.

Application

Vision et al. (2000) implemented the randomized mixed greedy algorithm in a software program called MapPop that allows the selection of samples from populations of 500 or less individuals within minutes. This software also computes the locations of markers that have been genotyped in the selected sample, relative to a user-supplied framework map. A software called IntiMap (Simon and Bonierbale 2003) was developed to create a Java interface to MapPop with the additional advantage of a graphical map output. We used this software to add other marker classes to the RFLP-based BCT framework map (Bonierbale et al. 1994). Thus, the BCT framework map was enriched by two types of microsatellite markers, i.e., simple sequence repeat (SSR), and expressed sequence tag-simple sequence repeat (EST-SSR), (Milbourne et al. 1998; Feingold et al. 2005), as well as by conservative ortholog set markers (COS) (Feinan et al. 2006). Our aim was to count on a common set of PCR-based markers for comparative mapping and functional analysis of genes. EST-SSR and COS are both PCR-based markers developed from EST databases. Since ESTs are from expressed genes, and particularly in the case of COS, these markers are expected to be useful for comparative mapping (Yu et al. 2004). Furthermore, as COS are single copy conserved orthologous genes from distantly related species, predictions of common gene functions may also be possible across wide phylogenetic distances.

5 HIGH- AND ULTRA HIGH-DENSITY MAPS

Earlier published maps have been of low to moderate density (i.e., average marker spacing > 5 cM) built on approximately 300 or fewer markers. These maps, though useful, have inherent and obvious limitations, that can be overcome if a greater number of markers are mapped at very close intervals throughout the genome (Tanksley et al. 1992). The idea of higher levels of mapping saturation gave rise to high density maps in which markers are spaced at average intervals less than 5 cM and their number goes beyond the thousand. These maps are key tools for several important applications in basic and applied research. The fact that high-density maps provide a greater probability that the entire genome is covered with molecular markers, assures that any gene of interest will be tightly linked to at least one molecular marker. This is important in breeding for the detection and characterization of loci underlying quantitative traits and for marker-assisted selection of desirable genes. Furthermore, these denser maps offer a starting point for chromosome walking toward map-based gene cloning.

Maps with this level of marker saturation have been developed in the last 20 years for several economically important crops such as tomato and potato (Tanksley et al. 1992; Haanstra et al. 1999), maize (Vuylsteke et al. 1999), wheat (Somers et al. 2004), rice (Harushima et al. 1998) and sorghum (Menz et al. 2002; Bowers et al. 2003) among others. Some of them are based on RFLPs, others on PCR-based markers i.e., AFLPs or SSRs; and some others on the combination of these marker types.

Though, as expected, levels of resolution differ between species, populations, marker class, and chromosomal regions, these maps have reached estimated average distances between markers of 0.23 to 8 cM and comprise in most cases, a range of approximately 1,000 to more than 2,000 mapped marker loci (Tanksley et al. 1992; Bowers et al. 2003; Somers et al. 2004).

Such maps have enabled the investigation of the meiotic behavior and the pattern of distribution of crossovers along chromosomes. High-density clustering of markers at centromeric and some telomeric areas have indicated low levels of meiotic recombination in these regions, which have been found to be associated with the presence of heterochromatin flanking regions (Tanksley et al. 1992; Harushima et al. 1998; Vuylsteke et al. 1999). Furthermore, heterogeneities in marker densities along maps have provided evidence of differences in recombination frequencies along the genome (Tanksley et al. 1992).

High-density maps have also offered the opportunity to perform structural, functional, and evolutionary genomic studies that could not have been possible with previous less dense maps. The illustration of structural differences between species, such as that of the five arm inversions between potato and tomato demonstrated by marker order inversion, is one of the most striking contributions of high-density maps. These paracentric inversions have suggested the movement of genetic loci from lower to higher recombination regions or vice versa, with the consequential change in the evolutionary outlook for those loci (Tanksley et al. 1992). Another example is the probable chromosome structural rearrangements between sorghum species as suggested by differences in the abundance of dominant genetic markers and lack of homolog hypomethylated regions. Likewise, incongruities found in the alignment of maize and sorghum genomes suggested chromosomal rearrangements that might have occurred since maize-sorghum divergence (Bowers et al. 2003). It is important to point out that comparisons between taxa were possible by mapping markers derived from heterologous DNA clones.

5.1 Construction of High-Density Linkage Maps

High-density maps are constructed using the same criteria, algorithms and software described throughout this chapter. A distinctive feature of the process worth mentioning is that a framework map is first constructed at a conservative LOD value, using only one marker from each set of cosegregating markers, and all remaining markers are then assigned at a more relaxed LOD score. The chromosomal affiliation of each linkage group is performed by identifying markers of known chromosomal position based on previously published map.

Alternatively, some high-density maps have been generated by joining independent genetic maps based on the same or a combination of different marker classes (Haanstra et al. 1999; Somers et al. 2004). In this case, PCR-based markers, which have steadily gained importance for genetic map construction, have been mapped to formerly developed RFLP maps in order to provide consensus maps across populations within crops. Among these PCR-based markers, AFLPs have become the marker of preference in the construction of high-density maps because of its ability to generate large numbers of markers in a relatively short time (Vos et al. 1995; Vuylsteke et al. 1999). The main requirement for the integration of distinct maps is a minimum number of common markers. Stam (1993) developed a computerized mathematical procedure to join linkage maps taking into account the variation in the precision of estimates of recombination frequencies between data sets and in information type i.e., population type (F_2, BC, RIL) and size. Combining recombination frequencies from different experiments assume identical underlying linkage in all populations, i.e., that the true recombination frequencies are the same in these experiments. The approach is based on assigning weights to the recombination frequencies available for a given pair of markers to then replace all by a single one. The statistical procedure to assign these weights assume that recombination frequency estimates from distinct populations were obtained from binomial samples, which allow the calculation of hypothetical binomial sample sizes that would each have yielded the same LOD value. These hypothetical samples sizes are used as weights to calculate the joint estimate and LOD value, as these weights assigned to distinct estimates of a certain recombination frequency correspond to the amount of information comprised in each estimate. The estimates of pairwise recombination frequencies are obtained by maximum likelihood. The integrated map is then sequentially built-up in a manner by which at each step, a numerical search is performed for the best fitting linear arrangement. Finally, the weighted least squares method is used for the estimation of map distances for a given order of the markers. This method is suitable for distances that can only be estimated indirectly, which occur when working with data from distinct sources in which several recombination frequency estimates are not available.

The procedure was implemented in the author's developed program JoinMap (Stam 1993; Stam and van Ooijen 1995). This program can process raw data (coded genotypes for segregating markers) of various types (F_2s, BC, RILs) and optionally if available, independent estimates of recombination (accompanied by their standard errors) to construct integrated linkage maps. In the case of outbreeders in which F_1 offspring may segregate as a backcross for one marker and as an F_2 type for another, an indication of the segregation type should be supplied for each marker. Furthermore, if information is available on the ordering of certain subsets of markers (e.g. from multipoint tests of closely linked markers), these can also be supplied as a list of fixed sequences in addition to the raw data. In this case, the program will produce an ordering which is not contradictory to any of these known sequences. JoinMap also provides a goodness-of-fit criterion for the integrated map built, corresponding to the two mapping functions (Kosambi's or Haldane's) as an indication to which the data as a whole fit best. An alternative approach for constructing high-density linkage maps in those species already represented by low-density genetic maps, is the application of Vision's selective mapping described earlier.

5.2 Ultra High-Density Maps (UHD-maps)

UHD maps are genome-wide saturated genetic maps made up of thousands instead of hundreds of markers, aiming at achieving the coverage of each and every linkage group of an organism. These maps offer additional advantages and facilities to those of high density maps, though this latter may attain some of them to a lesser extent. For example, high-density maps can only facilitate map-based cloning if the gene of interest happens to fall in a saturated region, while ultra high-density maps virtually guarantee that any gene of interest will be tightly linked to a marker, by their feature of global saturation of the genome. Moreover, smaller distances between markers in ultra high-density maps can facilitate chromosome landing by anchoring genomic clones or contigs, provided that the insert length of a BAC library is larger than the average marker distance. Lastly, ultra high-density maps accomplish the genetic anchoring of a physical map that will culminate in a sequence-ready minimal tiling path of BAC contigs of specific chromosomal region (Klein et al. 2000; van Os et al. 2006). Succeeding in the development of a genetically anchored physical map requires several thousands of BAC contigs to be anchored to the genetic map, a task that requires the identification of mapped markers in fingerprints of BAC pools.

UHD-maps have also enabled researchers to confirm and postulate hypothesis concerning the meiotic behavior of a genome. Thus, non-random patterns of marker distribution have provided insights into the positions of putative recombination hot spots and centromeric regions. Significant clustering of markers on the centromeres confirms the suppression of recombination in these regions, while gaps have been postulated to be associated with recombination hot spot or fixation of the genome (lack of heterozygocity). Marker saturation has also demonstrated the occurrence of more than one chiasma per chromosome arm at least in potato, indicating that there is no point for the assumption of absolute chiasma interference (van Os et al. 2006).

Construction of UHD maps

Developers of methods for the construction of UHD maps faced several problems. One of these was the limited calculation speed of available mapping software for dealing with the huge quantity of data contained in ultra dense mapping data sets. Another issue was the presence, as in all mapping studies, of errors, inconsistencies, and incomplete data (missing observations or incomplete genotype information, i.e., dominant markers), which even in low frequency, cause ordering ambiguities, inflation of map distances, and computing intensive calculations. Hence UHD map construction required the implementation of a time-efficient criterion and an efficient search algorithm to tackle numerous constraints and find the optimal map order. This entailed the development of new software much faster and less sensitive to missing and erroneous data. In the process of developing an ultra dense meiotic linkage map of potato, Isidore et al. (2003) experienced and thoroughly analyzed these difficulties, and systematically devised new approaches. Their 10,000 marker UHD potato map is the densest map based on meiotic recombination yet obtained in any species (van Os et al. 2006).

These authors also developed the recombination bin mapping concept in their UHD map construction approach. They defined a cosegregation bin as a multiply marked chromosomal segment that results from a very large number of co-segregating markers. They referred to the consensus segregation pattern of all those markers in a bin as a bin signature. Since the number of recombination events in a population and not the number of markers, defines the maximum number of bins in a chromosome, UHD maps are frameworks of ordered bins in which all recombination events in the mapping population are identified. Adjacent bins should be separated by a single recombination event. However, all theoretical bins cannot be identified directly from the data, due to situations such as chromosomal segments being either 'identical by descent' or simply physically small. In this case, segregation data from the adjacent filled bins provide enough information to calculate the minimum number of intervening recombination events, allowing empty bins to be inserted between the full ones. This calculating process is performed until each chromosome is represented as a linear string of bins, each separated by a single recombination event. A careful inspection of Figure 8 in which the recombination bin mapping concept is represented, allow us to observe two recombinations between the same two marker loci (Dd and Ee) resulting in the insertion of an empty bin between the CD and the EFGHI bins. According to this recombination bin mapping concept, one can figure out how erroneous data may create mapping conflicts. An erroneous data point appears as an artifactual double recombinant (singleton), and hence introduces two false recombination events. The term singleton in this context indicates the misclassification of a marker phenotype, i.e., a single locus in one plant that appears to have recombined with both its directly neighboring loci (Nilsson et al. 1993). Isidore et al. (2003) illustrated the significance of this, exemplifying that with a marker scoring accuracy of 99%, a single chromosome consisting of 1,000 markers in a population of 100 individuals can potentially introduce 2,000 false recombination events. The consequence is the generation of inflated maps and potentially erroneous marker orders. It is impossible to distinguish between singletons that are due to scoring errors, local DNA inversions, or methylation polymorphism from those occurred by double recombinations.

van Os et al. (2005a) proposed and applied as an optimization criterion to define the best map, the minimum number of recombination events (COUNT). The number of recombination events is easily obtained in a data set of marker segregation data, by counting the number of recombinants per locus pair, and for a given sequence of loci, by adding over adjacent loci. COUNT as other optimization criteria, such as SARF and likelihood, gives the same optimal order for perfect data. Instead, when dealing with incomplete data sets because of missing observations or dominance in an F_2 mapping population, the expected number of crossovers based on the maximum likelihood (ML) estimate of recombination frequency between the two loci is applied. The advantage of COUNT over other optimization criteria is that the COUNT function becomes close to the full likelihood, since it uses observable recombination events for the part of the data which has complete information, and maximum likelihood estimates for the data that are incomplete.

van Os et al. (2005a, b) combined COUNT with a heuristic search algorithm to search for an optimum map order, and implemented them in a program, called RECORD (Recombination Counting and Ordering). The search algorithm uses elements of branch-and bound with local reshuffling, producing optimal orderings without intensive calculations. The branch and bound like algorithm works by constructing a sequence stepwise, starting with a randomly chosen pair of markers, by adding one marker at a time and determining at each step the marker's best position. Then by reshuffling, a window of increasing size is allowed to move from head to tail of the linkage group, and for every position of this window, the sub-sequence within it is inverted, and the resulting COUNT-value calculated. Since the authors proved using simulated data with missing values, that stepwise assembling depends on the order in which markers were added to the sequence, the seriation principal was adopted. This entails the addition at each step, of the marker that is closest to the one at the current head or tail. The procedure is repeated a number of times in order to select the best optimum (best sequence of markers) from the replicated assemblages. The authors also implemented an additional procedure to this software in order to test the certainty of the best sequence found. The procedure searches starting from the best optimum, for 'almost equivalent sequences' -those that induces a pre set additional number of cossovers-, using the same search procedure described above. Then, from the set of sequences that fall within the range of admissible values of COUNT, the distribution of positions for each locus is recorded, and used to provide a quick impression of the local certainty of the sequence by simple inspection.

One important feature of RECORD is that it produces orders (bin orders) and not map positions in centiMorgans as classical mapping softwares. The reason for this is that locus ordering results are more important in UHD maps, since the resolution is primarily dictated by the size of the mapping population, usually not surpassing 1.0-0.25 cM, and under this order of magnitude, exact map distances do not make much sense as their standard error tends to exceed these values.

RECORD has proven to perform well when dealing with marker-dense regions, missing values and relatively high scoring errors. However, scoring errors lead RECORD to ordering ambiguities only when they occur near recombination events. This is because, for the artifactual double recombinant or singleton in question, it is not possible to determine which of the two markers between which a recombination actually occurred (as evidenced by the detection of true recombinants at this point), contains the error. The sensitivity of softwares to this type of errors is of great importance because as the percentage of errors increases, they are more likely to fall in these locations. Furthermore, the penalty for a typing error has been proved to be roughly fivefold than that for a missing observation (van Os et al. 2005a). Taking these constraints into account, the same authors developed a statistical procedure, that identifies and removes singletons (by replacing them with missing values) from data sets (van Os et al. 2005b). This procedure has been implemented in their software called SMOOTH. This program calculates the difference between the observed and predicted values of data points based on data points of neighboring loci in a given marker order. This is perfomed by first calculating the probability that each data point of a segregating marker locus is true on the basis of the genotype of flanking markers, that is for instance, 15 flanking data points on either side, with the nearest data points being given a higher weighting. A threshold for singleton removal can be set. SMOOTH is applied in conjunction with RECORD by cyclically reiterating the process of marker ordering and singleton removal. Initially, a strict probability threshold (p <0.01) is used to remove the least well supported data points and then this threshold is released by p = 0.01 through each reiterative cycle of marker order recalculation and removing of conflicting data points, until no further poorly supported inconsistent data points (i.e., singletons) can be identified. Removing these conflicting data points may occassionally cause adjacent bins to merge (the equivalent of removing a recombination event from the population). As opposed to classical mapping softwares in which an increasing amount of errors become difficult to handle in denser maps, SMOOTH takes advantage of the redundancy in high-density data sets. In high density data sets, it is more likely to count with a greater amount of neighboring data points at close genetic distance to calculate the veracity of a data point at a segregating marker locus than in low density data sets.

RECORD is able to handle only data sets of first generation backcross populations. Thus, for outbreds, such as potato, van Os et al. (2005a) performed several modifications of the raw data: marker data have to be split into three sets on the basis of markers segregating in the maternal, paternal, and in both parents (bridge markers). Within the separate parental data sets, the linkage phase of each marker has to be assessed. To perform this task, they suggest using the 'Quick-And-Dirty' mapping module of JoinMap 2.0 software package (Stam and van Ooijen 1995) in order to calculate marker orders, and then use this information to determine linkage phases by hand, based on the neighboring markers. Thus, all markers that are in repulsion should be converted to coupling phase. The data obtained in this way are comparable to two separate BC_1 populations for each parent or two-way pseudo-testcross (Grattapaglia and Sederoff 1994). Bridge markers are mapped based on the information provided by the parental bin maps. This is performed by assigning these bridge markers into putative bridge bins obtained by superimposing telomeric maternal and paternal bin signatures.

6   LINKAGE MAPS IN POLYPLOIDS

Construction of linkage maps in polyploids is more challenging than that in diploids due to the complex nature of multiple alleles in polyploids. Therefore, many studies on constructing linkage maps have focused on diploid relatives of polyploids, such as in potato (Bonierbale et al. 1988; Gebhardt et al. 1989; van Eck et al. 1994); and alfalfa (Brummer et al. 1993; Diwan et al. 2000; Kaló et al. 2000). However, not all polyploids have a diploid relative available in nature and genetic analyses of these polyploids must be carried out in the polyploid form. On the other hand, though mapping using diploid relatives have been a good alternative for some polyploids, the management of breeding programs in these species is actually carried out at the polyploid level. Therefore linkage maps at high ploidy levels are mandatory, especially if the purpose is to study quantitative trait loci (QTL) of economically important traits, as these maps offer the opportunity to detect higher order interactions between alleles at QTLs.

Several mapping strategies have been developed for mapping in polyploids. For allopolyploids derived from combination of distinct genomes and subsequent chromosome doubling (Soltis and Soltis 2000), statistical methods developed for molecular linkage mapping by estimating recombination fractions between different loci in diploid species (Lander and Green 1987) will also apply. However these methods cannot be used in autopolyploids that are formed due to the chromosome doubling of the same genome or by fusion of unreduced gametes (Soltis and Soltis 2000). Autopolyploids may undergo either bivalent (two chromosomes pair) or multivalent pairing (more than two chromosomes pair) or both, at meiosis, in which a gene has more than one possible partner (or set of partners). However, most of the available statistical methods for autopolyploid linkage analysis assume bivalent pairings (Wu et al. 1992; Hackett et al. 1998; Ripol et al. 1999; Luo et al. 2000, 2001). Theoretical analysis and statistical methods assuming multivalent pairings are under explored because of the complexity of polysomic inheritance. Luo et al. (2006) have presented a method that models quadrivalent pairings as well as a mixture of bivalent and quadrivalent pairings. Early mapping strategies were based on single dose restriction fragments, that is, those markers that segregate 1:1 as proposed by Ritter et al. (1990) for diploid parents, and subsequently extended to tetraploid species by Wu et al. (1992) and applied for crops such as sugarcane (*Saccharum officinarum* L.) (Da Silva et al. 1993) and alfalfa (Brouwer and Osborn 1999).

As such, linkage analysis of autopolyploids has been largely based on the use of single dose dominant markers (e.g., AFLPs and RAPDs) that occur in parents with genotype *Mmmm* where *M* stands for the dominant allele and *m* for the absent or null allele. A parent with such a single dose of the dominant allele is called the simplex. Yu and Pauls (1993) and Da Silva et al. (1993) extended this method to linkage analysis between single (*Mmmm*) and double dose (duplex) markers (*MMmm*) and between duplex markers, both present in one parent and absent in the other. All this required derivation of the expected phenotype frequencies formulae for coupling and repulsion linkages, as well as the maximum likelihood equations for the estimation of recombination fractions.

Both of these derivations can be followed in Hackett et al. (1998), for several combinations of linkages (simplex-simplex, e.g., MN/mn/mn/mn; duplex-simplex, e.g., Mn/MN/mn/mn, or Mn/Mn/mN/mn and duplex-duplex, e.g., MN/MN/mn/mn, MN/Mn/mN/mn or Mn/Mn/mN/mN) for an autotetraploid species, where N is a dominant allele at a linked locus, slash separates homologous chromosomes, m and n are absent or null alleles; and M and N absent in the other parent, i.e., mn/mn/mn/mn. Likewise, Meyer et al. (1998) report formulae for various marker pair configurations of biparental simplex markers, such as simplex (Mn/mn/mn/mn) x double simplex, also called simplex-simplex (MN/mn/mn/mn or Mn/mN/mn/mn); or double simplex (MN/mn/mn/mn or Mn/mN/mn/mn) x double simplex (MN/mn/mn/mn or Mn/mN/mn/mn).

The accuracy of linkage estimates depends on the type of markers involved. The simplex-simplex coupling pairs (MN/mn/mn/mn) are most reliable, followed by double simplex x double simplex coupling (MN/mn/mn/mn x MN/mn/mn/mn), duplex-duplex coupling (MN/MN/mn/mn), simplex x double simplex coupling (Mn/mn/mn/mn x MN/mn/mn/mn); and simplex-duplex (both coupling- MN/mN/mn/mn-, and repulsion- Mn/mN/mN/mn)- whereas the simplex-simplex, and duplex-duplex repulsion or mixed pairs, and other repulsion configurations are least reliable (Meyer et al. 1998). All estimations assumed chromosomes to pair at random, forming bivalents; and absence of double reduction. Double reduction is a phenomenon in which sister chromatids stay together in the same gamete as a result of homologous chromosomes forming a multivalent, causing systematic segregation distortion and complex segregation pattern (Darlington 1929; De Winton and Haldane 1931; Mather 1936; Fisher 1947).

Based on all the above considerations, a first linkage map in tetraploid potato which integrates simplex, duplex, and double–simplex AFLP markers were constructed by Meyer et al. (1998). Mapping took advantage of JoinMap to group and order markers from a set of pairwise recombination fractions and LOD scores previously estimated in another program. Thus, homologous cosegregation groups were constructed by JoinMap using estimates of recombination fractions and LOD scores between pairs of solely simplex markers, whereas the combined map, that merged the homologues into unique linkage groups, was performed by using the complete set of recombination fractions between all types of marker pairs. Duplex markers were used to identify and assemble homologous cosegregation groups into a single group by exploiting their linkage to simplex markers, which in turn allowed determination of the location of its alleles on the respective homologous groups by simple inspection. On the other hand, double simplex markers were useful as bridging markers to identify homologies between parental linkage groups. This first map also showed that even though AFLPs provide the volume of information required for this mapping approach, masking of polymorphism by marker dosage significantly reduces the number of individual AFLP alleles that can be scored. Moreover, when linked in repulsion, dominant simplex markers do not provide reliable estimates of recombination frequencies. Therefore, researchers have recommended the use of co-dominant and highly polymorphic markers, such as SSRs, to increase the power of polyploid mapping (Milbourne et al. 1997; Meyer et al. 1998). SSR markers offer the additional advantage of a means to align intra- and inter-parental maps.

A main concern in polyploid linkage analysis has been to achieve a feasible and accurate prediction of parent genotypes from their phenotype, currently scored as a gel band pattern. This information is essential for distinguishing recombinant and parental genotypic classes. The challenges of efficiently distinguishing allele dosage from gel band patterns, dealing with the complexities of distorted segregation patterns (due to double reduction), and the effect of null alleles on the efficiency of genotype prediction, were confronted from a mathematical angle. Luo et al. (2000) took these complexities into account in their development of a theoretical model based on Bayes's theorem, and a novel computational methodology to reconstruct the genotypes of a pair of parents using their phenotypes together with the segregation information on their progenies' phenotypes, observed at an SSR locus in tetraploid populations.

The conditional probabilities of all possible parental genotypes consistent with their phenotype banding patterns are computationally calculated in three steps. In the first step, all the possible genotypic configurations corresponding to the band pattern of the parents are determined. For example, if a parental phenotype shows two bands, there may be four corresponding genotypes: (1200), (1120), (1220), (1122), (1222), (1112), where 1 and 2 represent different alleles, and 0 denotes a null allele. Then, the gamete distributions for each parental genotype are solved considering different probabilities of double reduction, and the progeny genotypic distribution assembled by random union among parent gamete pools for each possible pair of parental genotypic configurations. In the second step, the corresponding phenotypic distribution for the progeny genotypic distribution of each of the predicted parental genotypic configurations is derived from relating all possible progeny genotypes to their phenotype banding patterns. Here, for each of the predicted parental genotypic configurations, the expected phenotypic classes and their frequencies are calculated. In the third step, observed and expected phenotypic classes are compared, and if some are not present then the given pair of parental genotypic configuration together with the given probability of double reduction is rejected. Otherwise, the likelihood of the observed offspring phenotypes is calculated assuming they are random samples from a multinomial distribution with frequencies of their corresponding marker classes and observed sample sizes, at any fixed value of double reduction coefficient between 0 and 1/6 (Fisher and Mather 1943). The most likely pair of parental genotypes is determined by examining the likelihood function over all possible parental genotypic pair configuration. A goodness of fit test highlights loci where the offspring data do not fit the expected frequencies, allowing alternative hypothesis such as multilocus markers or a mistyped parental banding pattern to be investigated. Besides providing a direct test for double reduction detection, the model considers the calculation of the maximum likelihood estimate of the coefficient of double reduction.

Once the most probable parental genotypic configurations or set of possible pair configurations (if more than one is consistent with all the phenotypic data), tetrasomic linkage analysis may proceed. Luo et al. (2001) outlined the steps in the construction of tetraploid maps assuming bivalent pairing as follows:

1. Detection of linkage between pair of marker loci using a two-way contingency table and testing for independent segregation by Pearson's Chi square.

2. Partitioning of loci into linkage groups by means of a cluster analysis based on the significance of the test of independent segregation, so that markers that segregate independently of others will be on different linkage groups. For this analysis, the authors proposed the transformation of the significance levels (only those from 0 to 0.05 that indicate a more likely linkage between loci) to a measure of distance. Accordingly, the closest markers will be clustered together, and marker loci can be partitioned into linkage groups by inspection of the dendrogram. The authors applied two cluster analyses in order to compare dendrograms, and thus decide on the partition of marker loci into linkage groups. Comparing dendrograms from different clustering methods can help to avoid the inadvertent combination of large groups. The author applied the nearest-neighbor cluster that adds a marker to a cluster according to its distance from the closest marker in the cluster, and the average cluster distance (Anderson 1973).

3. Estimation of recombination frequency and LOD scores for pairs of markers within each linkage group. Estimation of recombination frequencies requires first the calculation of the conditional distribution of the offspring phenotypes at every two linked loci for the most likely parental genotypes obtained as described above. Luo et al. (2001) provide the general formula to calculate the frequencies of all possible gametic genotypes formed at two linked loci of given parental genotypes, taking into account the three gamete classes – nonrecombinants, single recombinants and double recombinants- generated from the three equally-likely pairs of bivalents that may occur during meiosis, i.e., $A_iB_i/A_jB_j//$ A_kB_k/A_lB_l, $A_iB_i/A_kB_k//A_jB_j/A_lB_l$, $A_iB_i/A_lB_l//A_jB_j/A_kB_k$, where A and B are two linked loci, subscripts represent alleles and $//$ distinguishes pair of homologous chromosome. A general formula is also provided to calculate the frequency of zygote genotypes obtained after random union between all possible gametes generated from any pair of parental genotypes. Finally, the phenotypic distribution of the offspring is obtained by combining the probabilities of those genotypes from the offspring genotype distribution that result in the same phenotype. Computer algorithms and routines were developed by the authors to perform these calculations and to work out all possible linkage

phase configurations for any given pair of parental genotype configurations. Linkage phase configuration of the parental genotypes is an essential requirement for the estimation of recombination frequency, and correct construction of homologous linkage groups. Having the joint expected phenotypic distribution of the offspring based on the parental genotypes and their possible linkage phases, recombination frequencies can be estimated by maximum likelihood. Luo et al. (2001) built the likelihood equation of recombination frequency and possible two locus-linkage phases for parents and solved it applying the expectation-maximization (EM) algorithm (Dempster et al. 1977). Here, the two-locus linkage phase configurations of the parents for which the maximum is the highest is taken as the maximum likelihood estimate of the parental linkage phase, and its corresponding value of recombination frequency as its maximum-likelihood estimate. LOD score values are also calculated for all possible phases.

4. Ordering the markers within each linkage group. Once the recombination frequency and LOD score for each pair of markers is available, ordering of markers and calculation of map distances can proceed. An approach that has worked efficiently to calculate the best order was the optimization by 'simulation annealing' of the least squares criterion for estimation of multilocus map distances developed by Stam (1993) (Hackett et al. 2003). An initial order can be obtained by this method. Then, by simulation annealing, random changes are generated (a segment of the order taken at random is set in a reverse direction or transferred to a random location) leading to a decrease or an increase in the least squares criterion with respect to the previous order. The probability of accepting a change is a function of the difference between the least squares criteria of the old and new order and a parameter referred as the 'temperature'. Changes leading to a decrease in the criterion, i.e., those by which the least squares criteria of the new orders are less or equal to the old ones, are always accepted, whereas those that lead to an increase are accepted with a probability that decreases slowly according to the 'temperature'. The set of possible orders is explored by generating a number of random changes at each of the different temperatures. A cooling factor α is applied to reduce the temperature, so as to decrease the probability of accepting a change that increases the least square criterion, and further changes are generated.

The fact that the least square criterion uses the information from all pairs of markers (Stam 1993) is particularly advantageous in an autotetraploid cross, where pairs of markers vary considerably in their information content (Luo et al. 2001). Moreover, this criterion is less sensitive to allele typing errors than multipoint likelihoods (Shields et al. 1991). Low levels of missing values (error rate of 2%) have been proven to have little effect on the estimation of the map applying the simulation annealing approach (Hackett et al. 2003).

5. Reconstruction of the allelic linkage phases of the parental genotypes at the marker loci. Once the best fitting map is obtained, the last step is to reconstruct the parental linkage phases to distinguish the homologous linkage groups. Luo et al. (2001) proposed an intuitive algorithm to predict the multilocus parental linkage phase on the basis of the range of likelihood values of the alternative linkage phases obtained in the two-locus analysis. Here, the phase of the marker pair with the largest log-likelihood difference between its most likely and second most likely linkage phase is reconstructed first. For example, if the largest log-likelihood difference value were 18.72 for a marker pair M_2 and M_3, with the following phase, BO/CB/AC/EA, the allelic linkage phase of the parent would be:

$$
\begin{array}{c}
\text{Markers} \\
\begin{array}{ccccc}
M_2 & B & C & A & E \\
M_3 & O & B & C & A
\end{array}
\end{array}
$$

Then further markers are placed relative to this pair, placing markers with large log-likelihood difference values before those with smaller ones. For the same example, if the next largest log-likelihood difference value were 15.19 for a marker pair M_1 and M_2, with the following phase CB/AC/BA/BE, then marker M_1 would be placed as follows

$$
\begin{array}{c}
\text{Markers} \\
\begin{array}{ccccc}
M_1 & C & A & B & B \\
M_2 & B & C & A & E \\
M_3 & O & B & C & A
\end{array}
\end{array}
$$

The authors indicated that contradictions may arise between the phase of two markers estimated directly and the phase estimated when each of the pair is referred to a third marker. In such a case, they would reject an overall configuration for a pair with large log-likelihood difference value, but accept the overall configuration if the value were close to zero. The inferred linkage phases should be checked to be the most likely ones for all pairs with a substantial difference (d > 3) in their log-likelihoods. For pairs where the inferred phase is not the most likely, estimates of the recombination frequencies and LOD scores should be compared, and if necessary the linkage map should be recalculated on the basis of the inferred phase.

Computer routines developed for each of the steps described above, including that developed by Luo et al. (2000) for the inference of parental genotypes were implemented in the software TetraploidMap developed by Hackett and Luo (2003) which is freely available from the website of Bioinformatics and Statistics Scotland at http://www.bioss.ac.uk. TetraploidMap handles both co-dominant and dominant markers, in all possible configurations, and takes the presence of null alleles into account in the analysis. A population size of at least 150 full sibs can be used although larger numbers –say 250- have been recommended for a better chance of identifying homologous linkage groups and achieving reliable orders. It is important to keep in mind that the power to detect linkage decreases, and the standard error of the estimated recombination frequency increases, as the marker separation increases or the population size decreases (Hackett et al. 1998).

We applied TetraploidMap to develop an AFLP-based linkage map from a dihaploid population derived from the haploidization of a tetraploid potato cultivated accession of *Solanum tuberosum* subsp. *andigena*. The map comprised 442 AFLP markers and three SSRs, with a total length of 1,074.8 cM. A single dominant gene for resistance to potato leafroll virus (PLRV) segregating in double dose (duplex) from the resistant maternal parent was mapped on two homologues of potato chromosome 5 (Velásquez et al. 2007).

7 CONCLUDING REMARKS

Genetic mapping has been successfully employed in gene discovery, identification of markers linked to important traits and the elucidation of their genetic control. These maps have provided frameworks and bridges for QTL mapping, giving scientists the possibility to locate genes in genomes and in germplasm and to predict how QTLs and even traits differ and interact.

Common sets of markers genetically anchored in these frameworks have allowed a wide range of investigations in structural and evolutionary genomics. These studies have shed light on species differentiation and variation on genome organization.

Currently, genetic maps are constructed for physical mapping and development of integrated physical-genetic maps. They will contribute enormous efficiencies in the extension and applicability of emerging genome sequence data.

A unique feature of genetic mapping that cannot be provided by physical maps is the information it provides on the distinctive recombination patterns along the genome that have allowed the identification of recombination hotspots and recombinationally suppressed regions. This information has permitted studies on the distribution of linkage disequilibrium over genome regions, which is the basis of allele-trait association studies in both QTL mapping and association mapping.

We believe that genetic mapping will continue to be an important activity in the coming years for the reasons given above, and the fact that a large number of species remain as yet unrepresented by any of the organisms mapped so far.

References

Allard RW (1956) Formulas and tables to facilitate the calculation of recombination values in heredity. Hilgardia 24(10): 235-278

Anderson MR (1973) Cluster Analysis for Applications. Academic Press, New York.

Bonierbale MW, Plaisted RL, Tanksley SD (1988) RFLP maps based on a common set of Clones reveal modes of chromosomal evolution in potato and tomato. Genetics 120: 1095-1103

Bonierbale MW, Plaisted RL, Pineda O, Tanksley SD (1994) QTL analysis of trichome-mediated insect resistance in potato. Theor Appl Genet 87: 973-987

Bowers JE, Abbey C, Anderson S, Chang C, Draye X, Hoppe A, Jessup R, Lemke C, Lennington J, Li Z, Lin Y, Liu S, Luo L, Marler BS, Ming R, Mitchell S, Qiang D, Reischmann K, Schulze SR, Skinner DN, Wang Y, Kresovich S, Schertz KF, Paterson A (2003) A high-density genetic recombination map of sequence-tagged sites for Sorghum, as a framework for comparative structural and evolutionary genomics of tropical grain and grasses. Genetics 165: 367-386

Brouwer DJ, Osborn TC (1999) A molecular marker linkage map of tetraploid alfalfa (*Medicago sativa* L.). Theor Appl Genet 99: 1194-1200

Brummer EC, Bouton JH, Kochert G (1993) Development of an RFLP map in diploid alfalfa. Theor Appl Genet 86: 329-332

Crane F, Crane YM (2005) A nearest-neighboring-end algorithm for genetic mapping. Bioinformatics 21(8): 1579-1591

Darlington CD (1929) Chromosome behaviour and structural hybridity in the Tradescantiae. J Genet 21: 207-286

Da Silva JAG, Sorrells ME, Burnquist WL, Tanksley SD (1993) RFLP linkage map and genome analysis of *Saccharum spontaneum*. Genome 36: 782-792

De Winton D, Haldane JBS (1931) Linkage in the tetraploid *Primula sinensis*. J Genet 24: 121-144

Dempster AP, Laird NM, Rubin DB (1977) Maximum likelihood from incomplete data via EM algorithm (with discussion). J R Stat Soc Ser B 39: 1-38

Diwan N, Bouton JH, Kochert G, Cregan PB (2000) Mapping of simple sequence repeat (SSR) DNA markers in diploid and tetraploid alfalfa. Theor Appl Genet 101: 165-172

Doerge RW (1996) Constructing genetic maps by rapid chain delineation. J Quant Trait Loci 2: article 6. http://probe.nalusda.gov:8000/otherdocs/jqtl

Doucleff M, Jin Y, Gao F, Riaz S, Krivanek AF, Walker MA (2004) A genetic linkage map of grape utilizing *Vitis rupestris* and *Vitis arizonica*. Theor Appl Genet 109: 1178-1187

Dvorák J, Luo M-C, Yang Z-L (1998) Restriction fragment length polymorphism and divergence in the genomic regions of high and low recombination in self-fertilizing and cross-fertilizing *Aegilops* species. Genetics 148: 423-434

Feingold S, Lloyd J, Norero N, Bonierbale M, Lorenzen J (2005) Mapping and characterization of new EST-derived microsatellites for potato (*Solanum tuberosum* L.). Theor Appl Genet 111: 456-466

Fisher RA (1947) The theory of linkage in polysomic inheritance. Phil Trans Roy Soc Lond B 233: 55-87

Fisher RA, Mather K (1943) The inheritance of style length in *Lythrum salicaria*. Ann Eugen 12: 1-23

Gebhardt C, Ritter E, Debener T, Schachtschabel U, Walkemeier B, Uhrig H, Salamini F (1989) RFLP analysis and linkage mapping in *Solanum tuberosum*. Theor Appl Genet 78: 65-75

Gebhardt C, Ritter E, Barone A, Debener T, Walkemeier B, Schachtschabel U, Kaufmann H, Thompson RD, Bonierbale MW, Ganal MW, Tanksley SD, Salamini F (1991) RFLP maps of potato and their alignment with the homoeologous tomato genome. Theor Appl Genet 83: 49-57

Ghislain M, Trognitz B, Herrera MR, Solis J, Casallo G, Vásquez C, Hurtado O, Castillo R, Portal L, Orrillo M (2001) Genetic loci associated with field resistance to late blight in offspring of *Solanum phureja* and *S. tuberosum* grown under short day conditions. Theor Appl Genet 103: 433-442

Grattapaglia D, Sederoff RR (1994) Genetic linkage map of *Eucalyptus grandis* and *E. urophylla* using a pseudo-testcross strategy and RAPD markers. Genetics 137: 1121-1137

Haanstra JPW, Wye C, Verbakel H, Meijer-Dekens F, van den Berg P, Odinot P, van Heusden AW, Tanksley SD, Lindhout P, Peleman J (1999) An integrated high-density RFLP-AFLP map of tomato based on *Lycopersicum esculentum* x *L pennellii* F_2 populations. Theor Appl Genet 99: 254-271

Hackett CA, Bradshaw JE, Meyer RC, McNicol JW, Milbourne D, Waugh R (1998) Linkage analysis in tetraploid species: a simulation study. Genet Res Camb 71: 143-154

Hackett CA, Luo ZW (2003) TetraploidMap: Construction of a linkage map in autotetraploid species. J Hered 94(4): 358-359

Hackett CA, Pande B, Bryan GJ (2003) Constructing linkage maps in autotetraploid species using simulated annealing. Theor Appl Genet 106: 1107-1115

Haldane JBS (1919) The combination of linkage values, and the calculation of distance between the loci of linked factors. J Genet 8: 299-309

Harushima Y, Yano M, Shomura A, Sato M, Shimano T, Kuboki Y, Yamamoto T, Lin SY, Antonio BA, Parco A, Kajiya H, Huang N, Yamamoto K, Nagamura Y, Kurata N, Khush GS, Sasaki T (1998) A high-density rice genetic linkage map with 2275 markers using a single F_2 population. Genetics 148: 479-494

Hemmat M, Weeden NF, Maganaris AG, Lawson DM (1994) Molecular marker linkage map for apple. J Hered 85: 4-11

Isidore E, van Os H, Andrzejewski S, Bakker J, Barrena I, Bryan GJ, Caromel B, van Eck HJ, Ghareeb B, De Jong W, van Koert P, Lefebvre V, Melbourne D, Ritter E, Douppe van der Voort J, Rousselle-Bourgeois F, van Vliet J, Waugh R (2003) Toward a marker-dense meiotic map of the potato genome: Lessons from linkage group I. Genetics 165: 2107-2116

Jansen J, De Jong AG, van Ooijen (2001) Constructing dense genetic linkage maps. Theor Appl Genet 102: 1113-1122

Kaló P, Endre G, Zimányi L, Csanádi G, Kiss GB (2000) Construction of an improved linkage map of diploid alfalfa (*Medicago sativa*). Theor Appl Genet 100: 641-657

Klein PE, Klein RR, Cartinhour SW, Ulanch PE, Dong J, Obert JA, Morishige DT, Schlueter SD, Childs KL, Ale M, Mullet JE (2000) A high-throughput AFLP-based method for constructing integrated genetic and physical maps: progress toward a sorghum genetic map. Genome Res 10: 789-807

Kosambi DD (1944) The estimation of map distances from recombination values. Ann Eugen 12: 172-175

Lander E, Green P (1987) Construction of multilocus genetic linkage maps in humans. Proc Natl Acad Sci USA 84: 2363-2367

Lander E, Green P, Abrahamson J, Barlow A, Daley M, Lincoln S, Newburg L (1987) MAPMAKER: An interactive computer package for constructing primary genetic linkage maps of experimental and natural populations. Genomics 1: 174-181

Liu BH, Knapp SJ (1990) GMENDEL: A program for Mendelian segregation and linkage analysis of individual or multiple progeny populations using log-likelihood ratios. J Hered 81: 407

Luo ZW, Hackett CA, Bradshaw JE, McNicol JW, Milbourne D (2000) Predicting parental genotypes and gene segregation for tetrasomic inheritance. Theor Appl Genet 100: 1067-1073

Luo ZW, Hackett CA, Bradshaw JE, McNicol JW, Milbourne D (2001) Construction of a genetic linkage map in tetraploid species using molecular markers. Genetics 157: 1369-1385

Luo ZW, Zhang Z, Leach L, Zhang RM, Bradshaw JE, Kearsey MJ (2006) Constructing Genetic linkage maps under a tetrasomic model. Genetics 172: 2635-2645

Mather K (1936) Segregation and linkage in autotetraploids. J Genet 32: 287-314

Menz MA, Klein RR, Mullet JE, Obert JA, Unruh NC, Klein PE (2002) A high-density genetic map of *Sorghum bicolor* (L.) Moench based on 2960 AFLP®, RFLP, SSR markers. Plant Mol Biol 48: 483-499

Meyer RC, Milbourne D, Hackett CA, Bradshaw JE, McNicol JW, Waugh R (1998) Linkage analysis in tetraploid potato and association of markers with quantitative resistance to late blight (*Phytophtora infestans*). Mol Gen Genet 259: 150-160

Milbourne D, Meyer RC, Bradshaw JE, Baird E, Bonar N, Provan J, Powell W, Waugh R (1997) Comparision of PCR-based marker systems for the análisis of genetic Relationships in cultivated potato. Mol Breed 3: 127-136

Milbourne D, Meyer RC, Collins AJ, Ramsay LD, Gebhardt C, Waugh R (1998) Isolation, characterization and mapping of simple sequence repeats loci in potato. Mol Gen Genet 259: 233-245

Nilsson N-O, Sall T, Bengtsson BO (1993) Chiasma and recombination data in plants: Are they compatible? Trends Genet 9(10): 344-348

Ripol MI, Churchill GA, da Silva JAG, Sorrels M (1999) Statistical aspects of genetic mapping in autopolyploids. Gene 235: 31-41

Ritter E, Gebhardt C, Salamini F (1990) Estimation of recombination frequencies and construction of RFLP linkage maps in plants from crosses between heterozygous parents. Genetics 125: 645-654

Schiex T, Gaspin C (1997) CARTHAGENE: constructing and joining maximum likelihood genetic maps. Proc Intl Conf of Intelligent Syst for Mol Biol 5: 258-267

Shields DC, Collins A, Buetow KH, Morton NE (1991) Error filtration, interference and human linkage map. Proc Natl Acad Sci USA 88: 6501-6505

Simon R, Bonierbale MW (2003) IntiMap - A Program for Comparing and Publishing Genomic Maps. In: Plant and Animal Genome XI Conf, San Diego, CA, USA. http://www.intl-pag.org/pag/11/abstracts/P10c_P894_XI.html

Sobral BWS, Honeycutt RJ (1993) High output genetic mapping of polyploids using PCR-generated markers. Theor Appl Genet 86: 105-112

Soltis PS, Soltis DE (2000) The role of genetic and genomic attributes in the success of polyploids Proc Natl Acad Sci USA 97: 7051-7057

Somers DJ, Isaac P, Edwards K (2004) A high-density microsatellite consensus map for bread wheat (*Triticum aestivum* L.). Theor Appl Genet 109: 1105-1114

Stam P (1993) Construction of integrated-linkage maps by means of a new computer packages:JoinMap. Plant J 3(5): 739-744

Stam P, van Ooijen J (1995) JoinMap™ version 2.0: Software for the calculation of genetic linkage maps. CPRO-DLO, Wageningen, The Netherlands

Tanksley, SD, Ganal MW, Prince JP, de Vicente MC, Bonierbale MW, Broun P, Fulton TM, Giovannoni JJ, Grandillo S, Martin GB, Messeguer R, Miller JC, Miller L, Paterson AH, Pineda O, Roder MS, Wing RA, Wu W, Young ND (1992) High density molecular linkage map of the potato and tomato genomes. Genetics 132: 1141-1160

Thompson EA (1987) Crossover counts and likelihood in multipoint linkage analysis. IMA J Math Appl Med Biol 4: 93-108

van Eck HJ, Jacobs JM, Stam P, Ton J, Stiekema WJ, Jacobsen E (1994) Multiple alleles for tuber shape in diploid potato detected by qualitative and quantitative genetic analysis using RFLPs. Genetics 137: 303-309

van Os H (2005) The construction of an ultra-dense genetic linkage map of potato. PhD Thesis. Wageningen Univ, The Netherlands. ISBN 90-8504-221-6. http://library.wur.nl/wda/dissertations/dis3777.pdf

van Os H, Andrzejewski S, Bakker E, Barrena I, Bryan GJ, Caromel B, Ghareeb B, Isidore E, de Jong W, van Koert P, Lefebvre V, Milbourne D, Ritter E, Rouppe van der Voort JNAM, Rousselle-Bourgeois F, van Vliet J, Waugh R, Visser RGF, Bakker J, van Eck HJ (2006) Construction of a 10,000-marker ultradense genetic recombination map of potato: Providing a framework for accelerated gene isolation and a genomewide physical map. Genetics 173: 1075-1087

van Os H, Stam P, Visser RGF, van Eck HJ (2005a) RECORD: a novel method for ordering loci on a genetic linkage map. Theor Appl Genet 112(1): 30-40

van Os H, Stam P, Visser RGF, van Eck HJ (2005b) SMOOTH: a statistical method for successful removal of genotyping errors from high-density genetic linkage data. Theor Appl Genet 112(1): 187-194

Velásquez AC, Mihovilovich E, Bonierbale M (2007) Genetic characterization and mapping of major gene resistance to potato leafroll virus in *Solanum tuberosum* ssp. *andigena*. Theor Appl Genet 114: 1051-1058

Vision TJ, Brown DG, Shmoys DB, Durres RT, Tanksley SD (2000) Selective Mapping: A strategy for optimizing the construction of high-density linkage maps. Genetics 155: 407-420

Vos P, Hogers R, Bleeker M, Reijans M, van de Lee T, Hornes M, Frijters A, Pot J, Peleman J, Kuiper M, Zabeau M (1995) AFLP: new technique for DNA fingerprinting. Nucl Acids Res 23: 4407-44014

Vuysteke M, Mank R, Antonise R, Bastiaans E, Senior ML, Stuber CW, Melchinger AE, Lubberstedt T, Xia XC, Stam P, Zabeau M, Kuiper M (1999) Two high-density AFLP® linkage maps of *Zea mays* L.: analysis of distribution of AFLP markers. Theor Appl Genet 99: 921-935

Watanabe KN (1994) Molecular genetics. In: JE Bradshaw and GR Mackay (eds) Potato Genetics. CAB International, Cambridge University Press, UK, pp 213-235

Wu KK, Burnquist W, Sorrells ME, Tew TL, Moore PH, Tanksley SD (1992) The detection and estimation of linkage in polyploids using single-dose restriction fragments. Theor Appl Genet 83: 294-300

Yu J-K, Dake TM, Singh S, Benscher D, Li W, Gill B, Sorrells ME (2004) Development and mapping of EST-derived simple sequence repeat markers for hexaploid wheat. Genome 47: 805-818

Yu KF, Pauls KP (1993) Segregation of random amplified polymorphic DNA markers and strategies for molecular mapping in tetraploid alfalfa. Genome 36: 844-851

5 | Mapping and Tagging of Genes Controlling Simple-inherited Traits

Ilan Paran[1*] and Ilan Levin[2]

[1]Department of Plant Genetics and Breeding, Agricultural Research Organization, The Volcani Center, P. O. Box 6, Bet Dagan, 50250, Israel

[2]Institute of Plant Sciences, Agricultural Research Organization, The Volcani Center, P. O. Box 6, Bet Dagan, 50250, Israel

*Corresponding author: iparan@volcani.agri.gov.il

1 INTRODUCTION

1.1 Simple-Inherited Traits

Genetics, the study of inheritance, explains why offspring both resemble and differ from their parents. Offspring and parents may display a range of identical and diverse characteristics, which are usually referred to by geneticists and breeders as traits. Eye-color in humans and petal-color in flowering plants are characteristics commonly used as examples for highly visible traits. However, a certain individual organism may be characterized by myriad characteristics, or traits, many of which may not be easily seen, scored or measured. These countless traits form the characteristic phenotype of that individual organism. Plant geneticists and breeders seek to identify a desired trait and propagate it into a certain population, thus creating a uniform population in which all individuals carry the same desired trait or phenotype. This work of creation probably led the British scientist William Bateson in a personal letter to Adam Sedgwick, dated April 18, 1905, to offer the term 'genetics' from the Greek word *genno* (γεννώ), meaning "I give birth". Bateson first used this term publicly at the Third International Conference on Genetics (London, England) in 1906.

Given that many desired traits are not easily noticeable; the identification of such traits is sometimes considered to be an art. In many instances such identification also requires special instrumentations and dedicated technologies. Good examples of such traits are sugar and carotenoid content of vegetables and fruits, metabolites often associated with taste and functional qualities of these agricultural products, respectively. Once scored, quantified or measured geneticists and breeders can rank individuals within a certain population according to levels of expression of a desired phenotype and select for individuals with outstanding performances or traits. Furthermore, we seek to transmit or introgress these valuable traits into the different populations by breeding, i.e. a time-consuming repeated process of crosses and selection. The time requirement and cost of a breeding process is highly dependent upon the visual characteristics of the target traits, the number of genes that control their phenotypic characteristics, whether they can be measured in early developmental stage, and whether the desired traits are associated with undesired characteristics.

Traditionally, traits are divided into two categories: 1. Qualitative or Mendelian traits, and 2. Quantitative traits. Qualitative traits are those traits controlled by one or two (and rarely a few) genes, none of which are significantly influenced by the environment. The effect of each gene is typically 'large' and discernable in nature and overall, will result in discrete, observable, phenotypic classes (Figure 1; Tables 1 and 2). For

Table 1 Examples of simple-inherited traits, encoded by a single gene already cloned, in selected crop plants.

Plant	Trait	Gene symbol	Reference
Tomato	Pseudomonas resistance	PTO	Martin et al. 1993
	High pigment	HP-1	Mustilli et al. 1999
	Ripening-inhibitor	RIN	Vrebalov et al. 2002
	High pigment	HP-2	Lieberman et al. 2004
Potato	Late blight resistance	R3a	Huang et al. 2005
Maize	Vernalization	VRN1	Yan et al. 2003
	Barren stalk1	BA1	Gallavotti et al. 2004
	Vernalization	VRN2	Yan et al. 2004
	Teosinte glume architecture	TGA1	Wang et al. 2005
	Inflorescence architecture	RA2	Bortiri et al. 2006b
Barley	Stem rust-resistance	RPG1	Brueggeman et al. 2002

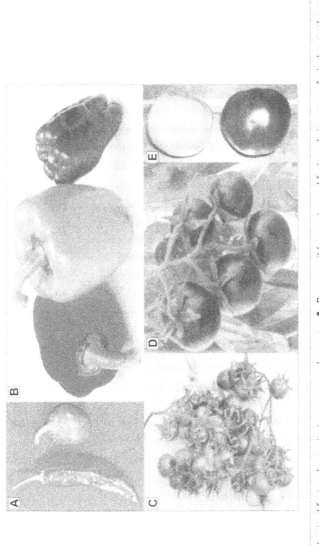

Fig. 1 Simple-inherited fruit color traits in pepper and tomato. **A.** Pepper wild type ripe red fruit and ripe green fruit due to the recessive mutation *chlorophyll retainer* that inhibits chlorophyll degradation during ripening. **B.** Green immature pepper fruit, ripe red fruit and ripe yellow fruit due to the recessive mutation *y* caused by inhibition of the synthesis of the red pigments. **C.** Tomato fruits of mutant plants carrying the dominant *Aubergine (Abg)* mutation characterized by accumulation of anthocyanins in fruit epidermis. **D.** Tomato fruits of mutant plants carrying the dominant *Anthocyanin fruit (Aft)* mutation characterized by accumulation of anthocyanins in fruit peel and upper epidermis. **E.** Wild type mature green tomato fruit (upper fruit) and mature green fruit carrying the recessive *hp-2^{dg}* mutation, also known as *darg green (dg*, lower fruit).

Table 2 Examples of morphological traits encoded by single genes mapped to the tomato genome (http://tgrc.ucdavis.edu/).

Symbol	Name	Phenotype	Generated by	Chromosome
a	anthocyaninless	Stems and leaves always lack anthocyanin	Spontaneous mutation	11
B	β-carotene	High β-carotene, low lycopene in ripe fruit	Spontaneous mutation	6
c	potato leaf	Number of leaf segments reduced	Spontaneous mutation	6
d	dwarf	All parts foreshortened; leaves dark and rugose	Spontaneous mutation	2
e	entire	Fewer fused leaf segments; mid-vein distorted	Spontaneous mutation	4
f	fasciated fruit	Fruits many-loculed	Spontaneous mutation	11
glau	glaucescens	Leaves shortened, dull green to yellowish gray-green; small upright habit	Radiation induced	8
h	hairs absent	Large trichomes absent except on hypocotyl and at growing point; incompletely dominant	Spontaneous mutation	10
ic	inclinata	Small bush; short internodes, leaves; older leaves epinastic, yellowish	Radiation induced	–
j	jointless	Jointless pedicels; proliferates inflorescence	Spontaneous mutation	11
Lpg	Lapageria	Leaves small, dark green, glossy; flowers campanulate; homozygote sterile	Spontaneous mutation	1
ms-02	male-sterile-2	Anthers pale, shrunken; no pollen	Spontaneous mutation	2
n	nipple-tip	Nipple tips at stylar end of fruits	Spontaneous mutation	5

qualitative traits, the individual's phenotype is usually a clear representation of its genotype. These traits are also known as simple-inherited traits. The genetic discipline that focuses on qualitative traits is known as Mendelian genetics. Traits such as seed color, seed shape, pod shape, pod color and petal color observed by Gregor Mendel among pea plants are good examples of such traits. These traits enabled Mendel to formulate the two basic laws of genetics: The law of segregation and the law of independent assortment. These two laws, formulated between 1856 and 1863, are still fundamental in predicting the mode of genetic inheritance of traits.

Unlike qualitative traits, quantitative traits are influenced by multiple genes (~ three genes or more) as well as the effects of the environment. These traits display continuous phenotypic classes and are also known as polygenic or complex traits. Examples of such traits are: yield, fruit size, and cotton fiber quality in plants; growth rate, meat quality, and milk production in animals; height, weight and blood pressure in humans. Methodologies are currently available that can dissect the genetic basis of quantitative traits into distinct genes that can be eventually identified and cloned, thus bridging the gap between qualitative and quantitative traits. Good examples of such gene identification are *FW2.2* controlling fruit size (Frary et al. 2000), and *LIN5* and *AGPL1* controlling fruit sugar content in tomato (Schaffer et al. 2000; Fridman et al. 2004).

This chapter focuses on mapping and tagging of simple-inherited traits. Many of these traits share profound importance from both practical as well as scientific perspectives. Throughout this chapter we used many sources of information, most of which are cited herein below. It should be however noted that we have relied quite extensively on several recent comprehensive reviews that summarize in good order the history of genetic mapping and several of the principle techniques involved in such activity (Harper and Cande 2000; Paterson 2002; Jenkins 2003; Peters et al. 2003; Alonso-Blanco et al. 2005; Bortiri et al. 2006a).

1.2   The Early History of Genetic Maps

Chromosomes were first described cytologically in the 1870s. While it was suspected for many years, Bridges determined conclusively that chromosomes carry the genetic material in 1916 (Bridges 1916; Sturtevant 1965). The first genetic map, describing position and order of genes along the chromosome, however, was created earlier when Sturtevant developed a map of the *Drosophila* sex chromosome (Sturtevant 1913, 1965).

A genetic map can be defined as an abstract map of chromosomal loci, based on recombinant frequencies. The foundations for genetic mapping were laid by Thomas Hunt Morgan and his contemporaries.

Following the discovery of partial linkage between two characters in *Lathyrus odoratus* (Bateson et al. 1905), Morgan confirmed the existence of this phenomenon in segregating populations of *Drosophila melanogaster* (Morgan 1911a, b). Since the genes for the characters he was using were both sex linked, Morgan concluded that there must be exchange of material between homologous chromosomes, in accordance with an earlier hypothesis of chromosome exchange formulated by De Vries (1903). Morgan and Cattell (1912) proposed the term 'crossing over' to describe this physical exchange, which we now accept as being manifested cytologically as a chiasma- a cross-shaped structure commonly observed between nonsister chromatids during meiosis. Morgan's student, Arthur Sturtevant, realized the potential of this observation by pointing out that the proportion of crossovers could be used as an index of the distance between any two genes. He named this proportion centiMorgan (cM) for his mentor T. H. Morgan. One cM is defined as the genetic distance between two loci with a recombination frequency of 1% and is also often called a 'map unit'. These pioneering studies enabled the construction of the first linkage map (Sturtevant 1913), and paved the way for genetic mapping in many different species.

Several features are required to place a gene on a genetic map: two different alleles of the gene, two different alleles of a second gene to map relative to, and an opportunity for meiotic recombination. The latter is achieved by constructing segregating population, usually Filial$_2$ (F$_2$) or backcross (BC) populations, originating between two phonotypically or genetically divergent parental lines. Genetic maps report the linear order of genes and the amount of recombination between linked genes. They do not contain information on physical distance, neither cytological distance nor number of DNA base pairs between markers.

1.3 The Evolution of Genetic Maps

Earlier maps were based on simple-inherited traits, showing visible and discrete variation in a population (Figure 2). Color and morphological traits that show clear segregation patterns in populations generated by cross-breeding were among these traits (Table 2). Initially, such traits were identified as abnormal or unusual phenotypes spontaneously appearing among pure-bred lines or population. Later on mutant phenotypes were induced, mostly in plants, either by radiation, including: fast neutron and gamma radiation, or by chemical treatments, including: ethyleneamine and ethyl-methane sulfonate (EMS). Today, transgenic plants generated by over-expression or silencing of specific genes are also considered as mutant accessions when they show a

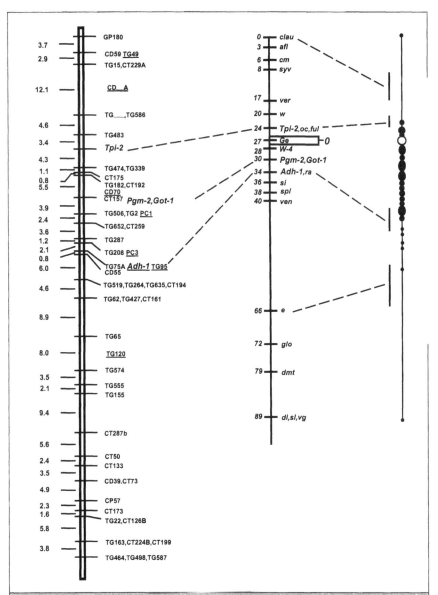

Fig. 2 Molecular linkage map of the tomato chromosome 4 (left) in comparison with its classical map (center) and cytological (pachytene) map (right). Positions of markers from the classical map on both the molecular and cytological maps are shown by dashed lines. Positions of markers from classical map on pachytene chromosome are based on deletion mapping (Khush and Rick 1968). Maps were taken from Tanksley et al. (1992), and presented with the kind permission of SD Tanksley.

particular phenotype that is different from their azygous control counterparts.

Five distinctive morphological traits: dwarf plant habit, potato leaf, peach (fuzzy) fruits, yellow fruit flesh and colorless fruit epidermis, identified among tomato accessions, and cited as early as 1905, are good examples of such traits (Halsted et al. 1905). Monogenic inheritance for these traits was demonstrated later on, in addition to two supplementary traits: lutescent foliage and pyriform fruit shape (Price and Drinkard 1908). The first linkage study in tomato, which ranks as one of the classical organism for such studies, can be traced back to Jones (1917), who reinterpreted data of Hedrick and Booth (1907) on the co-segregation of dwarfness (d) and elongated (ovate) fruit shape (o) as the consequence of linkage between them. Lindstrom followed with an intensive study of linkage on chromosome 2, utilizing Jones' markers in addition to peach fruit (p) and compound inflorescence (s) (Rick 1991). Throughout the years, about 1,200 such traits were mapped or assigned to the tomato 12 chromosomes (Stevens and Rick 1986). However, since then another 3417 induced mutations have been cataloged (Menda et al. 2004), and a collection of those have been curated *(http://zamir.sgn.cornell.edu/ mutants/)*. Among them are most of the previously described phenotypes from the monogenic mutant collection of the Tomato Genetics Resource Center *(http://tgrc.ucdavis.edu/)*, and over a thousand new mutants, with multiple alleles per locus. This rather new, ambitious and successful attempt to saturate the tomato genome with induced mutations indicated that some organs such as leaves are more prone to alterations as others (Menda et al. 2004). This study is also a comprehensive resource study describing well how saturated muta-tional studies should be carried out and how phenotypes could be categorized. In this study a total of 13,000 M_2 families, derived from EMS and fast-neutron mutagenesis, were visually phenotyped in the field and categorized into a morphological catalog that includes 15 primary and 48 secondary categories. These categories, presented in Table 3, are a good demonstration of the type and nature of mutant morphological traits that can be discovered in plant species.

First generation maps based on qualitative mutant traits are still limited in terms of size, density, and further progress. These limitations are due to the fact that in order to map two distinct traits, one needs to construct a specific segregating population between two parents that differ in the phenotypic expression of these two traits. At times two parents could be found that display variation in several such traits, enabling their mapping in a single segregating population, still the need to construct many different populations hampered rapid progress in

Table 3 Tomato mutants phenotypic catalog (Menda et al. 2004).

Major category	Sub category
Seed	Germination
	Seedling lethality
	Slow germination
Plant size	Extremely small
	Small plant
	Large plant
Plant habit	Internode length
	Branching
	Aborted growth
	Other plant habit
Leaf morphology	Leaf width
	Leaf size
	Leaf complexity
	Leaf texture
	Other leaf development
Leaf color	Purple leaf
	White leaf
	Yellow leaf
	Yellow-green leaf
	Dull green/gray leaf
	Dark green leaf
	Variegation
Flowering	Flowering timing
Inflorescence	Inflorescence structure
Flower morphology	Flower homeotic mutation
	Flower organ size
	Flower organ width
	Other flower morphology
Flower color	White flower
	Pale yellow flower
	Strong yellow flower
Fruit size	Small fruit
	Large fruit
Fruit morphology	Long fruit
	Rounded fruit
	Other fruit morphology
Fruit color	Yellow fruit
	Orange fruit
	Dark red fruit
	Epidermis
	Green fruit

Contd.

Contd.

Fruit ripening	Early ripening
	Late ripening
Sterility	Partial sterility
	Full sterility
Disease and stress response	Necrosis
	Wilting
	Other disease response

establishing denser linkage maps in plant species. Another reason for the delayed progress in genetic mapping, was deleterious effects that the expression of several mutant phenotypes could have when incorporated into a single stock.

The advent of molecular markers, based on protein and DNA molecules, advanced substantially the size and density of molecular maps. This was primarily due to the fact that two genetic stocks may harbor many DNA and protein sequence polymorphisms that are phenotypically neutral. Therefore, protein and DNA polymorphisms may be identified even between phenotypically identical genotypes. This is a significant advantage compared to traditional phenotypic markers, which led the way to the creation of dense genetic linkage maps.

The first molecular markers used were isozymes. Isozymes are copies of an enzyme that differ slightly in amino acid sequence. They are often detected by electrophoresis of proteins in a gel followed by identification of the position of the protein based on stains that are deposited as a consequence of catalytic activity of the enzyme. Homozygous individuals will have a single form of the enzyme, while heterozygous individuals will usually have 2 forms of the enzyme, and in some cases 3 forms, if the enzyme functions also as a dimmer. The key for mapping is that the equivalent enzyme forms from two genetic lines and/or species may migrate to different position on the gel and thus can be distinguished. For example, two alleles of alcohol dehydrogenase could both perform the correct enzymatic function, but the electrophoretic mobility of the two may differ. Therefore, two alleles would not migrate to the same location in a starch gel. The procedure used to identify isozyme variation is simple. A crude protein extract is made from some tissue sources, usually leaves. The extracts are next separated by electrophoresis in a starch gel. The gel is then placed in a solution that contains reagents required for the enzymatic activity of the enzyme you are monitoring. In addition, the solution contains a dye that the enzyme can catalyze into a color reagent that stains the protein. In this manner, allelic variants of the protein can be visualized in a gel.

An example for the use of a combination of morphological and isozyme markers is the double tagging of male-sterility gene in tomato (Tanksley and Zamir 1988). Male-sterility genes are useful for hybrid seed production, eliminating the need to emasculate the female parent. Several nuclear male-sterility genes have been identified in tomato but their use suffers from two drawbacks. First, because male-sterility is recessive, its transfer in backcross programs requires progeny testing. Second, once the gene has been transferred into the parent line, it will continue to segregate which requires rouging fertile plants before pollination. To overcome these problems, a male-sterility gene *ms-10* was placed in *cis* linkage configuration with the co-dominant isozyme marker *Prx-2* (at 0.5 cM distance) and the recessive morphological marker *aa* (at 5 cM distance) which results in the absence of anthocyanine in the seedling hypocotyls. The use of the co-dominant *Prx-2* marker allows selection of heterozygous plants during backcrossing, eliminating the need of progeny testing, while the use of the recessive morphological marker allows rouging of fertile plants at the seedling stage. Additional examples of isozyme loci associated with agriculturally important traits are listed in Table 4.

Table 4 Examples of isozyme loci associated with agriculturally important traits.

Trait	Species or plant	Isozyme locus	Reference
Resistance to Watermelon Mosaic Virus 2	*Cucurbita ecuadorensis*	*Aldo-p*	Weeden et al. 1984
Chilling tolerance	*Glycine max (L.) Merr.*	*Apx1*	Funatsuki et al. 2003
Resistance to Fusarium wilt race 1	*Pisum sativum L.*	*Lap-1*	Grajal-Martin and Muehlbauer 2002
Resistance to root-knot nematode	*Solanum lycopersicum L.*	*Asp-1*	Rick and Tanksley 1983
Plant habit	*Cicer arietinum L.*	*Pgd-c*	Kazan et al. 1993
Photoperiod sensitivity	*Oryza sativa L.*	*Pgi-2*	Mackill et al. 1993

Several drawbacks should be noted with regard to isozyme markers. First, the number of isozyme loci that can be scored is limited. To date, only 40-50 reagent systems have been developed that permit the staining of a particular protein in starch. Furthermore, not all of these reagent systems work efficiently with all plant species. Therefore, for many species only 15-20 loci can be mapped. A second drawback is tissue variability. Some isozymes are better expressed in certain tissues such as

polymorphisms (RFLPs) and the technique of fluorescence in situ hybridization (FISH). Cytogenetic maps reveal intriguing insights into the structure-function relationship of chromosomes, especially in regard to chromosome function during meiosis (Harper and Cande 2000; Jenkins 2003). They do not however report the exact absolute physical-distance among markers or genes. Such absolute physical-distance should be measured as the number of DNA base pairs between loci, reveled by DNA sequencing technologies (Maxam and Gilbert 1977; Sanger et al. 1977).

Traditionally, genome sequence data have not been included on genetic, cytological or cytogenetic maps. Maps of DNA sequences are physical maps, and show the position of DNA motifs or DNA marker sequence relative to an absolute scale in base pairs. For example, a map of overlapping clones forming a contig is a physical map. Within the framework of the plant genome sequencing efforts, these data have already been included, thus creating a new type of map that integrates cytogenetic data with DNA sequence information (Harper and Cande 2000; Jenkins 2003).

1.5 Generating Cytogenetic Maps

In general, two approaches have been used to integrate genetic and cytological maps (Harper and Cande 2000; Jenkins 2003). The first, and much more widely-used method, is to determine the location of genetically mapped markers relative to chromosome breakpoints. Chromosomal rearrangements such as deficiencies, duplications, inversions and translocations all contain breakpoints. Using gene order information from the genetic map, groups of genes, which are lost or gained on either side of a breakpoint are determined, and two markers closely flanking the breakpoint can be identified. This establishes the position of the breakpoint on the genetic map. This is then correlated with the physical position of the breakpoint as determined by squashes of meiotic prophase or mitotic metaphase chromosomes. A variety of chromosomal rearrangements have been used for this type of breakpoints analysis including, but not limited to, inversions, A–A translocations (reciprocal translocations between two different chromosomes of the normal complement), A–B translocations (also called TB translocations; reciprocal translocations between one chromosome of the normal or A complement and one supernumerary B chromosome), secondary trisomics (one pair of homologous chromosomes plus one iso-chromosome representing only one chromosome arm), tertiary trisomics (one pair of homologous chromosomes plus one translocation chromosome) or terminal or interstitial deficiencies (chromosomes missing a terminal or internal part, respectively) (Appels et al. 1998). Breakpoints can also be generated by irradiation (e.g., Khush and Rick 1968) or by special properties of the species, such as the gametocidal factor on certain alien chromosomes introduced into wheat (Endo and Gill 1996; Nasuda et al. 1998). Breakpoint analysis has been used to generate cytogenetic maps in tomato (Khush and Rick 1968), maize (Coe 1993), wheat (Gill et al. 1996), barley (Kunzel et al. 2000), rye (Alonso-Blanco et al. 1994), pea (Hall et al. 1995; Hall et al. 1997), rice (Singh et al. 1996a, b; Delos Reyes et al. 1998) and other plants.

A second way to integrate maps is by direct hybridization of genetically mapped sequences onto chromosomes. This is done by DNA–DNA in situ hybridization techniques. The first in situ hybridization experiments used radioactive probes and photographic emulsion for detection (Gall and Pardue 1969; John et al. 1969). The preferred choice at present is fluorescent detection and mapping of hybridization sites, since fluorescence is safer, faster, and has the capability of mapping several DNA sequences simultaneously and at high resolution. This fluorescence in situ hybridization (FISH) technique and its applications have already been described in several manuals (Wilkinson 1992; Leitch et al. 1994; Schwarzacher and Heslop-Harrison 2000). Efforts to integrate cytological and genetic maps have been made using FISH with repetitive and single-copy probes (Gustafson et al. 1990; Lehfer et al. 1993; Leitch and Heslop-Harrison 1993; Pedersen et al. 1995; Pedersen and Linde-Laursen 1995; Jiang et al. 1996; Ren et al. 1997; Fuchs et al. 1998). For example, Pedersen et al. (1995) determined the cytological position of ten loci by FISH in barley. This allowed anchoring parts of the genetic map onto the barley mitotic metaphase cytological map. As with the deletion breakpoint mapping, they found these loci to reside at the distal chromosome tips, showing the increase of recombination towards the tips. In tomato, Peterson et al. (1999) used as FISH probes three single/low copy genomic lambda clones containing RFLP probe sequences that genetically mapped to chromosome 11. To their surprise, they discovered a different arrangement of markers compared to the order in the genetic map. This illustrates an advantage of FISH mapping; an unambiguous gene order can be determined. The three RFLP markers were located on the euchromatic arms, and not in the centric heterochromatin. The longest possible genetic distance between the three RFLPs comprised 35% of the genetic map, yet physically they occupy only the distal 19% of the chromosome.

2 IDENTIFICATION OF MARKERS LINKED TO TRAITS OF INTEREST

In order to identify molecular markers linked to simple-inherited traits, segregation analysis should be performed in F_2 or other types of populations constructed from parents that differ at alternate alleles of the trait and are polymorphic at the DNA level. However, when *a priori* information about the location of genes controlling the trait does not exit, the task of finding linkage could be formidable especially for species with large number of chromosomes. Therefore, to alleviate the problem, methods that utilize special genetic stocks have been developed that include the use of near isogenic lines (NILs) and bulked segregant analysis (BSA). These methods described in the following paragraphs allowed tagging molecular markers to many traits in different plant species; some examples were listed by Mohan et al. (1997).

2.1 Use of NILs to Identify Linkage with Molecular Markers

NILs are produced by repeated backcrossing of a donor parent, characterized by a unique or mutant phenotype with a recurrent parent characterized by a common phenotype often referred to as wild type. An initial F_1 is crossed to the recurrent parent to produce BC_1 progeny. BC_1 progenies (in case of dominant mutation) or BC_1S_1 progenies (in case of recessive mutation) carrying the mutant phenotype are crossed again to the recurrent parent. This process usually continues for at least 5-6 backcross generations, consequently reducing the sequence divergence between the donor and recurrent parents and limiting it to the region spanning the gene of interest (Fehr 1987; Muehlbauer et al. 1988). The result of such a backcrossing program is that NILs share common genetic background throughout the genome except for a region containing the locus that controls the selected trait (Figure 3). Models were developed to describe the proportion of the donor parent genome at the background of the recurrent genome parent as a function of chromosome length (Stam and Zeven 1981). For example, the length of the donor parent segment in BC_6 is estimated to be 32 cM for a chromosome of 100 cM, assuming no other chromosomes are carrying donor-parent segments. In practice, the size of the introgressed segment may vary in different sets of NILs. For example, the size of the introgressed region containing the *Tm-2a* gene in different NILs of tomato varied between 4 cM to 51 cM even after 11 backcross generations (Young and Tanksley 1989). In addition, unlinked

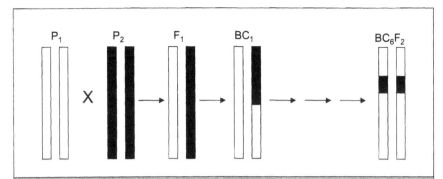

Fig. 3 Construction of near-isogenic lines (NILs). The F_1 results from crossing the recurrent parent (P_1) and the donor parent (P_2). The F_1 is backcrossed to P_1 to produce BC_1 progenies and the mutant phenotype is selected among the progenies. Backcrossing continues until BC_6. After this generation, selfing (BC_6F_2) results in the production of a homozygous line identical to the recurrent parent except for a segment from the donor parent (represented as black bar) containing the selected locus.

chromosome segments may also be introgressed, causing ambiguities in data interpretation.

NILs for many traits have been produced in breeding programs for various crops such as soybean and tomato (Bernard 1976; Maxon Smith and Ritchie 1983). However, the major limitation for using these NILs for identification of linked markers is the low level of DNA polymorphism between the recurrent and donor parents' DNA. Therefore, the use of NILs was applied mainly for tagging disease resistance genes because these genes are often introgressed from wild species that exhibit a high level of DNA polymorphism relative to the recurrent parent DNA (Young et al. 1988; Hinze et al. 1991; Klein-Lankhorst et al. 1991; Martin et al. 1991; Paran et al. 1991).

Polymorphism between NILs visualized using DNA markers indicates a putative linkage between the marker and the target gene. However, linkage must be confirmed by analysis of segregating progenies because false positive polymorphisms can arise due to the occurrence of additional unlinked donor segments in the background of the recurrent parent genome. Reduction of false positive markers can be achieved by screening independent sets of NILs because the chances that the same unlinked segments are introgressed to different NILs is small (Paran et al. 1991).

NILs screening is best suited for high throughput PCR markers such as RAPD or AFLP. However, single locus markers such as RFLP or SSR can also be used. In order to speed up the process of screening the NILs, pools of 5-8 RFLP clones were hybridized together allowing a significant reduction in the number of hybridizations (Young et al. 1988). Similarly, SSR markers can be multiplexed in a single reaction.

2.2 Use of Bulked Segregant Analysis (BSA) to Identify Linkage with Molecular Markers

The major limitation of using NILs for tagging target genes is the long time required for their production. A method, bulked segregant analysis (BSA) was developed that alleviates this problem by comparing pools of DNA from individuals that segregate for the target trait in a single cross (Giovannoni et al. 1991; Michelmore et al. 1991; Figure 4). Each pool consists of individuals that have an identical genotype at the locus of interest and mixed genotypes at non-target loci. The two pools are identical at alternate alleles of the target locus. In the case of disease resistance, one pool consists of resistant individuals while the second pool consists of susceptible individuals. Ideally each pool should consist of homozygous individuals only, but this requires progeny testing. A typical population used for BSA analysis is an F_2. DNA is extracted from

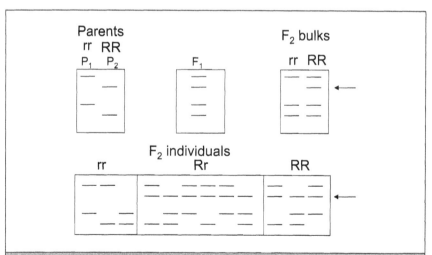

Fig. 4 Bulked segregant analysis for tagging a disease resistance gene. The four bands of the two parents (two per parent) represent four dominant loci. The bulks are derived from homozygous resistant F_2 (RR) and from homozygous susceptible F_2 (rr). The locus marked by an arrow is polymorphic between the bulks because of linkage to the R gene as illustrated by its segregation in the F_2. The other three loci are unlinked to the R gene and are therefore monomorphic between the bulks.

each individual progeny and selfed F_3 seeds are used for progeny testing. If progeny testing is not feasible, due to fertility problems that often occurred in interspecific crosses, F_2 DNA can also be used to construct the bulks. The two pools of DNA are screened with DNA markers and the polymorphic markers are considered as putatively linked to the target gene. Estimation of recombination frequency between the markers and the target gene should be done by genotyping the individual progenies in the segregating population. The advantage of using only homozygotes to construct the bulks is that markers can be identified in both *cis* and *trans* configurations relative to the desired allele at the target locus. If homozygotes cannot be distinguished from heterozygotes, only markers in *cis* configuration can be identified. An added advantage of the BSA method over the use of NILs is that the detection of polymorphic but unlinked markers (false positives) is minimized in BSA compared to use of NIL. The frequency of false positives is dependent on the size of the bulks. If sufficiently large bulks are used (at least 10 individuals per bulk), the probability of false positives is very low (Mackay and Caligari 2000).

BSA was applied in many plant species using mainly RAPD and AFLP markers. Some examples include identification of linked markers to the resistance gene *Cf-9* in tomato (Thomas et al. 1995), mapping of nodulation genes in pea (Schneider et al. 2002), identification of AFLP markers linked to resistance gene for parasitism in cowpea (Quedraogo et al. 2001), mapping stalk rot resistance gene in maize (Yang et al. 2004) and mapping resistance gene to pod weevil in bean (Blair et al. 2006). In the study designed to identify AFLP markers linked to the *Cf-9*, a gene that confers resistance to *Cladosporium fulvum*, 728 primer pairs were used. These primer pairs amplified approximately 42,000 loci, of which three cosegregated with the resistance gene. As the bulks were constructed from F_2 DNA, only *cis* markers could be identified. Further reduction in the number of potentially informative markers occurs because of lack of DNA polymorphism between the parents. Assuming the frequencies of *cis* and *trans* linked markers are equal, and an average of 50% polymorphism between the parents, about one quarter of the screened markers was informative in this study.

2.3 Targeting Disease Resistance Genes by NBS Profiling

The recent cloning of plant resistance genes from several plant species revealed sequence conservation at specific motifs, nucleotide-binding site (NBS) and leucine-rich repeat (LRR) required for their function (Belkhadir et al. 2004). Plants contain numerous NBS-containing genes, e.g., 149 genes in Arabidopsis, that are often organized in clusters (Pan et al. 2000; Meyers et al. 2003). The conservation of the NBS motif has been exploited to specifically isolate putative resistance genes containing this motif by NBS profiling (van der Linden et al. 2004). The method involves digestion of genomic DNA with a single restriction enzyme, ligation of adapter and PCR amplification of NBS containing fragments by using a combination of adapter and NBS degenerate primers. The PCR products are then separated and visualized similar to AFLP analysis. This method was applied to several plant species such as potato, tomato, lettuce, barley and apple. In apple, 43 NBS-containing fragments were isolated, 23 of them revealed sequence conservation with known resistance genes and most were mapped in the vicinity of known resistance genes (Calenge et al. 2005), indicating the usefulness of the method to specifically tag resistance genes.

3 CONVERSION OF DOMINANT TO CO-DOMINANT MARKERS

RAPD and AFLP primers amplify multiple loci; screening hundreds and even thousands of loci is feasible in a short time. Therefore, RAPD and AFLP are often used to tag target genes by methods such as NIL or BSA screening. However, for use in marker-assisted selection these markers have disadvantages of being muti-allelic and dominant. AFLP markers are also expensive and laborious for large-scale screening of a single locus. In order to address these problems, there is a need to convert these multi-allelic and dominant markers to single co-dominant locus markers. A method termed sequence characterized amplified region (SCAR) was developed to convert RAPD markers to single locus markers (Paran and Michelmore 1993). In this method a specific amplified fragment of a multi-allelic PCR amplification product is excised from a gel, cloned and sequenced. The sequence is used to design specific primers that amplify the two alleles of a single locus from parents of a mapping population. Conversion of AFLP is more cumbersome than conversion of RAPD markers to single locus markers. Detailed protocols for the conversion of AFLP markers were outlined by Brugmans et al. (2003).

Polymorphism can be searched for by sequence comparison of the PCR products from the two parents. When an insertion-deletion (InDel) mutation differentiates the DNA samples, the size difference can be also visualized in gel electrophoresis. If a single nucleotide polymorphism (SNP) exits between the two samples, it can be scored after restriction enzyme digestion (if the SNP is recognized by a restriction enzyme) or by other methods of SNP detection (Kwok and Chen 2003). SCAR markers are similar to CAPS (cleaved amplified polymorphic sequence) which are based on amplification of DNA sequence and its cleavage by a restriction enzyme that differentiates the two alleles (Konieczny and Ausubel 1993). An elegant and simple technique to detect SNP that does not create unique restriction site is the creation of a recognition site for restriction enzyme by introducing mismatches in the primers used in the PCR (Neff et al. 1988; Michaels and Amasino 1998).

An example for tagging and mapping a target gene using NILs and BSA as well as development of specific PCR markers is based on the work of Zhang and Stommel (2000, 2001) on the *Beta* (*B*) gene in tomato and is described in detail in the following paragraphs. Red tomatoes accumulate the carotenoid lycopene as the major pigment in the mature fruit. Mutation at the *B* locus results in an orange-colored tomato because of accumulation of high level of beta-carotene at the expense of lycopene and is inherited as a single incomplete dominant gene. NILs that differ at *B* were developed, in which *B* was introgressed from the wild species *Lycopersicum hirsutum*, thus providing ample DNA polymorphism between wild type cultivated tomato (*L. esculentum* cv. Rutgers) and its isogenic mutant line. For BSA, an interspecific F_2 population was constructed from a cross of red-fruited *L. esculentum* and orange-fruited *L. cheesmanii*. Two bulks were constructed by combining equal amount of DNA from each of seven red-fruited and seven orange-fruited F_2 plants.

The NILs were screened with RAPD (1,018 primers) and AFLP primers (64 primer combinations). Only two RAPD and three AFLP products were found as polymorphic between the NILs, however, none was linked to the *B* gene as revealed by segregation in the F_2 population. The inability to find linked markers to *B* using the NILs may indicate that the *L. hirsutum* introgression containing *B* is small, while other non-targeted introgressions may exist throughout the genome. RAPD and AFLP screening of the bulks by BSA analysis resulted in four RAPD and three AFLP polymorphic products. Out of these markers, two RAPD loci (OPAR18$_{1100}$ and UBC792$_{830}$) were found as linked to B (Figure 5A). The rest of the markers were unlinked; this resulted because of the small number of plants in each bulk, which increases the probability of false positives. To map the *B*-linked markers on the tomato molecular linkage map, the PCR products were amplified from a set of 50 tomato introgression lines, each containing a marker-defined segment from the wild species *L. pennellii* (Eshed and Zamir 1995). Comparison of the polymorphism pattern between the parents and the introgression lines enabled mapping the *B*-linked markers to chromosome 6.

In order to develop specific PCR markers useful for marker-assisted selection, the dominant RAPD markers $OPAR18_{1100}$ and $UBC792_{830}$ were cloned, characterized and converted to co-dominant CAPS markers. For cloning, the two amplified fragments were purified from the gel, cloned and sequenced. On the basis of the DNA sequence, long oligonucleotide primers (22-24 bp) were designed such that each primer contained the original 10 bases of the RAPD primer plus the next 12 or 14 bases.

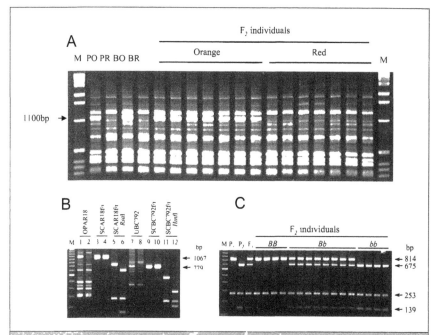

Fig. 5 Development of RAPD and CAPS markers linked to the B locus. **A.** RAPD marker $OPAR18_{1100}$ linked to the B locus. PO and PR are orange and red-fruited parents, respectively. BO and BR are bulks derived from orange and red-fruited F_2, respectively. **B.** CAPS markers derived from two RAPD markers. Lanes 1 and 2 and lanes 7 and 8 are the RAPD products of OPAR18 and UBC792, respectively amplified from two parents. Lanes 3 and 4 and lanes 9 and 10 are products of the two parents generated after cloning of $OPAR18_{1100}$ and $UBC792_{830}$ and amplification with specific PCR primers. Lanes 5 and 6 and lanes 11 and 12 are CAPS markers generated by cleaving the specific PCR products with restriction enzymes. **C.** Linkage of the co-dominant CAPS marker derived from the RAPD marker $OPAR18_{1100}$ to B. P_1, P_2 and F_1 are the parents of the mapping population and their F_1. BB, Bb and bb are F_2 individuals segregating for B. M- molecular DNA ladder. Figure 5A is taken from Zhang and Stommel 2000. Figure 5B and 5C are taken from Zhang and Stommel 2001 and presented with the kind permission of JR Stommel.

Amplification of the parents used in the cross for BSA analysis resulted in equal sized PCR products. Sequence comparison of the PCR products amplified from the two parents revealed restriction site polymorphisms that enabled the development of CAPS markers (Figure 5B). The CAPS markers were tested and verified for linkage with B in the F_2 population (Figure 5C).

4 USE OF GENOMICS TOOLS FOR MAPPING

The completion of whole-genome sequences of the model plants Arabidopsis and rice and the ongoing sequencing projects for several additional crop plants in conjunction with improving DNA array technologies and bioinformatics tools provide new opportunities to rapidly generate a large number of molecular markers for gene mapping. The main approach for developing high-density and high-throughput markers is hybridization of genomic DNA to oligonucleotide array that contains thousands of short probes usually up to 100 bp long. Current technologies permit printing up to 6,000,000 oligos (features) per chip (Mockler and Ecker 2005). These ultra-density tilling arrays can contain whole-genome oligonucleotides arranged at different designs depending on the length of the oligonucleotides and the degree of overlap between them. So far, the Affymetrix ATH1 GeneChip expression array has been used to detect polymorphism at the DNA level in Arabidopsis. Detection of polymorphism is based on differential hybridization intensities of DNA with perfect match (high intensity) to the array feature compared to DNA with mismatch sequence (low intensity). The difference in hybridization intensity is interpreted as sequence polymorphism referred to as single feature polymorphism (SFP) that can be used as a molecular marker. As the ATH1 GeneChip array contains oligonucleotides representing approximately 20,000 genes, it is possible to assay simultaneously multiple SFPs throughout the genome. Initial experiments using Arabidopsis and barley chips, allowed identification of approximately 4,000 and 10,500 SFPs between two accessions of Arabidopsis and barley, respectively (Borevitz et al. 2003; Rostoks et al. 2005). Sequence verification of the barley SFPs revealed that only 67% contained single nucleotide polymorphism (SNP). The high rate of false positive SFPs could be explained by other non-sequence related mechanisms that result in differential hybridization intensities such as alternative splicing and polyadenylation. Detailed protocols for use of arrays in mapping are provided by Borevitz (2005).

A combination of array mapping and bulked segregant analysis can provide a robust and fast method to map genes controlling simple-inherited traits. This technique was applied to map several developmental EMS mutations in Arabidopsis (Hazen et al. 2005). In case the mutation is caused by deletion, direct comparison of the hybridization patterns between mutant and wild type strains can map the mutation to a small interval without the need to construct bulks from segregating population. Using this approach, the flowering time deletion mutations *fkf1* and *cry2-1* were delineated to missing 185 features covering 77 kb and 19 features covering 7 kb genomic regions in chromosome 1, respectively (Hazen et al. 2005). With future whole genome tilling arrays, the resolution power of the array mapping approach will be much greater.

5   STATISTICAL CONSIDERATIONS

As we have already noted, construction of linkage maps based on molecular markers is an important tool in genetic analysis. These maps are useful for the localization of genes underlying quantitative and qualitative traits, and thus aid map-based gene cloning. Commonly used programs like JOINMAP (Stam 1993), MAPMAKER (Lander et al. 1987), LINKAGE-1 (Suiter et al. 1983), and GMENDEL (Echt et al. 1992) are very suitable for low-density genetic linkage map construction. However, molecular marker systems like AFLP and array-mapping allow that many markers can be generated in a short time. This leads to the construction of highly saturated maps to enable fine-scale genetic mapping and the anchoring of physical maps (Klein et al. 2000; Jansen et al. 2001).

In principle, statistical analysis methods for gene mapping originated in counting recombinant and non-recombinant offspring, but have now progressed to sophisticated approaches for the incorporation of markers into highly saturated or high-density maps. Basically, these highly saturated maps can be constructed with the same software as genetic linkage maps of normal density, such as MAPMAKER and JOINMAP. However, these methods have difficulty in solving the increasing ordering ambiguities in denser maps (Lincoln and Lander 1992; Jansen et al. 2001; van Os et al. 2005). This is due to the fact that denser maps have more loci than normal maps, but rely more or less on the same amount of recombination events observed among offspring genotypes. The limited number in recombination events may lead to increased probability that two markers are assigned as linked while in fact they are not, and incorrect localization and ordering of markers along the map. Methods are being therefore proposed that can cope with increased density of markers in maps (Jansen et al. 2001; van Os et al. 2005, 2006).

This section outlines the major principles of map construction. Detailed statistical considerations of genetic mapping were described in Chapter 4. Constructing a typical linkage map starts by calculation of pairwise recombination frequencies among markers. The corresponding LOD values, first proposed by Morton in 1955 (Morton 1955), are also calculated. The LOD score is indicative for the likelihood of linkage; LOD means the logarithm of odds, the 'odds' being the ratio of the probability that two loci are linked with a given recombination value over the probability that the two are not linked. A LOD score of 3.0 or more is generally taken to indicate that two gene loci are close to each other on the chromosome. A LOD score of 3.0 means the odds are a thousand to one in favor of genetic linkage. In other words: a critical LOD of 3.0 or more will, in general, prevent incorrect assignment of markers to the same linkage group. LOD score decreases with increasing recombination value; it increases with increasing sample size, expressing that, e.g. an estimate of 0.3 from a sample of 40 is less informative than an estimate of 0.3 from a sample of 100. LOD values can thus be seen as a measure of linkage information in the data (Stam 1993; Dawn Teare and Barrett 2005).

Constructing a linkage map, using computer programs like JOINMAP runs through the following basic steps (Stam 1993; van Ooijen and Voorrips 2001):

1. Create a data file: A raw data set consists of coded genotypes for all markers that are segregating in the mapping population. Mapping populations may vary, including: F_2s, backcrosses, recombinant inbred lines (RILs), haploids, doubled haploids, and full-sib families of cross-pollinating species (CP).

2. Read data.

3. Calculate pairwise recombination frequencies and LOD scores for all pairs of markers: JOINMAP uses every piece of linkage information, i.e., all the available pairwise estimates of recombination, weighed by the corresponding LOD scores. Thus, recombination percentages near to 50% are assigned low weights and, consequently, contribute little to the combined estimate of the distance covering such pairs. Nevertheless, recombination frequencies near to 50% are informative; knowing that two markers are at a large distance is helpful in finding the best fitting order on a multi-locus map (Weide et al. 1993).

4. Establish linkage groups.

5. Sequential build-up of the map: this step involves a 'reshuffling' of all map positions to find, by trial and error, the best fitting order within each linkage group.

The mapping procedure is basically a process of building a map by adding loci one by one, starting from the most informative pair of loci. For each added locus the best position is searched and a goodness-of-fit measure is calculated. When the goodness-of-fit reduces too sharply, or when the locus gives rise to negative distances, the locus is removed. This is continued until all loci have been handled once. Subsequently, all loci previously removed are attempted to be added to the map a second time. This can be successful since the map will contain more loci than at the first attempt. But it may also be unsuccessful again because of a sharp reduction in the goodness-of-fit or the appearance of negative distances, so that a locus will be removed once again. Following that, all loci previously removed are attempted to be added to the map for a final time, now ignoring the requirements of maximum allowed reduction in goodness-of-fit and no negative distances (van Ooijen and Voorrips 2001).

6 APPLICATION OF GENETIC MAPPING

DNA markers tightly linked to traits or phenotypes of economical importance are known to be an excellent tool to expedite breeding, using marker-assisted selection. In particular when such markers display complete linkage disequilibrium with genes responsible for a desired phenotype. When DNA markers are based on the gene sequences that encode proteins that clearly cause such phenotypes, these gene-sequences can be further used to study its developmental expression profile under certain conditions. These sequences can also be used in transgenic modifications to decipher the role of the nucleotide and/or amino-acid sequence in establishing a particular phenotype or directly constitute a phenotype of interest. In this sense, genetic maps proved to be an excellent tool in identifying genes that cause discrete or quantitative phenotypes by map-based cloning (Tanksley et al. 1995; Jander et al. 2002; Alonso-Blanco et al. 2005). Moreover, it is also possible to use maps of one species to isolate an orthologous region in another, provided that the linear sequence and repertoire of markers is sufficiently well conserved between the two species being compared (see for example, Delseny et al. 2001). An impressive application of comparative genetic mapping is found in the Poaceae, where cereal genomes have been aligned using common DNA markers. These studies have shown that the genomes of at least 10 divergent cereals, plus a growing number of their wild relatives, can be described in terms of only 30 rice linkage blocks (Moore et al. 1995; Gale and Devos 1998; Devos and Gale 2000). In other words, despite 50–70 million years of evolution and divergence (Kellogg 1998), a single ancestral genome is still visible in the modern cereals and grasses (Jenkins 2003).

Although synteny is indisputable within the Poaceae, and in other groups of plants such as the Solanaceae (Tanksley et al. 1988; Tanksley et al. 1992; Kellogg 1998) and the Cruciferae (Kowalski et al. 1994; Lagercrantz 1998), the question remains as to what extent the structure of genomes is conserved at the microsyntenic level. Answers to this question have important implications for the utility of comparative genomics for isolating genes from one species using the map of another. The advances made in high throughput DNA sequencing enabled the creation of the ultimate map comprising markers of adjacent base-pairs (Jenkins 2003) facilitated the comparison of specific, well-defined regions of the genome with unprecedented resolution. For example, studies of orthologous segments over 30 kb in the cereals maize, sorghum, rice, and barley have shown that there is conservation of gene structure, but that microsynteny comprises small genomic rearrangements (Chen et al. 1997; Chen et al. 1998; Tikhonov et al. 1999; Dubcovsky et al. 2001). A more recent thorough comparison of larger orthologous sequences in four cereal genomes has shown microcolinearity of four common genes near the *Vrn1* locus, but has also highlighted sequence divergence and different lineages of genome evolution (Ramakrishna et al. 2002). It was therefore concluded that there is potential for gene isolation in this group using comparative physical maps, but its utility is likely to be governed by phylogenetic proximity (Ramakrishna et al. 2002; Jenkins 2003). However, additional comparative microsynteny studies should be perused because they may potentially highlight the functional role of such sequence divergence in terms of gene regulation rather then gene identification.

References

Alonso-Blanco C, Goicoechea PG, Roca A, Alvarez E, Giraldez R (1994) Genetic mapping of cytological and isozyme markers on chromosome 1R, 3R, 4R and 6R of rye. Theor Appl Genet 88: 208-214

Alonso-Blanco C, Mendez-Vigo B, Koornneef M (2005) From phenotypic to molecular polymorphisms involved in naturally occurring variation of plant development. Intl J Dev Biol 49: 717-732

Appels R, Morris R, Gill BS, May CE (1998) Chromosome Biology. Kluwer Acad Publ, Boston, USA

Bateson W, Saunders ER, Punnett RC (1905) Experimental studies in the physiology of heredity. Rep Evol Comm Roy Soc 2: 1-55, 80-99

Belkhadir Y, Subramaniam R, Dangl JL (2004) Plant disease resistance protein signaling: NBS-LRR proteins and their partners. Curr Opin Plant Biol 7: 391-399

Bernard RL (1976) United States national germplasm collections. In: Hill LD (ed) World Soybean Research. Interstate Printers and Publ, Danville, IL, USA, pp 286-289

Blair MW, Munoz C, Garza R, Cardona C (2006) Molecular mapping of genes for resistance to the bean pod weevil (*Apion godmani* Wagner) in common bean. Theor Appl Genet 112: 913-923

Borevitz JO (2005) Array genoming and mapping. In: Salinas J, Sanchez-Serrano JJ (eds) Arabidopsis Protocols. Edn 2. Humana Press, Totowa, NJ, USA

Borevitz JO, Liang D, Plouffe D, Chang HS, Zhu T, Weigel D, Berry CC, Winzeler E, Chory J (2003) Large-scale identification of single-feature polymorphisms in complex genomes. Genome Res 13: 513-523

Bortiri E, Jackson D, Hake S (2006a) Advances in maize genomics: the emergence of positional cloning. Curr Opin Plant Biol 9: 164-71

Bortiri E, Chuck G, Vollbrecht E, Rocheford TF, Martienssen R, Hake S (2006b) *ramosa2* encodes a LATERAL ORGAN BOUNDARY domain protein that determines the fate of stem cells in branch meristems of maize. Plant Cell 18: 574-585

Botstein D, White RL, Skolnick M, Davis RW (1980) Construction of a genetic linkage map in man using restriction fragment length polymorphisms. Am J Hum Genet 32: 314-331

Bridges CB (1916) Nondisjunction as proof of the chromosome theory of heredity. Genetics 1: 1-52, 107-163

Brueggeman R, Rostoks N, Kudrna D, Kilian A, Han F, Chen J, Druka A, Steffenson B, Kleinhofs A (2002) The barley stem rust-resistance gene Rpg1 is a novel disease-resistance gene with homology to receptor kinases. Proc Natl Acad Sci USA 99: 9328-9333

Brugmans B, van der Hulst RGM, Visser RGF, Lindhout P, van Eck HJ (2003) A new and versatile method for the successful conversion of AFLP markers into simple single locus markers. Nucl Acids Res 31: e55

Calenge F, van der Linden CG, van de Weg E, Schouten HJ, van Arkel G, Denance C, Durel C-E (2005) Resistance gene analogues identified through the NBS-profiling method map close to major genes and QTL for disease resistance in apple. Theo Appl Genet 110: 660-668

Chen M, SanMiguel P, de Oliveira AC, Woo S-S, Zhang H, Wing RA, Bennetzen JL (1997) Microcolinearity in *sh2*-homologous regions of the maize, rice and sorghum genomes. Proc Natl Acad Sci USA 94: 3431-3435

Chen M, SanMiguel P, Bennetzen JL (1998) Sequence organization and conservation in *sh2/a1*-homologous regions of sorghum and rice. Genetics 148: 435-443

Coe E (1993) In: Coe E (ed) Maize Genetics Cooperation Newsletter. Department of Agronomy and US Department of Agriculture, University of Missouri, Columbia, Mo, USA, pp 157-166

Dawn Teare M, Barrett JH (2005) Genetic linkage studies. Lancet 366: 1036-1044

Delos Reyes BG, Khush GS, Brar DS (1998) Chromosomal location of eight isozyme loci in rice using primary trisomics and monosomic alien addition lines. J Hered 89: 164-168

Delseny M, Salses J, Cooke R, Sallaud C, Regad F, Lagoda P, Guiderdoni E, Ventelon M, Brugidou C, Ghesquiere A (2001) Rice genomics: present and future. Plant Physiol Biochem 39: 323-334

Devos KM, Gale MD (2000) Genome relationships: the grass model in current research. Plant Cell 12: 637-646

De Vries H (1903) Fertilization and hybridization. In: Intracellular Pangenesis including a paper on Fertilization and Hybridization. Open Court Publ, Chicago, IL, USA, pp 217-263 [Translated from Befruchtung und Barstardierung]

Dubcovsky J, Ramakrishna W, SanMiguel PJ, Busso CS, Yan L, Shiloff BA, Bennetzen JL (2001) Comparative sequence analysis of collinear barley and rice BACs. Plant Physiol 125: 1342-1353

Echt C, Knapp S, Liu B-H (1992) Genome mapping with non-inbred crosses using GMendel 2.0. Maize Genet Coop Newsl 66: 27-29

Endo TR, Gill BS (1996) The deletion stocks of common wheat. J Hered 87: 295-307

Eshed Y, Zamir D (1995) An introgression line population of *Lycopersicon pennellii* in the cultivated tomato enables the identification and fine mapping of yield-associated QTL. Genetics 141: 1147-1162

Fehr WR (1987) Principles of Cultivar Development. Vol 1. Theory and Technique. Macmillan Publ, NY, USA

Frary A, Nesbitt TC, Grandillo S, Knaap E, Cong B, Liu J, Meller J, Elber R, Alpert KB, Tanksley SD (2000) *fw2.2*: a quantitative trait locus key to the evolution of tomato fruit size. Science 289: 71-72

Fridman E, Carrari F, Liu YS, Fernie AR, Zamir D (2004) Zooming in on a quantitative trait for tomato yield using interspecific introgressions. Science 305: 1786-1789

Fuchs J, Kuehne M, Schubert I (1998) Assignment of linkage groups to pea chromosomes after karyotyping and gene mapping by fluorescent in situ hybridization. Chromosoma 107: 272-276

Funatsuki H, Kurosaki H, Murakami T, Matsuba S, Kawaguchi K, Yumoto S, Sato Y (2003) Deficiency of a cytosolic ascorbate peroxidase associated with chilling tolerance in soybean. Theor Appl Genet 106: 494-502

Gale MD, Devos KM (1998) Comparative genetics in the grasses. Proc Natl Acad Sci USA 95: 1971-1974

Gall JG, Pardue ML (1969) Formation and detection of RNA-DNA hybrid molecules in cytological preparations. Proc Natl Acad Sci USA 69: 378-383

Gallavotti A, Zhao Q, Kyozuka J, Meeley RB, Ritter MK, Doebley JF, Pe ME, Schmidt RJ (2004) The role of *barren stalk1* in the architecture of maize. Nature 432: 630-635

Gill KS, Gill BS, Endo TR, Taylor T (1996) Identification and high-density mapping of gene-rich regions in chromosome group 1 of wheat. Genetics 144: 1883-1891

Giovannoni JJ, Wing RA, Ganal MA, Tanksley SD (1991) Isolation of molecular markers from specific chromosomal intervals using DNA pools from existing mapping populations. Nucl Acids Res 19: 6553-6558

Grajal-Martin MJ, Muehlbauer FJ (2002) Genomic location of the *Fw* gene for resistance to Fusarium wilt race 1 in peas. J Hered 93: 291-293

Gustafson JP, Butler E, McIntyre CL (1990) Physical mapping of a low-copy DNA sequence in rye (*Secale cereale* L.). Proc Natl Acad Sci USA 87: 1899-1902

Hall KJ, Ellis THN, Parker JS (1995) Examining concordance between cytological and genetic maps of the pea genome. Chrom Res 3: 99-100

Hall KJ, Parker JS, Ellis THN, Turner L, Knox MR, Hofer JMI, Lu J, Ferrandiz C, Hunter PJ, Taylor JD, Baird K (1997) The relationship between genetic and cytogenetic maps of pea. II. Physical maps of linkage mapping populations. Genome 40: 755-769

Halsted BD, Owen EJ, Shaw JK (1905) Experiments with tomatoes. New Jersey Agri Exp Stn Annu Rep 26: 447-477

Harper LC, Cande WZ (2000) Mapping a new frontier; development of integrated cytogenetic maps in plants. Funct Integr Genom 1: 89-98

Hazen SP, Borevitz JO, Harmon FG, Pruneda-Paz JL, Schultz TF, Yanovsky MJ, Liljegren SJ, Ecker JR, Kay SA (2005) Rapid array mapping of circadian clock and developmental mutations in Arabidopsis. Plant Physiol 138: 990-997

Hedrick UP, Booth NO (1907) Mendelian characters in tomatoes. Proc Am Soc Hort Sci 5: 19-24

Hinze K, Thompson RD, Ritter E, Salamini F, Schultze-Lefert P (1991) Restriction fragment length polymorphism-mediated targeting of the *ml-o* resistance locus in barley (*Hordeum vulgare*). Proc Natl Acad Sci USA 88: 3691-3695

Huang S, van der Vossen EA, Kuang H, Vleeshouwers VG, Zhang N, Borm TJ, van Eck HJ, Baker B, Jacobsen E, Visser RG (2005) Comparative genomics enabled the isolation of the R3a late blight resistance gene in potato. Plant J 42: 251-261

Jander G, Norris SR, Rounsley SD, Bush DF, Levin IM, Last RL (2002) *Arabidopsis* map-based cloning in the post-genome era. Plant Physiol 129: 440-450

Jansen J, de Jong AG, van Ooijen JW (2001) Constructing dense genetic linkage maps. Theor Appl Genet 102: 1113-1122

Jenkins G (2003) Unfolding large-scale maps. Genome 46: 947-952

Jiang J, Hulbert SH, Gill BS, Ward DC (1996) Interphase fluorescence in situ hybridization mapping: a physical mapping strategy for plant species with large complex genomes. Mol Gen Genet 252: 497-502

John HA, Birnstiel ML, Jones KW (1969) RNA-DNA hybrids at the cytological level. Nature (London) 223: 582-587

Jones DF (1917) Linkage in *Lycopersicum*. Am Nat 51: 608-621

Jones CM, Mes P, Myers JR (2003) Characterization and inheritance of the *Anthocyanin fruit (Aft)* tomato. J Hered 94: 449-456

Kazan K, Muehlbauer FJ, Weeden NF, Ladizinsky G (1993) Inheritance and linkage relationships of morphological and isozyme loci in chickpea (*Cicer arietinum* L.). Theor Appl Genet 86: 417-426

Kellogg EA (1998) Relationships of cereal crops and other grasses. Proc Natl Acad Sci USA 95: 2005-2010

Khush G, Rick C (1968) Cytogenetic analysis of the tomato genome by means of induced deficiencies. Chromosoma 23: 452-484

Klein-lankhorst R, Rietveld P, Machiels B, Verkerk R, Weide R, Gebhardt C, Koornneef M, Zabel P (1991) RFLP markers linked to the root knot nematode resistance gene *Mi* in tomato. Theor Appl Genet 81: 661-667

Klein PE, Klein RR, Cartinhour SW, Ulanch PE, Dong J, Obert JA, Morishige DT, Schlueter SD, Childs KL, Ale M, Mullet JE (2000) A high-throughput AFLP-based method for constructing integrated genetic and physical maps: progress towards a sorghum genome map. Genome Res 10: 789-807

Konieczny A, Ausubel FM (1993) A procedure for mapping *Arabidopsis* mutations using co-dominant ecotype-specific markers. Plant J 4: 403-410

Kowalski S, Lan T, Feldmann K, Paterson A (1994) Comparative mapping of *Arabidopsis thaliana* and *Brassica oleracea* chromosomes reveals islands of conserved organization. Genetics 138: 499-510

Kwok PW, Chen X (2003) Detection of single nucleotide polymorphisms. Curr Iss Mol Biol 5: 43-60

Kunzel G, Korzum L, Meister A (2000) Cytologically integrated physical restriction fragment length polymorphism maps for the barley genome based on translocation breakpoints. Genetics 154: 397-412

Lagercrantz U (1998) Comparative mapping between *Arabidopsis thaliana* and *Brassica nigra* indicates that *Brassica* genomes have evolved through extensive genome replication accompanied by chromosome fusions and frequent rearrangements. Genetics 150: 1217-1228

Lander ES, Green P, Abrahamson J, Barlow A, Daly MJ, Lincoln SE, Newburg L (1987) MAPMAKER: an interactive computer package for constructing primary genetic linkage maps of experimental and natural populations. Genomics 1: 174-181

Lehfer H, Busch W, Martin R, Hermann RG (1993) Localization of the B-hordein locus on barley chromosomes using fluorescence in situ hybridization. Chromosoma 102: 428-432

Leitch IJ, Heslop-Harrison JS (1993) Physical mapping of four sites of 5S rDNA sequences and one site of the α-amylase-2 gene in barley (*Hordeum vulgare*). Genome 36: 517-523

Leitch AR, Schwarzacher T, Jackson D, Leitch IJ (1994) *In situ* Hybridization. BIOS Scientific Publ, Oxford, UK

Lieberman M, Segev O, Gilboa N, Lalazar A, Levin I (2004) The tomato homolog of the gene encoding UV-damaged DNA binding protein 1 (DDB1)

underlined as the gene that causes the *high pigment-1* mutant phenotype. Theor Appl Genet 108: 1574-1581

Lincoln SE, Lander ES (1992) Systematic detection of errors in genetic linkage data. Genomics 14: 604-610

Mackay IJ, Caligari PDS (2000) Efficiencies of F_2 and backcross generation for bulked segregant analysis using dominant markers. Crop Sci 40: 626-630

Mackill DJ, Salam MA, Wang ZY, Tanksley SD (1993) A major photoperiod-sensitivity gene tagged with RFLP and isozyme markers in rice. Theor Appl Genet 85: 536-540

Martin GB, Williams GK, Tanksley SD (1991) Rapid identification of markers linked to a *Pseudomonas* resistance gene in tomato by using random primers and near-isogenic lines. Proc Natl Acad Sci USA 88: 2336-2340

Martin GB, Brommonschenkel SH, Chunwongse J, Frary A, Ganal MW, Spivey R, Wu T, Earle ED, Tanksley SD (1993) Map-based cloning of a protein kinase gene conferring disease resistance in tomato. Science 262: 1432-1436

Maxam AM, Gilbert W (1977) A new method for sequencing DNA. Proc Natl Acad Sci USA 74: 560-564

Maxon Smith JW, Ritchie DB (1983) A collection of near-isogenic lines of tomato: research tool of the future? Plant Mol Biol Rep 1: 41-45

Menda N, Semel Y, Peled D, Eshed Y, Zamir D (2004) In silico screening of a saturated mutation library of tomato. Plant J 38: 861-872

Meyers BC, Kozik A, Griego A, Kuang H, Michelmore RW (2003) Genome-wide analysis of NBS-LRR-encoding genes in *Arabidopsis*. Plant Cell 4: 809-834

Michaels SD, Amasino RM (1998) A robust method for detecting single-nucleotide changes as polymorphic markers by PCR. Plant J 14: 381-385

Michelmore RW, Paran I, Kesseli RV (1991) Identification of markers linked to disease resistance genes by bulked segregant analysis: A rapid method to detect markers in specific genomic regions by using segregating populations. Proc Natl Acad Sci USA 88: 9828-9832

Mockler TC, Ecker JR (2005) Applications of DNA tiling arrays for whole-genome analysis. Genomics 85: 1-15

Mohan M, Nair S, Bhagwat A, Krishna TG, Yano M, Bhatia CR, Sasaki T (1997) Genome mapping, molecular markers and marker-assisted selection in crop plants. Mol Breed 3: 87-103

Moore G, Devos KM, Wang Z, Gale MD (1995) Grasses, line up and form a circle. Curr Biol 5: 737-739

Morgan TH (1911a) The application of the conception of pure lines to sex-limited inheritance and to sexual dimorphism. Am Nat 45: 65-78

Morgan TH (1911b) An attempt to analyze the constitution of the chromosomes on the basis of sex-limited inheritance in *Drosophila*. J Exp Zool 11: 365-414

Morgan TH, Cattell E (1912) Data for the study of sex-linked inheritance in *Drosophila*. J Exp Zool 13: 79-101

Morton NE (1955) Sequential tests for the detection of linkage. Am J Hum Genet 7: 277-318

Muehlbauer GJ, Specht JE, Thomas-Compton MA, Staswick PE, Bernard RL (1988) Near-isogenic lines- A potential resource in the integration of conventional and molecular marker linkage maps. Crop Sci 28: 729-735

Mustilli AC, Fenzi F, Ciliento R, Alfano F, Bowler C (1999) Phenotype of the tomato *high pigment-2* mutant is caused by a mutation in the tomato homolog of *DEETIOLATED1*. Plant Cell 11: 145-157

Nasuda S, Friebe B, Gill BS (1998) Gametocidal genes induce chromosome breakage in the interphase prior to the first mitotic cell division of the male genotype in wheat. Genetics 149: 1115-1124

Neff MM, Neff JD, Chory J, Pepper AE (1998) dCAPS, a simple technique for the genetic analysis of single nucleotide polymorphisms: experimental applications in *Arabidopsis thaliana* genetics. Plant J 14: 387-392

Pan Q, Liu Y–S, Budai-Hadrian O, Sela M, Carmel-Goren L, Zamir D, Fluhr R (2000) Comparative genetics of nucleotide binding site-leucine rich repeat resistance gene homologues in the genomes of two dicotyledons: Tomato and Arabidopsis. Genetics 155: 309-322

Paran I, Kesseli RV, Michelmore RW (1991) Identification of restriction fragment length polymorphism and random amplified polymorphic DNA markers linked to downy mildew resistance genes in lettuce using near-isogenic lines. Genome 34: 1021-1027

Paran I, Michelmore RW (1993) Development of reliable PCR-based markers linked to downy mildew resistance genes in lettuce. Theor Appl Genet 85: 985-993

Paterson AH (2002) What has QTL mapping taught us about plant domestication? New Phytol 154: 591-608

Pedersen C, Giese H, Linde-Laursen I (1995) Towards an integration of the physical and the genetic chromosome maps of barley by in situ hybridization. Hereditas 123: 77-88

Pedersen C, Linde-Laursen I (1995) The relationship between physical and genetic distances at the *Hor1* and *Hor2* loci of barley estimated by two-colour fluorescent in situ hybridization. Theor Appl Genet 91: 941-946

Peters JL, Cnudde F, Gerats T (2003) Forward genetics and map-based cloning approaches. Trends Plant Sci 8: 484-491

Peterson DG, Lapitan NLV, Stack SM (1999) Localization of single- and low-copy sequences on tomato synaptonemal complex spreads using fluorescence in situ hybridization (FISH). Genetics 152: 427-439

Price HL, Drinkard AW Jr (1908) Inheritance in tomato hybrids. Virginia Agri Exp Stn Bull 177: 1-53

Quedraogo JT, Maheshwari V, Berner DK, St-Pierre C-A, Belzile F, Timko MP (2001) Identification of AFLP markers linked to resistance to cowpea (*Vigna unguiculata* L.) to parasitism by *Striga gesnerioides*. Theor Appl Genet 102: 1029-1036

Ramakrishna W, Dubcovsky J, Park Y-J, Busso C, Emberton J, SanMiguel P, Bennetzen JL (2002) Different types and rates of genome evolution detected by comparative sequence analysis of orthologous segments from four cereal genomes. Genetics 162: 1389-1400

Ren N, Song YC, Bi XZ, Ding Y, Liu LH (1997) The physical location of genes *cdc2* and *prh1* in maize (*Zea mays* L.). Hereditas 126: 211-217

Rick CM, Tanksley SD (1983) Isozyme monitoring of genetic variation in *Lycopersicon*. Isozymes Curr Top Biol Med Res 11: 269-284

Rick CM (1991) Tomato paste: a concentrated review of genetic highlights from the beginnings to the advent of molecular genetics. Genetics 128: 1-5

Rostoks N, Borevitz JO, Hedley PE, Russell J, Mudie S, Morris J, Cardle L, Marshall DF, Waugh R (2005) Single-feature polymorphism discovery in the barley transcriptome. Genome Biol 6: R54

Sanger F, Nicklen S, Coulson AR (1997) DNA sequencing with chain terminating inhibitors. Proc Natl Acad Sci USA 74: 5463-5467

Schaffer AA, Levin I, Ogus I, Petreikov M, Cincarevsky F, Yeselson E, Shen S, Gilboa N, Bar M (2000) ADP-glucose pyrophosphorylase activity and starch accumulation in immature tomato fruit: the effect of a *Lycopersicon hirsutum*-derived introgression encoding for the large subunit. Plant Sci 152: 135-144

Schneider A, Walker S, Sagan M, Duc G, Ellis T, Downie J (2002) Mapping of the nodulation loci *sym9* and *sym10* of pea (*Pisum sativum* L.). Theor Appl Genet 104: 1312-1316

Schwarzacher T, Heslop-Harrison P (2000) Practical *In Situ* Hybridization. BIOS Scientific Publ, Oxford, UK

Singh K, Ishii T, Parco A, Huang N, Brar DS, Khush GS (1996a) Centromere mapping and orientation of the molecular linkage map of rice (*Oryza sativa* L.). Proc Natl Acad Sci USA 93: 6163-6168

Singh K, Multani DS, Khush GS (1996b) Secondary trisomics and telotrisomics of rice: origin, characterization, and use in determining the orientation of chromosome map. Genetics 143: 517-529

Stam P, Zeven AC (1981) The theoretical proportion of the donor genome in near-isogenic lines of self-fertilizers bred by backcrossing. Euphytica 30: 227-238

Stam P (1993) Construction of integrated genetic linkage maps by means of a new computer package: JOINMAP. Plant J 3: 739-744

Stevens MA, Rick CM (1986) Genetics and breeding. In: Atherton JG, Rudich J (eds) The Tomato Crop. Chapman and Hall, New York, USA, pp 35-109

Sturtevant AH (1913) The linear arrangement of six sex-linked factors in *Drosophila*, as shown by their mode of association. J Exp Zool 14: 43-59

Sturtevant AH (1965) A History of Genetics. Harper and Row, New York, USA

Suiter KA, Wendel JF, Case JS (1983) LINKAGE-1: a PASCAL computer program for the detection and analysis of genetic linkage. J Hered 74: 203-204

Tanksley SD, Zamir D (1988) Double tagging of a male-sterile gene in tomato using a morphological and enzymatic marker gene. HortScience 23: 387-388

Tanksley SD, Bernatzky R, Lapitan N, Prince J (1988) Conservation of gene repertoire but not gene order in pepper and tomato. Proc Natl Acad Sci USA 85: 6419-6423

Tanksley SD, Ganal M, Prince J, de Vicente M, Bonierbale M, Broun P, Fulton TM, Giovannoni JJ, Grandillo S, Martin GB, Messeguer R, Miller JC, Miller L, Paterson AH, Pineda O, Röder MS, Wing RA, Wu W, Young ND (1992) High density molecular maps of the tomato and potato genomes. Genetics 132: 1141-1160

Tanksley SD, Ganal M, Martin GB (1995) Chromosome landing - a paradigm for map-based cloning in plants with large genomes. Trends Genet 11: 63-68

Thomas CM, Vos P, Zabeau M, Jones DA, Norcott KA, Chadwick BP, Jones DG (1995) Identification of amplified restriction fragment polymorphism (AFLP) markers tightly linked to the tomato Cf-9 gene for resistance to *Cladosporium fulvum*. Plant J 8: 785-794

Tikhonov AP, SanMiguel PJ, Nakajima Y, Gorenstein NM, Bennetzen JL, Avramova Z (1999) Colinearity and its exceptions in orthologous *adh* regions of maize and sorghum. Proc Natl Acad Sci USA 96: 7409-7414

van der Linden CG, Wouters DC, Mihalka V, Kochieva EZ, Smulders MJM, Vosman B (2004) Efficient targeting of plant disease resistance loci using NBS profiling. Theor Appl Genet 109: 384-393

van Os H, Stam P, Visser RG, van Eck HJ (2005) SMOOTH: a statistical method for successful removal of genotyping errors from high-density genetic linkage data. Theor Appl Genet 112: 187-194

van Os H, Andrzejewski S, Bakker E, Barrena I, Bryan GJ, Caromel B, Ghareeb B, Isidore E, de Jong W, van Koert P, Lefebvre V, Milbourne D, Ritter E, van der Voort JN, Rousselle-Bourgeois F, van Vliet J, Waugh R, Visser RG, Bakker J, van Eck HJ (2006) Construction of a 10,000-marker ultradense genetic recombination map of potato: providing a framework for accelerated gene isolation and a genomewide physical map. Genetics 173: 1075-1087

van Ooijen JW, Voorrips RE (2001) JoinMap® 3.0, Software for the calculation of genetic linkage maps. Plant Research International, Wageningen, The Netherlands

Vrebalov J, Ruezinsky D, Padmanabhan V, White R, Medrano D, Drake R, Schuch W, Giovannoni J (2002) A MADS-box gene necessary for fruit ripening at the tomato ripening-inhibitor (*rin*) locus. Science 296: 275-276

Wang H, Nussbaum-Wagler T, Li B, Zhao Q, Vigouroux Y, Faller M, Bomblies K, Lukens L, Doebley JF (2005) The origin of the naked grains of maize. Nature 436: 714-719

Weeden NF, Robinson RW, Ignart F (1984) Linkage between an isozyme locus and one of the genes controlling resistance to watermelon mosaic virus 2 in *Cucurbita ecuadorensis*. Cucurbit Genet Coop Rep 7: 86-87

Weide R, van Wordragen MF, Lankhorst RK, Verkerk R, Hanhart C, Liharska T, Pap E, Stam P, Zabel P, Koornneef M (1993) Integration of the classical and molecular linkage maps of tomato chromosome 6. Genetics 135: 1175-1186

Wilkinson DG (1992) *In Situ* Hybridization: A Practical Approach. IRL Press, Oxford, UK

Yan L, Loukoianov A, Tranquili G, Helguera M, Fahima T, Dubcovsky J (2003) Positional cloning of the wheat vernalization gene *VRN1*. Proc Natl Acad Sci USA 100: 6263-6268

Yan L, Loukoianov A, Blechl A, Tranquilli G, Ramakrishna W, SanMiguel P, Bennetzen JL, Echenique V, Dubcovsky J (2004) The wheat *VRN2* gene is a flowering repressor down regulated by vernalization. Science 303: 1640-1644

Yang DE, Zhang CL, Zhang DS, Jin DM, Weng ML, Chen SJ, Nguyen H, Wang B (2004) Genetic analysis and molecular mapping of maize (*Zea mays* L.) stalk rot resistance gene *Rfg1*. Theor Appl Genet 108: 706-711

Young ND, Zamir D, Ganal MA, Tanksley SD (1988) Use of isogenic lines and simultaneous probing to identify DNA markers tightly linked to the *Tm-2a* gene in tomato. Genetics 120: 579-585

Young ND, Tanksley SD (1989) RFLP analysis of the size of chromosomal segments retained around the *Tm-2* locus of tomato during backcross breeding. Theor Appl Genet 77: 353-359

Zhang Y, Stommel JR (2000) RAPD and AFLP tagging and mapping of *Beta* (*B*) and *Beta* modifier (*Mo$_B$*), two genes which influence beta-carotene accumulation in fruit of tomato (*Lycopersicon esculentum* Mill.). Theor Appl Genet 100: 368-375

Zhang Y, Stommel JR (2001) Development of SCAR and CAPS markers linked to the *Beta* gene in tomato. Crop Sci 41: 1602-1608

6 Genetic Mapping of Quantitative Trait Loci

Yanru Zeng[1], Jiahan Li[2], Chenguang Wang[2], Myron M. Chang[3], Runqing Yang[4] and Rongling Wu[1,2,4*]

[1]School of Forestry and Biotechnology, Zhejiang Forestry University, Lin'an, Zhejiang 311300, People's Republic of China

[2]Department of Statistics, University of Florida, Gainesville, FL 32611, USA

[3]Department of Epidemiology and Health Policy Research, University of Florida, Gainesville, FL 32611, USA

[4]School of Agriculture and Biology, Shanghai Jiaotong University, Shanghai 200240, People's Republic of China

*Corresponding author: rwu@stat.ufl.edu

This chapter outlines the basic genetic and statistical principle of quantitative trait locus (QTL) mapping. Interval mapping and its extensions which are commonly used are described. An example in rice is used to illustrate the procedure of QTL mapping. Genetic mapping holds a great promise to understand the genetic architecture of quantitative traits important to agriculture.

1 INTRODUCTION

Genetics has been thought to play a dominant role in explaining fundamental biological issues in life sciences, given that almost every biological phenomenon involves a genetic component. As a result of this, there is a pressing need for studying the genetic architecture of biological traits, especially those that vary in continuous patterns. The past two decades have witnessed an unprecedented growth in our ability to dissect these so-called quantitative traits into individual loci at the molecular level in agricultural genetics, evolutionary biology (Paterson et al. 1988; Wu 1998; Walsh 2001; Paterson 2006) and human genetics aimed to detect genes for complex human diseases (Peltonen and McKusick 2001). Quantitative traits are thought to be controlled by multiple genes, each with a small effect and segregating according to Mendel's laws, and can also be affected by the environment to varying degrees (Lynch and Walsh 1998). According to this argument established by Fisher (1918), the observed phenotype of a quantitative trait (y) can be expressed as a linear combination of genetic (g), environmental (e) and genotype $\times$ environment interaction effects, i.e.,

$$y = \mu + g + e + g \times e + \varepsilon, \tag{1}$$

where μ is the population mean and ε is the residual error. Depending on different purposes of plant breeding, the environmental effect can be due to different climates or locations (Piepho 2000), which are usually called 'macroenvironments' in light of their evident varying patterns (Wu 1997). The macroenvironment effect can be discrete, like location, or continuous, such as temperature, moisture and nutrient, in nature (Via et al. 1995). The residual error is due to stochastic fluctuations, i.e., 'microenvironments' (Wu 1997). Under the prerequisite that the families or genotypes studied have multiple replicates in space, statistical approaches based on regression models and analysis of variance have been available to estimate the variances due to the genetic, environment and residual effects, as well as genotype $\times$ environment interaction effects. Heritability, defined as the proportion of the genetic variance to the total phenotypic variance, is then estimated to measure the contribution of genetic factors to quantitative variation.

With the development of modern molecular markers, classical quantitative genetics has been developed to a point at which individual genetic loci underlying a quantitative trait, called quantitative trait loci (QTL), can be mapped and identified (Lynch and Walsh 1998). By dissecting a quantitative trait into a total of m possible QTLs each segregating in a Mendelian ratio, the phenotypic value of the trait can be expressed as

$$y = \mu + \sum_{r=1}^{m} g_r + \sum_{r \neq s}^{m} (g_r \times g_s) + e + \sum_{r=1}^{m} (g_r \times e) + \sum_{r \neq s}^{m} (g_r \times g_s \times e) + \varepsilon \tag{2}$$

where $\sum_{r=1}^{m}$ and $\sum_{r \neq s}^{m}$ denote the summations associated with the main and epistatic genetic effects, respectively, among the m QTLs estimated from a linkage map constructed by molecular markers.

Equation (2) presents a general statistical model for QTL mapping (Lander and Botstein 1989; Kao and Zeng 2002). The detection of the underlying QTL for a quantitative trait is based on a segregating population of progeny derived from crossing genotypes containing different alleles at phenotypically important loci. The crossed parents should be adequately divergent to identify discrete molecular markers that sample the genome at sufficiently dense intervals. Statistical principles and methods for mapping QTL with the linkage map constructed from genotyped molecular markers have been well established (Lander and Botstein 1989; Piepho 2000; Kao and Zeng 2002; Ma et al. 2002; Wu and Lin 2006). In practice, QTL mapping approaches have been instrumental for the characterization and discovery of thousands of QTLs responsible for a variety of traits in plants, animals and humans. In a recent study, Li et al. (2006) were able to characterize the molecular basis of the reduction of grain shattering – a fundamental selection process for rice domestication – at the detected QTL. Many other examples for the success of QTL mapping include the positional cloning of QTL responsible for fruit size and shape in tomato (Frary et al. 2000) and for branch, florescence and grain architecture in maize (Doebley et al. 1997; Gallavotti et al. 2004; Wang et al. 2005). In this chapter, we purport to outline the basic tenet of QTL mapping with different experimental designs and review several commonly used statistical methods for mapping quantitative traits. Examples are used to demonstrate the utilization of various methods. We will not intend to give a complete review of the detection of QTL for various quantitative traits in plants by genetic mapping.

2 NATURE OF QUANTITATIVE VARIATION

Most quantitative traits are determined by a web of many interacting loci and by an array of environmental factors (Falconer and Mackay 1996). The traditional polygenic theory of quantitative traits (Mather 1943) envisaged a fairly large number of loci, each with relatively small and equal effects, acting in a largely additive way. Over the years it has indeed been observed that a quantitative trait may display complicated genetic architecture (Mackay 2001; Anholt and Mackay 2004; Carlborg and Haley 2004), described below:

1. It may be controlled by a fairly large number of loci; for example, of the order of 50, according to the work of Shrimpton and Robertson (1988a, b);

2. Genes act in ways which may be additive, dominant, epistatic and interactive with environmental factors;

3. The magnitude of the effect produced by each locus can vary considerably;

4. The same genes may affect different phenotypic traits through pleiotropic effects;

5. The genes affecting the trait may be distributed over the genome at random or in a certain pattern.

With the use of genetic mapping to analyze quantitative traits, increasing evidence has been observed for the third point, which suggests that typically a small number of loci account for a very large fraction of the variation in the trait. For this reason, the traditional polygenic model may be replaced by a new oligogenic model in which a small number of major genes each with a large effect, combined with many minor genes each with a small effect, determine the genetic variation of a quantitative trait (see Mackay 1996 for an excellent review). According to the oligogenic model, the distribution of genetic effects may be approximated by a geometric series (Lande and Thompson 1990). When incorporated into a QTL mapping model, such an approximation can significantly increase the power of QTL mapping and the precision of parameter estimation.

3 MAPPING POPULATIONS

The central idea of QTL mapping is based on the co-segregation of different but linked genes, which is reflected in terms of the co-transmission of the genes from a parental to progeny population. Thus, a segregating progeny population is necessary to detect the linkage of different genes. Below, we introduce several mapping populations that have proven powerful for QTL mapping in plants.

3.1 Controlled Crosses

Inbred pedigrees, such as the backcross and F_2, initiated with two contrasting lines, establish the principle behind the linkage analysis of markers and, therefore, the genetic mapping of QTL (Paterson et al. 1988). Linkage analysis is based on the occurrence of recombination events between genetic loci (measured by the recombination fraction) when gametes are formed and transmitted from parents to offspring. By estimating the recombination fraction between markers and QTL, the genomic location of the QTL that affects the variation of a quantitative trait can be determined. Other inbred pedigrees used for QTL mapping in plants include the doubled haploid (DH) population by doubling chromosomes of the F_1 from two inbred lines, and recombinant inbred lines (RIL) derived either by repeated selfing or by repeated sibling (brother-sister) mating from the offspring of the F_1 cross between two inbred lines. Genetic models for QTL mapping in the DH population are equivalent to the backcross, whereas a special derivation is needed for QTL mapping in the RIL population (Wu et al. 2007).

In forest trees, long-generation intervals and outcrossing nature prevent the generation of inbred lines and, therefore, of any advanced cross. However, because these species are highly heterozygous, their cross of one generation (F_1) often displays substantial segregation and has many different types of segregation. Some loci may have four different alleles between the crossing parents, generating four genotype classes in the progeny. Many others may also follow the F_2 pattern in a 1:2:1 ratio (called intercross loci) and the backcross pattern in a 1:1 ratio (called testcross loci) (Lu et al. 2004). Using the testcross markers, i.e., those that are segregating in one parent but not in the other, Grattapaglia and Sederoff (1994) proposed a so-called pseudo-testcross strategy for linkage mapping in a controlled cross between two outbred parents. Although it only makes use of a portion of markers from the genome, this strategy provides a simple way for genetic mapping and has been widely utilized in practical mapping projects for outcrossing species (Cervera et al. 2001; Yin et al. 2004). More recently, Lin et al. (2003) have proposed a general statistical model for simultaneously estimating the QTL-marker linkage phase, QTL location and QTL effects in an outcrossed family. This QTL mapping model takes into account uncertainties about the number of alleles across the genome and the linkage phase among different loci, and can be reduced to the pseudo-testcross strategy.

3.2 Structured Pedigrees

In some plant breeding programs, multiple related families are often used in order to accumulate a sufficient number of progeny for linkage analysis and QTL mapping. For such a family-based pedigree, among-member relationships are defined by the probability of identical by descent (IBD). A random-effect model based on IBD sharing can be developed to estimate the genetic variance explained by a QTL. This is in contrast to the fixed-effect model that estimated the genetic effects of a QTL for an unrelated controlled cross (Xu and Atchley 1995). One advantage of the random model lies in its power to estimate the genetic variance and covariance for any QTL with an arbitrary number of alleles inferred by any types of molecular markers with varying numbers of alleles. This advantage is particularly important for outcrossing species, like forest trees, with high heterozygosity.

3.3 Natural Populations

For some species, genetic mapping can only rely on a collection of unrelated individuals sampled from a natural population (Wu and Zeng 2001; Li and McKeand 2004). In this case, mapping is based on linkage disequilibrium (LD). As a particular allele at a marker locus tends to co-segregate with one allelic variant of the gene of interest, provided the marker and gene are very closely linked, LD mapping can potentially be used to map QTL to very small regions (Wall and Pritchard 2003). In order to perform efficient LD mapping, markers must be mapped at a density compatible with the distances that LD extends in the population.

There is a special group of land plants, called gymnosperm, which include pine, fir, spruce and ginkgo. Genetic mapping for the gymnosperm can be based on the haploid megagametophyte. The megagametophyte, with the genotype identical to the maternal gamete, surrounds the embryo in the mature seed and supplies initial nutrients during seed to germination. As the megagametophyte is genetically equivalent to a haploid progeny, any heterozygous locus in the seed parent will always segregate 1:1 in the megagametophytes (Wilcox et al. 1996; Wu 1999) regardless of the pollen contribution, if segregation distortion does not occur. As a result, dominant markers derived from haploid megagametophytes are as informative as co-dominant markers. Attempts have been made to employ the megagametophyte to construct genetic linkage maps by collecting PCR-based dominant markers from the progeny of a heterozygous tree.

3.4 Joint linkage and linkage disequilibrium mapping

Although LD mapping has tremendous potential to fine map QTL for a dynamic trait, it is limited in practice because the association between a marker and QTL is also affected by evolutionary forces, such as mutation, drift, selection and admixture (Lynch and Walsh 1998). This disadvantage can be overcome by a mapping strategy that combines linkage and LD (Wu and Zeng 2001; Wu et al. 2002). Linkage and LD mapping has been integrated into a unifying framework to take advantages of both approaches in genome-wise scan (linkage mapping) and fine-structured estimates of QTL positions (LD mapping).

4 STATISTICAL MODELS, METHODS AND ALGORITHMS

4.1 Models

The main task of QTL mapping is to map the locations of individual QTLs and estimate their effects on the phenotype of a quantitative trait. If each individual in a mapping population has a known QTL genotype, simple analysis methods such as the t test or analysis of variance (ANOVA) can be directly used to associate QTL genotypes with the phenotypes. However, in practice, QTL genotypes cannot be observed and rather they should be inferred from observed marker genotypes based on the co-segregation between the QTL and markers in a specific segregating population. To clearly describe statistical algorithms for QTL mapping, we will first define the models that relate QTL information with observed data.

4.2 The QTL Model

The simplest case is that the genotypes of QTL can be observed for all individuals. Suppose there are two inbred lines that are homozygous for alternative alleles, Q and q, at each locus, respectively. These two contrasting inbred lines are crossed to generate the heterozygous F_1, from which different kinds of mapping populations are derived, such as the backcross (BC), doubled haploids (DH), recombinant inbred lines (RIL) or F_2. For the BC, DH and RIL populations, there are two segregating QTL genotypes at each locus and, thus, only the additive effect of the QTL can be estimated. For the F_2, in which three genotypes are segregating at a locus, both the additive and dominant effects of the QTL can be estimated.

The statistical model for the phenotypic value of individual i at a quantitative trait affected by a QTL can be expressed, for different designs, as

$$y_i = \begin{cases} \mu + x_i a + e_i & \text{for BC, DH and RIL} \\ \mu + x_i a + z_i d + e_i & \text{for the } F_2 \end{cases} \tag{3}$$

where μ is the overall mean, a and/or d are the additive and dominant effects of a QTL, respectively, x_i and/or z_i are the indicator variables that are related to the additive and dominant effects by specifying the QTL genotypes, which are defined, for the F_2, as

$$x_i = \begin{cases} 1 & \text{for QTL genotype } QQ \\ -1 & \text{for QTL genotpye } qq \end{cases}$$

and

$$z_i = \begin{cases} 1 & \text{for QTL genotype } Qq \\ 0 & \text{for QTL genotpye } QQ \text{ or } qq \end{cases}$$

and e_i is the residual error that is assumed to follow a normal distribution $N(0, \sigma^2)$. By estimating and testing a and/or d, QTL mapping attempts to identify the underlying QTL for the trait.

4.3 The Marker Model

The genotypes of a QTL cannot be observed, but they are related with marker genotypes that are observable, with the degree depending on how strongly the QTL and markers are physically linked on the same region of a chromosome. The practical data for a QTL mapping project include two sets, marker and phenotype, which are collected for each individual in a mapping population.

Although the markers are not a QTL, they are sometimes associated with the phenotypes to provide approximate estimates of the effects of the underlying QTL. The statistical model for the phenotypic value of individual i at a marker can be expressed as

$$y_i = \begin{cases} \mu + x_i^* a^* \, e_i & \text{for BC, DH and RIL} \\ \mu + x_i^* a^* + z_i^* d^* + e, & \text{for the } F_2 \end{cases} \tag{4}$$

where a^* and/or d^* are the additive and dominant effects associated with a marker (with alternative alleles M and m), respectively, x_i^* and/or z_i^* are the indicator variables that are related to the marker additive and dominant effects by specifying the marker genotypes. For the F_2, we have

$$x_i^* = \begin{cases} 1 & \text{for marker genotype } MM \\ -1 & \text{for marker genotype } mm \end{cases}$$

and

$$z_i^* = \begin{cases} 1 & \text{for marker genotype } Mm \\ 0 & \text{for marker genotype } MM \text{ or } mm \end{cases}$$

By estimating and testing a^* and/or d^*, whether there exists an underlying QTL linked with the marker can be inferred.

4.4   Marker-conditional QTL Model

Although the QTL model (3) provides precise estimation of QTL effects, this model is not realistic in practice because it is impossible to know the QTL genotypes. On the other hand, the marker model (4) makes use of observed marker data, but its estimation cannot precisely reflect the QTL effects. These two models should be integrated for more precise and realistic QTL mapping.

The statistical model for the phenotypic value of individual i at a QTL linked with a known marker can be expressed as

$$y_i = \begin{cases} \mu + \xi_i a + e_i & \text{for BC, DH and RIL} \\ \mu + \xi_i a + \zeta_i d + e_i & \text{for the } F_2 \end{cases} \tag{5}$$

where ξ_i (and/or ζ_i) is the variable that relates marker genotype x_i^* (and/or z_i^*) with QTL genotype x_i (and/or z_i). For example, in the F_2 they are defined as

$$\xi_i = \frac{1}{2} \left(\varpi_{2|i} - \varpi_{0|i} \right)$$

and

$$\zeta_i = \varpi_{1|i} - \frac{1}{2} \left(\varpi_{2|i} - \varpi_{0|i} \right),$$

where $\varpi_{2|i}$, $\varpi_{1|i}$ and $\varpi_{0|i}$ are the conditional probabilities of QTL genotypes QQ (2), Qq (1) and qq (0), conditional upon the observed marker genotype of individual i, respectively. These conditional probabilities can be derived in terms of the linkage between the QTL and markers (see below).

4.5   Methods and Algorithms

A variety of statistical methods have been developed for QTL mapping (Lynch and Walsh 1998). These methods can be classified as: t-tests and analysis of variance (ANOVA), least-squares analysis (LS), maximum-likelihood analysis (ML), and Bayesian analysis. They differ in computational requirements, efficiency in terms of extracting information, flexibility with regard to handling different data structures, and ability in mapping multiple QTL.

4.5.1 Marker Analysis

4.5.1.1 Testing

The existence of a QTL linked with a marker can be tested by a simple t test or ANOVA. In the BC, DH and RIL population, two classes of genotypes can be observed for each marker. For a specific marker, the difference between the two marker classes can be tested by a two-sample t test based on the hypotheses

$$H_0 : m_1 = m_0 \text{ vs. } H_1 : m_1 \neq m_0,$$

where m_1 and m_0 are the means for two different classes of marker genotypes. The t test statistic used to test for the significance of the difference between the two means is

$$t = \frac{m_1 - m_0}{\sqrt{s^2 \left(\dfrac{1}{n_1} + \dfrac{1}{n_0} \right)}}, \tag{6}$$

where s^2 is the pooled sampling variance given by

$$s^2 = \frac{(n_1 - 1)s_1^2 + (n_0 - 1)s_0^2}{n_1 + n_0 - 2},$$

with n_1, n_0 and s_1^2, s_0^2 being the sample sizes and variances in two different marker classes, respectively.

The null hypothesis H_0 will be rejected if the t test statistic calculated is larger than, or equal to, the critical value to be obtained from the t-distribution. If we denote the upper α critical point by $t_{(\alpha, v)}$, we would reject the hypothesis at $\alpha = 0.05$ if $t > t_{(0.025, v)}$, the two-tailed t value for the 0.05 significance level, with $v = n_1 + n_0 - 2$ degrees of freedom.

For an F_2 population, there are three different classes of marker genotypes at each marker, MM (2), Mm (1) and mm (0). Let n_2, n_1, n_0 and s_2^2, s_1^2, s_0^2 be the sample sizes and variances in three different marker classes of the F_2, respectively. To test the overall difference among the three genotypes, a traditional ANOVA can be used. The mean square due to the difference among the three marker genotypes reflects the degree to which the marker is associated with putative QTL for a particular trait, while the mean square due to the difference within the genotypes reflect the residual variance. The ratio of these two mean squares, i.e., the F-value, is a test statistic used to test for the significance of the difference among the three marker genotypes.

Table I Summary of ANOVA for the difference among three genotype groups in an F_2 population

Source of variation	df	Mean square	F-value	Expected mean square
Among marker genotypes	2	MS_1	MS_1/MS_2	$\sigma_e^2 + k\sigma_q^2$
Within marker genotypes	$n-3$	MS_2		σ_e^2

Note: $k = 3 \Big/ \left(\dfrac{1}{n_2} + \dfrac{1}{n_1} + \dfrac{1}{n_0} \right)$ is a harmonic mean, with n_2, n_1 and n_0 standing for sample sizes of the three different marker genotypes in an F_2 population.

The calculated F-value is compared with the critical value obtained from the F distribution, $F_{0.05,\,(2,\,n-3)}$. The genetic variance due to a significant marker can be estimated by equating the expected mean squares (Table 1) to the mean squares (MS) and solving the resulting equation:

$$\sigma_g^2 = \frac{MS_1 - MS_2}{k}. \tag{7}$$

The overall difference among the three marker genotypes in the F_2 population may be due to either the additive or dominant effect, or both. The significance of these two effects can also be tested by using the t test. To test the marker additive effect, we have the test statistic

$$t_1 = \frac{m_2 - m_0}{\sqrt{s^2 \left(\dfrac{1}{n_2} + \dfrac{1}{n_0} \right)}}, \tag{8}$$

where

$$s_2 = \frac{(n_2 - 1)\, s_2^2 + (n_0 - 1)\, s_0^2}{n_2 + n_0 - 2}.$$

The critical value for this test at the 0.05 significance level is $t_{(0.025,\, v)}$ with $v = n_2 + n_0 - 2$ degrees of freedom.

Similarly, to test the marker dominance effect, we have the test statistic

$$t_2 = \frac{m_1 - \dfrac{1}{2}(m_2 + m_0)}{\sqrt{s^2 \left(\dfrac{1}{4n_2} + \dfrac{1}{n_1} + \dfrac{1}{4n_0} \right)}}, \tag{9}$$

where

$$s_2 = \frac{(n_2 - 1)s_2^2 + (n_1 - 1)s_1^2 + (n_0 - 1)s_0^2}{n_2 + n_1 + n_0 - 3}$$

The critical value for this test at the 0.05 significance level is $t_{(0.025, v)}$ with $v = n_2 + n_1 + n_0 - 3$ degrees of freedom.

4.5.1.2 Linkage analysis

Consider a putative QTL linked to a marker, with a recombination fraction of r. The conditional expected genotypic values associated with each marker genotype are calculated from the conditional probabilities of the QTL genotypes given a marker genotype and from the genotypic values of different QTL genotypes. Given known marker gamete genotypes, M (1) and m (0), we can derive the conditional probabilities of two QTL gamete genotypes, Q (1) and q (0) for the F_1 as

Marker genotypic value	QTL genotypic value	
	μ_1	μ_0
m_1	$1 - r$	r
m_0	r	$1 - r$

The genetic values of these two QTL gamete genotypes can be denoted by

$$\mu_1 = \mu + \frac{1}{2}\, a, \text{ and } \mu_0 = \mu - \frac{1}{2}\, a,$$

respectively. For each marker genotype, two different QTL gamete genotypes are mixed, weighted by the conditional probabilities. Thus, the conditional expected genotypic values associated with different marker gamete genotypes can be calculated as

$$m_1 = (1 - r)\mu_1 + r\mu_0 = (1 - r)\left(\mu + \frac{1}{2}a\right) + r\left(\mu - \frac{1}{2}a\right) = \mu + \frac{1}{2}(1 - 2r)a$$

$$m_0 = r\mu_1 + (1 - r)\mu_0 = r\left(\mu + \frac{1}{2}a\right) + (1 - r)\left(\mu - \frac{1}{2}a\right) = \mu - \frac{1}{2}(1 - 2r)a$$

Thus, the difference of the two marker means is

$$\mu_1 - \mu_0 = (1 - 2r)a \tag{10}$$

If a is not significantly different from zero, the t test statistic based on (1) will be smaller than the critical value. In this sense, the t test can provide information about the significance of QTL effect. But a nonsignificant t value may also be due to non-linkage between the marker and QTL ($r = 0.5$). Therefore, the t test only gives a composite test for the QTL effect and QTL-marker linkage.

For an F_2 population, three marker means can be similarly derived by conditional probabilities expressed as

Marker genotypic value	QTL genotypic value		
	μ_2	μ_1	μ_0
m_2	$(1 - r)^2$	$2r(1 - r)$	r^2
m_1	$r(1 - r)$	$(1 - r)^2 + r^2$	$r(1 - r)$
m_0	r^2	$2r(1 - r)$	$(1 - r)^2$

and the assigned QTL genotype values. These marker means are written as

$$m_2 = (1 - r)^2 \, \mu_2 + 2r(1 - r) \, \mu_1 + r^2 \, \mu_0 = \mu + (1 - 2r)a + 2r(1 - r)d$$
$$m_1 = r(1 - r) \, \mu_2 + [(1 - r)^2 + r_2] \, \mu_1 + r(1 - r) \, \mu_0 = \mu + (1 - 2r + 2r^2)d$$
$$m_0 = r^2 \, \mu_2 + 2r(1 - r) \, \mu_1 + (1 - r)^2 \, \mu_0 = \mu - (1 - 2r)a + 2r(1 - r)d$$

The tests for the additive (a) from (8) and dominant effects (d) from (9) are equivalent to testing whether composite parameters

$$\frac{1}{2} (m_2 - m_0) = (1 - 2r)a$$

and

$$m_1 - \frac{1}{2} (m_2 + m_0) = (1 - 2r)^2 d$$

are equal to zero, respectively.

From the above analysis, although the t test and ANOVA can be used to test the significance of marker differences, they cannot separate QTL genotypic means and the recombination fraction between a single marker and a QTL. If the marker difference is significant, we still do not know whether this difference is due to a tight linkage (small r) between the marker and a QTL of small effect, or a loose linkage (large r) between the marker and a QTL of large effect. The two confounded parameters, QTL genetic value and the recombination fraction, can be separated using the approaches explained below.

4.5.2 Interval Mapping

The confounded effects of markers can be ruled out by assuming the underlying QTL within the interval bracketed by a pair of flanking markers. This so-called interval mapping can be analyzed by a regression method (Haley and Knott 1992) or maximum likelihood (ML, Lander and Botstein 1989). Although these two methods may provide similar estimates of QTL parameters, the ML method has several attractive statistical properties, such as consistency and asymptotic efficiency, and, therefore, it has great potential for the precise estimation of QTL parameters. Furthermore, the ML method has better interpretability than the regression model in terms of the genetic model, suggesting its applicability to practical genetic mapping problems.

4.5.2.1 Mixture model

The central idea of the ML-based interval mapping is to infer unknown QTL genotypes based on observed phenotypic and marker data via genetic and statistical models that connect the observed and unknown information. In statistics, this is a missing data problem, which can be solved within the mixture model framework (Lander and Botstein 1989). Since the first attempt to analyze a mixture model by Pearson (1894), mixture models have been used in an incredible range of applications (Titterington 1997). Pearson used a method of moments to estimate the parameters of a mixture of two univariate normal distributions, but mixture models are now commonly analyzed using maximum likelihood or Bayesian methods.

We consider a model in which data $y = (y_1, ..., y_n)$ are assumed to be independent observations from a mixture density with G components. This model is expressed as

$$y \sim p(y \mid \varpi, \phi, \eta) = \varpi_1 f_1(y \mid \phi_1, \eta) + ... + \varpi_1 f_G(y \mid \phi_G, \eta) \qquad (11)$$

where $\varpi = (\varpi_1, ..., \varpi_G)$ are the mixture proportions which are constrained to be non-negative and sum to unity; $\phi = (\phi_1, ..., \phi_G)$ are the component specific parameters, with ϕg being specific to component g; and η is a parameter which is common to all components.

Mixture models are typically used to model data where each observation is assumed to have arisen from one of G (G possibly unknown but finite) components, each component being modeled by a density from the parametric family f. The mixture proportions then represent the relative frequencies of occurrence of each group in the population, and the model provides a framework by which observations may be clustered together into groups for discrimination or classifications. For a mapping population, the QTL genotype an individual carries is unknown, but must be one and only one of all possible QTL genotypes. When known marker genotypes are used to predict QTL genotypes based on the linkage between the QTL and markers, the mixture proportions correspond to the conditional probabilities of QTL genotypes, conditional upon marker genotypes, in a particular population. As the marker genotype of each individual is known, we generally use $\varpi_{g|i}$ to denote the conditional probability of QTL genotype g given the marker genotype of individual i.

The density distribution of each component can be modeled by a parametric family. In QTL mapping, the density distribution is usually assumed to be normal because most quantitative traits are continuously distributed. For individual i that carries a QTL genotype g, the normal distribution density is expressed as

$$f_g(y_i \mid \phi_g, \eta) = \frac{1}{\sqrt{2\pi}\sigma} \exp\left[-\frac{(y_i - \mu_g)^2}{2\sigma^2} \right] \tag{12}$$

where μ_g is the mean value of the phenotypic trait corresponding to QTL genotype g, which is a common-specific parameter (ϕ_g), and σ^2 is the residual variance, which is the common parameter (η). These two parameters represent the magnitude of the genetic effect of a QTL and its relative contribution to the total phenotypic variance.

4.5.2.2 Conditional probabilities

The information about the QTL number and locations is reflected within conditional probabilities of unknown QTL genotypes given known marker genotypes. The derivation process of conditional probabilities varies, depending on different genetic settings. Suppose there is a QTL (with alleles Q and q) responsible for variation in a quantitative trait, bracketed by two flanking markers, **M** (with alleles M and m) and **N** (with alleles N and n). Let $r(\bar{r} = 1 - r)$, $r_1(\bar{r}_1 = 1 - r)$ and $r_2(\bar{r}_2 = 1 - r_2)$ denote the recombination fractions between the two markers, between marker **M** and the QTL and between the QTL and marker **N**, respectively. Crossing two parental inbred lines, $MMQQNN$ and $mmqqnn$, leads to a heterozygous F_1. The F_1 generate 8 different gametes at these markers and QTL, whose frequencies can be derived by assuming the independence of recombinant events between each pair of loci, expressed as

Gamete	MQN	MqN	MQn	Mqn	mQN	mqN	mQn	mqn
Frequency	$\frac{1}{2}\bar{r}_1\bar{r}_2$	$\frac{1}{2}r_1r_2$	$\frac{1}{2}\bar{r}_1r_2$	$\frac{1}{2}r_1\bar{r}_2$	$\frac{1}{2}r_1\bar{r}_2$	$\frac{1}{2}\bar{r}_1r_2$	$\frac{1}{2}r_1r_2$	$\frac{1}{2}\bar{r}_1\bar{r}_2$

The frequency of an observed marker gamete genotype is the sum of the frequencies of the same marker genotype containing different QTL gamete genotypes. For example, the frequency of MN or mn is

$$\frac{1}{2}\bar{r}_1\bar{r}_2 + \frac{1}{2}r_1r_2 = \frac{1}{2}\bar{r}$$

and of Mn or mN is

$$\frac{1}{2}\bar{r}_1r_2 + \frac{1}{2}r_1\bar{r}_2 = \frac{1}{2}r,$$

where $r = r_1 + r_2 - 2r_1r_2$. The conditional probability ($\varpi_{g|i}$) of a QTL genotype given the marker genotypes can be derived using Bayes' theorem. All together we have two QTL genotypes in the BC, DH and RIL population, each predicted by the four known marker genotypes. These conditional probabilities are tabulated in Table 2.

Table 2 Conditional probabilities of QTL gamete genotypes given the marker gamete interval genotype in the F_1, assuming the independence of recombination between two adjacent intervals

Interval marker		QTL genotype	
Genotype	Size	Q	q
MN	n_{11}	$\dfrac{\bar{r}_1\bar{r}_2}{1-r}$	$\dfrac{r_1r_2}{1-r}$
Mn	n_{10}	$\dfrac{\bar{r}_1r_2}{r}$	$\dfrac{r_1\bar{r}_2}{r}$
mN	n_{01}	$\dfrac{r_1\bar{r}_2}{r}$	$\dfrac{\bar{r}_1r_2}{r}$
mm	n_{00}	$\dfrac{r_1r_2}{1-r}$	$\dfrac{\bar{r}_1\bar{r}_2}{1-r}$

If a dense linkage map is used, the recombination fraction between any two flanking markers should be small and, thus, the recombination fraction between the QTL and one of the flanking markers (r_1 or r_2) is even smaller. In this case, the product r_1r_2 can be reasonably assumed to be zero so that $r_1 + r_2 = r$. This relation corresponds to the assumption of no double recombination between marker **M** and the QTL and between the QTL and marker **N**. Letting $\theta = r_1/r$, the conditional probabilities can be simplified as given in Table 3.

Table 3 Approximate conditional probabilities of QTL genotypes given the marker interval genotype in the backcross, assuming no double recombination

Interval marker		QTL genotype	
Genotype	Size	Q	q
MN	n_{11}	1	0
Mn	n_{10}	$1 - \theta$	θ
mN	n_{01}	θ	$1 - \theta$
mm	n_{00}	0	1

Note: The ratio $\theta = r_1/r$ and $1 - \theta = r_2/r$.

The conditional probabilities in Tables 2 and 3 can be used for the BC and DH population, but are not appropriate for the RIL population. RILs are derived either by repeated selfing or by repeated brother-sister mating of the progeny from an F_1 cross between two inbred lines. RILs are fixed, with homozygous genotypes 2 and 0, for all genes and can serve a permanent mapping population for multiple uses. Let R_1 and R_2 be the proportions of recombinant zygotes between marker **M** and the QTL and between the QTL and marker **N**, respectively. Martin and Hospital (2006) derived the conditional probabilities of QTL genotypes given the marker genotypes in terms of R_1 and R_2 in the selfing RIL population which are tabulated in Table 4. The relationship between R and r for two loci in a selfing RIL population has been derived by Haldane and Waddington (1931), expressed as

$$R_1 = \frac{2r_1}{1 + 2r_1}, \; r_1 = \frac{R_1}{2(1 - R_1)}$$

$$R_2 = \frac{2r_2}{1 + 2r_2}, \; r_2 = \frac{R_2}{2(1 - R_2)} \tag{13}$$

The random combinations among the four gametes of the F_1 lead to the F in which a total of nine marker interval genotypes each correspond to one of three possible QTL genotypes. Table 5 lists the joint genotype probabilities of two flanking markers and QTL in the F_2 population, as well as the joint genotype probabilities of the two markers. From these joint probabilities, conditional probabilities of the QTL genotypes given marker genotypes can be derived.

4.5.2.3 Likelihood and algorithm

Statistical methods have been available to estimate the parameters about the QTL number, locations and effects. These methods include the

Table 4 Conditional probabilities of QTL genotypes given marker genotypes in the selfing RIL population

Marker interval		QTL genotype	
Genotype	Size	QQ	q
MMNN	n_{22}	$1 - \dfrac{R_1R_2(3 - 2R_1 - 2R_2)}{2(1 - R_1)(1 - R_2)}$	$\dfrac{R_1R_2(3 - 2R_1 - 2R_2)}{2(1 - R_1)(1 - R_2)}$
MMnn	n_{20}	$1 - \dfrac{2R_1 - R_1R_2(3 + 2R_1 - 2R_2)}{2R_2 + R_1(2 - 6R_2)}$	$\dfrac{2R_1 - R_1R_2(3 + 2R_1 - 2R_2)}{2R_2 + R_1(2 - 6R_2)}$
mmNN	n_{02}	$\dfrac{2R_1 - R_1R_2(3 + 2R_1 - 2R_2)}{2R_2 + R_1(2 - 6R_2)}$	$1 - \dfrac{2R_1 - R_1R_2(3 + 2R_1 - 2R_2)}{2R_2 + R_1(2 - 6R_2)}$
mmnn	n_{00}	$\dfrac{R_1R_2(3 - 2R_1 - 2R_2)}{2(1 - R_1)(1 - R_2)}$	$1 - \dfrac{R_1R_2(3 - 2R_1 - 2R_2)}{2(1 - R_1)(1 - R_2)}$

Table 5 Joint marker-QTL-marker genotype frequencies in the F_2

Marker Interval			QTL Genotype		
Genotype	Frequency	Size	QQ	Qq	qq
MMNN	$\frac{1}{4}\bar{r}^2$	n_{22}	$\frac{1}{4}\bar{r}_1^2\bar{r}_2^2$	$\frac{1}{2}r_1r_2\bar{r}_1\bar{r}_2$	$\frac{1}{4}r_1^2r_2^2$
MMNn	$\frac{1}{2}r\bar{r}$	n_{21}	$\frac{1}{2}r_2\bar{r}_1^2\bar{r}_2$	$\frac{1}{2}r_1\bar{r}_1(r_2^2 + \bar{r}_2^2)$	$\frac{1}{2}r_1^2r_2\bar{r}_2$
MMnn	$\frac{1}{4}r^2$	n_{20}	$\frac{1}{4}\bar{r}_1^2r_2^2$	$\frac{1}{2}r_1r_2\bar{r}_1\bar{r}_2$	$\frac{1}{4}r_1^2\bar{r}_2^2$
MmNN	$\frac{1}{2}r\bar{r}$	n_{12}	$\frac{1}{2}r_1\bar{r}_1\bar{r}_2^2$	$\frac{1}{2}r_2\bar{r}_2(r_1^2 + \bar{r}_1^2)$	$\frac{1}{2}r_1\bar{r}_1r_2^2$
MmNn	$\frac{1}{2}(r^2 + \bar{r}^2)$	n_{11}	$r_1r_2\bar{r}_1\bar{r}_2$	$\frac{1}{2}(r_1^2 + \bar{r}_1^2)(r_2^2 + \bar{r}_2^2)$	$r_1r_2\bar{r}_1\bar{r}_2$
Mmnn	$\frac{1}{2}r\bar{r}$	n_{10}	$\frac{1}{2}r_1\bar{r}_1r_2^2$	$\frac{1}{2}r_2\bar{r}_2(r_1^2 + \bar{r}_1^2)$	$\frac{1}{2}r_1\bar{r}_1\bar{r}_2^2$
mmNN	$\frac{1}{4}r^2$	n_{02}	$\frac{1}{4}r_1^2\bar{r}_2^2$	$\frac{1}{2}r_1r_2\bar{r}_1\bar{r}_2$	$\frac{1}{4}\bar{r}_1^2r_2^2$
mmNn	$\frac{1}{2}r\bar{r}$	n_{01}	$\frac{1}{2}r_1^2r_2\bar{r}_2$	$\frac{1}{2}r_1\bar{r}_1(r_2^2 + \bar{r}_2^2)$	$\frac{1}{2}r_2\bar{r}_1^2\bar{r}_2$
mmnn	$\frac{1}{4}\bar{r}^2$	n_{00}	$\frac{1}{4}r_1^2r_2^2$	$\frac{1}{2}r_1r_2\bar{r}_1\bar{r}_2$	$\frac{1}{4}\bar{r}_1^2\bar{r}_2^2$

regression analysis using equation (5) (Haley and Knott 1992), and maximum likelihood and Bayesian approaches based on $p(y \mid \varpi, \phi, \eta)$ (equation 11). The advantages and disadvantages of these methods have been discussed in Kao (2000) and Mayer (2005). Here, we will focus on the maximum likelihood method, implemented with the EM algorithm. For Bayesian approaches implemented with the Markov chain Monte Carlo algorithm, the readers are referred to Satagopan et al. (1996) and Sillanpaa and Arjas (1998, 1999).

For simplicity, our description will be focused on a simple BC or DH design. By assuming that the distribution of a phenotypic trait follows a normal distribution, each of two possible genotype classes can be described by

$$f_1(y_i \mid \mu_1, \sigma^2) = N(\mu_1, \sigma^2) \text{ and } f_0(y_i \mid \mu_0, \sigma^2) = N(\mu_0, \sigma^2).$$

Thus, the likelihood of the phenotypic (y) and marker observations (M) can be written, in terms of a mixture model, as

$$L(\mu_1, \mu_2, \sigma^2, \theta \mid y, M) = \prod_{i=1}^{n} [\varpi_{1 \mid j} f_1(y_i \mid \mu_1, \sigma^2) + \varpi_{0 \mid j} f_0(y_i \mid \mu_0, \sigma^2)]. \quad (14)$$

with the conditional probabilities of QTL genotypes given marker genotypes being substituted by Tables 2 or 3. Assuming no double recombination, we have

$$L(\mu_1, \mu_2, \sigma^2, \theta \mid y, M) = \prod_{i=1}^{n_{11}} f_1(y_i \mid \mu_1, \sigma^2)$$

$$\times \prod_{i=1}^{n_{10}} [(1 - \theta) f_1(y_i \mid \mu_1, \sigma^2) + \theta f_0(y_i \mid \mu_0, \sigma^2)] \quad (15)$$

$$\times \prod_{i=1}^{n_{01}} [\theta f_1(y_i \mid \mu_1, \sigma^2) + (1 - \theta) f_0(y_i \mid \mu_0, \sigma^2)]$$

$$\times \prod_{i=1}^{n_{00}} f_0(y_i \mid \mu_0, \sigma^2)$$

The solution of equation (15) can be obtained by implementing the EM algorithm (Dempster et al. 1977; Meng and Rubin 1993). In the E step, we define the posterior probabilities of a QTL genotype given the marker genotype and phenotypic value of individual i as

$$P_{1|i} = \frac{(1-\theta)f_1(y_i|\mu_1,\sigma^2)}{(1-\theta)f_1(y_i|\mu_1,\sigma^2)+\theta\, f_0(y_i|\mu_0,\sigma^2)} \tag{16}$$

$$P_{0|i} = \frac{\theta\, f_0(y_i|\mu_0,\sigma^2)}{(1-\theta)f_1(y_i|\mu_1,\sigma^2)+\theta\, f_0(y_i|\mu_0,\sigma^2)}$$

with $P_{1|i} + P_{0|i} = 1$.

In the M step, a group of the following log-likelihood equations are solved that are derived by differentiating $\log L(\mu_1, \mu_2, \sigma^2, \theta\,|\,y, M)$ and letting the derivatives equal to zero:

$$\frac{\partial}{\partial\mu_1}\log L(\mu_1, \mu_2, \sigma^2, \theta\,|\,y, M) = 0 \Rightarrow \hat{\mu}_1 = \frac{\sum_{i=1}^{n_{11}} y_i + \sum_{i=1}^{n_{10}+n_{01}} P_{1|i}y_i}{n_{11} + \sum_{i=1}^{n_{10}+n_{01}} P_{1|i}} \tag{17}$$

$$\frac{\partial}{\partial\mu_0}\log L(\mu_1, \mu_2, \sigma^2, \theta\,|\,y, M) = 0 \Rightarrow \hat{\mu}_0 = \frac{\sum_{i=1}^{n_{10}+n_{01}} P_{0|i}y_i + \sum_{i=1}^{n_{00}} y_i}{\sum_{i=1}^{n_{10}+n_{01}} P_{0|i} + n_{00}} \tag{18}$$

$$\frac{\partial}{\partial\sigma^2}\log L(\mu_1, \mu_2, \sigma^2, \theta\,|\,y, M) = 0 \Rightarrow$$

$$\hat{\sigma}^2 = \frac{1}{n}\left(\sum_{i=1}^{n_{11}}(y_i-\hat{\mu}_1)^2 + \sum_{i=1}^{n_{10}+n_{01}}\left[P_{1|i}(y_i-\hat{\mu}_1)^2 + P_{0|i}(y_i-\hat{\mu}_0)^2\right] + \sum_{i=1}^{n_{00}}(y_i-\hat{\mu}_0)^2\right) \tag{19}$$

$$\frac{\partial}{\partial\theta}\log L = 0 \Rightarrow \hat{\theta} = \frac{\sum_{i=1}^{n_{10}} P_{0|i} + \sum_{i=1}^{n_{01}} P_{1|i}}{n_{10} + n_{01}} \tag{20}$$

The solutions for the unknown parameters are not in analytical form because each estimate depends on estimates of other parameters. Different from other numerical problems, each estimate is also a function of the posterior probability for each individual with an expected QTL genotype. The iteration between the E (equations 16) and M steps (equations 17 – 20) is repeated until the estimates are stable. The convergence of this iteration can be guaranteed. The stable estimates are regarded as the maximum likelihood estimates (MLEs) that will maximize the likelihood.

As genotypic values can be defined as $\mu_1 = \mu + a$ and $\mu_0 + \mu$ in the BC, DH or RIL design, the additive genetic effect of the QTL can be estimated as $\hat{a} = \hat{\mu}_1 - \hat{\mu}_0$. Based on the invariance property of maximum likelihood, the estimate of a in this way should be the MLE because $\hat{\mu}_1$ and $\hat{\mu}_0$ are the MLEs.

In practical computations, the QTL position parameter (θ) can be viewed as a fixed parameter by searching for a putative QTL at every 1 or 2 cM on a map interval bracketed by two markers throughout the entire linkage map. The likelihood values for a QTL at different map positions are displayed graphically to generate a likelihood map or profile. The genomic position that corresponds to a peak of the profile is the MLE of the QTL location.

4.5.2.4 Hypothesis testing

The genetic effect of a detected QTL should be tested for its significance. This can be done by formulating the following hypotheses:

$$H_0 : \mu_1 = \mu_0 = \mu \text{ vs. } H_1 : \mu_1 \neq \mu_0, \tag{21}$$

which is equivalent to the hypotheses

$$H_0 : a = 0 \text{ vs. } H_1 : a \neq 0,$$

Under hypothesis (21), we calculate the likelihood value. Let $L_0 = L(\tilde{\mu}, \tilde{\sigma}^2 \,|\, y)$ and $L_1 = L(\hat{\mu}_1, \hat{\mu}_0, \hat{\sigma}^2, \hat{\theta} \,|\, y, M)$ be the maximum likelihood value under the null (reduced model, there is no QTL) and alternative hypothesis (full model, there is a QTL), respectively. Note that under the null hypothesis the estimates are not dependent on the marker information and the QTL position is not identifiable. We then calculate the LOD score as a test statistic for testing the existence of a QTL,

$$LOD = \frac{\log_{10} L_1}{\log_{10} L_0}. \tag{22}$$

The LOD score is equivalent to the log-likelihood ratio (LR) test statistic expressed as

$$LR = -2\left(\frac{\log_e L_0}{\log_e L_1} \right) \tag{23}$$

with

$$LOD = 0.217LR \text{ or } LR = 4.608LOD$$

The determination of a critical threshold for the significance tests for declaring the existence of a QTL are a difficult, still controversial area. As hundreds of independent loci are associated with a trait, there is a need for statistical thresholds that account for multiple comparisons. Lander and Botstein (1989) and Lander and Kruglyak (1995) established a common threshold for all normally distributed traits, based on the length of genetic map and density of markers. Different approaches have now evolved for setting significance thresholds (e.g., Piepho 2001). An empirical approach based on permutation tests has been proposed to determine the threshold (Churchill and Doerge 1994) by re-shuffling the associations between the markers and phenotypic values of a trait across the offspring genotypes of a mapping population.

4.5.2.5 Example

Two inbred lines, semi-dwarf IR64 and tall Azucena, were crossed to generate an F_1 progeny population. By doubling haploid chromosomes of the gametes derived from the heterozygous F_1, a DH population of 123 lines were founded (Huang et al. 1997). Such a DH population is equivalent to a backcross population because its marker segregation follows 1:1. With 123 DH lines, Huang et al. genotyped a total of 175 polymorphic markers (including 146 RFLPs, 8 isozymes, 14 RAPDs and 12 cloned genes) to construct a linkage map (Figure 1), representing a good coverage of 12 rice chromosomes. The DH plants were phenotyped for various quantitative traits. In this example, the phenotype chosen for QTL mapping is plant height measured at 10 weeks after rice was transplanted to the field.

Interval mapping was used to scan for the existence and distribution of QTL for plant height throughout the entire genome. We incorporate two different models of the conditional probabilities of QTL genotypes given marker genotypes into the model for interval mapping. One model is based on the assumption of no double recombination, whereas the second model is based on the independence of recombination. The log-likelihood ratio (LR) test statistics were calculated by assuming a QTL at a fixed position between two flanking markers. An LR profile for the two models was drawn across all the linkage groups (Figure 2). The critical thresholds for the declaration of QTL were empirically determined on the basis of 1,000 permutations tests.

The no double recombination model identified three significant QTLs on chromosomes 4, 7 and 12 at the 0.05 significance level (Fig. 2). As expected, the favorable allele for plant height at each detected QTL is contributed by the tall Azucena parent (Table 6). Using the independence model, five QTLs were characterized on chromosomes 2, 3, 4, 6 and 7, at each of which the tall Azucena parent contributes a favorable allele (Fig. 2; Table 6). Each QTL detected using the two models explained a small proportion (0.02 – 0.20) of the phenotypic variance in plant height.

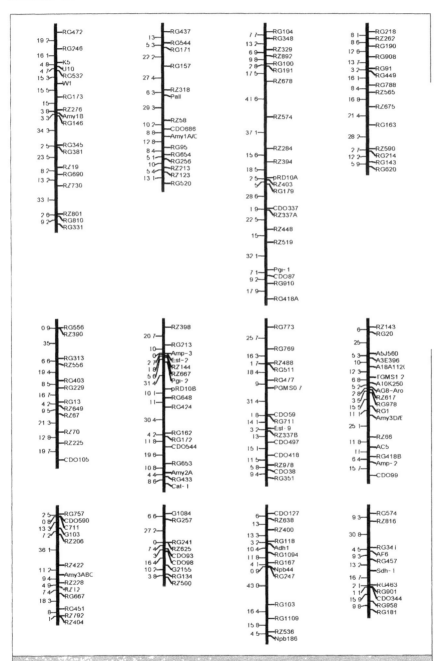

Fig. I Genetic linkage maps constructed from 135 RFLP and 40 isozyme and RAPD markers for 123 DH plants derived from the tall Azucena and short IR64 parents.

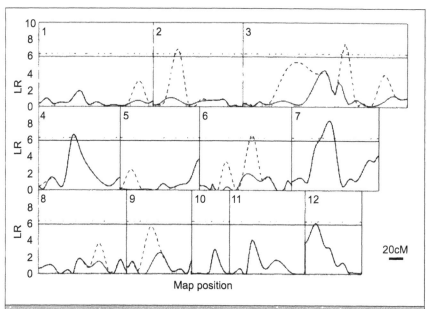

Fig. 2 The log-likelihood ratio profile across the genome under two different models assuming no double recombination (solid) and the independence of recombination (broken). The ticks on the x-axis indicate the positions of markers. The genome-wide critical thresholds, shown by the horizonal lines, were determined on the basis of 1,000 permutation tests.

Table 6 The MLEs of the QTL positions and effects for plant height in a DH population of rice under two different models by assuming the independence of recombination and no double recombination between two intervals, respectively

Chromosome	Marker interval	μ	a	R^2
Model assuming no double recombination				
4	RG788–RZ565	37.24	2.75	0.06
7	PGMS0.7–CDO59	34.21	2.90	0.06
12	RZ816–RG341	37.71	3.01	0.07
Model assuming the independence of recombination				
2	RG157–RZ318	40.70	8.37	0.20
3	RG179–CDO337	40.40	8.24	0.19
4	RG788–RZ565	37.24	2.75	0.02
6	RG424–RG162	40.13	8.08	0.19
7	PGMS0.7–CDO59	37.12	2.90	0.02

It is interesting to note the similarity and difference in QTL detection between the two types of models. The same QTLs on chromosome 4 and 7 were detected by the two models, whereas more QTLs were observed by the independence model than the no double recombination model (Figure 2). The characterization of different QTLs reveals that the co-segregation of markers and QTLs underlies different meiotic mechanisms at different genomic regions. This further suggests the importance of the use of different mechanistic models to draw a full picture of the genetic architecture of a complex trait in plants.

4.5.3   Composite and Multiple Interval Mapping

Lander and Botstein's (1989) interval method has an advantage for mapping QTL genome-wide by scanning for the position of a QTL throughout the genome. But this method can lead to biased estimates of QTL positions and effects when multiple QTLs occur on the same linkage group because it makes use of one single marker interval at a time and has no mechanism to alleviate the impact of other QTLs outside the interval. For this reason, if a real QTL is located nearby a marker interval with no QTL, interval mapping may still detect a 'ghost' QTL due to the linkage between the real QTL and the interval being tested (Martinez and Curnow 1992). Although a simultaneous search for multiple QTL on different intervals can overcome this problem, this will bring about new difficulties in parameter estimation and model identifiability.

A natural way for eliminating the influences of genetic background is attempted to remove this confounding information using co-variates or co-factors. This is the approach of *composite interval mapping*, which constructs test statistics by combining interval mapping on two flanking markers and multiple regression analysis with the other markers. This composite interval mapping strategy, proposed independently by Zeng (1994) and Jansen and Stam (1994), removes some of the interference from QTL outside the interval being tested. Zeng (1993) particularly demonstrates the advantages and disadvantages of composite interval mapping.

Both interval mapping and composite interval mapping model one QTL at a time. Although composite interval mapping that combines the idea of interval mapping and marker regression analysis can overcome the problem of multi-QTL linkage, it has less power to characterize the detailed genetic architecture of a quantitative trait. As a complex trait may be controlled by a number of QTLs, it is crucial to have a mapping approach that can model multiple QTLs simultaneously and identify and locate all the QTLs responsible for quantitative variation. Such an approach has been proposed by Zeng and colleagues, which is called multiple interval mapping (Kao et al. 1999; Zeng et al. 2000). Kao and Zeng (1997) have derived general formulae for obtaining maximum likelihood estimates for the positions and effects of multiple QTLs.

Multiple interval mapping models multiple QTLs in a way that QTLs can be directly controlled in the model through the simultaneous use of multiple marker intervals. It has proven more powerful and precise for estimating the positions and effects of QTLs than conventional interval mapping and composite interval mapping. In addition, by searching and mapping all possible QTLs in multiple marker intervals simultaneously, multiple interval mapping allows for the full estimation of the genetic architecture of a quantitative trait in terms of the number of the underlying QTLs, their genetic effects, pleiotropic effects and epistatic network among different QTLs. The area of research that is open to multiple interval mapping is the procedure for the model selection of multiple QTLs, their genomic positions and effects, which collectively provide the best fit of the data observed.

5 CONCLUDING REMARKS

Thanks to the recent development of molecular technologies and instruments, genetic mapping of quantitative trait loci has become a routine approach for studying the genetic control of complex phenotypes. Facing the increasing availability of genetic and genomic data for a variety of species, statistical analysis has become critical for giving precise explanations regarding genetic variation in quantitative traits occurring among species, populations, families and individuals.

In 1989, Lander and Botstein published a hallmark methodological paper for interval mapping that enables geneticists to detect and estimate individual genes or quantitative trait loci (QTL) that control the phenotype of a trait. Today, interval mapping has been a mainstream statistical tool for characterizing the genetic architecture of quantitative traits at the molecular level, leading to the discovery of thousands of QTL responsible for a variety of traits in plants, animals and humans. To be suitable for various practical applications, interval mapping has been extensively modified and extended during the past 15 years. A host of useful statistical methods for QTL mapping have been produced through collective efforts of statistical geneticists (Wu et al. 2007). In the next decades, these still-developing and improving methods will increasingly prove their values for revealing the genetic secrets behind the formation and development of complex phenotypes of economical and biological interests.

Acknowledgements

The preparation of this manuscript was partially supported by NSF grant (0540745) to R. W.

References

Anholt RR, Mackay TFC (2004) Quantitative genetic analysis of complex behaviours in *Drosophila*. Nat Rev Genet 5: 838-849

Carlborg O, Haley CS (2004) Epistasis: too often neglected in complex trait studies? Nat Rev Genet 5: 618-625

Cervera MT, Storme V, Ivens B, Gusmao J, Liu BH, Hostyn V, van Slycken J, van Montagu M, Boerjan W (2001) Dense genetic linkage maps of three Populus species (*Populus deltoides*, *P. nigra* and *P. trichocarpa*) based on AFLP and microsatellite markers. Genetics 158: 787-809

Churchill GA, Doerge RW (1994) Empirical threshold values for quantitative trait mapping. Genetics 138: 963-971

Dempster AP, Laird NM, Rubin DB (1977) Maximum likelihood from incomplete data via the EM algorithm. J Roy Stat Soc Ser B 39: 1-38

Doebley J, Stec A, Hubbard L (1997) The evolution of apical dominance in maize. Nature 386: 485-488

Falconer DS, Mackay TFC (1996) Introduction to Quantitative Genetics, 4th edn. Longman, New York, USA

Fisher RA (1918) The correlation between relatives on the supposition of Mendelian inheritance. Trans Roy Soc Edinb 52: 399-433

Frary A, Nesbitt TC, Frary A, Grandillo S, van der Knaap E, Cong B, Liu JP, Meller J, Elber R, Alpert KB, Tanksley SD (2000) fw2.2: A quantitative trait locus key to the evolution of tomato fruit size. Science 289: 85-88

Gallavotti A, Zhao Q, Kyozuka J, Meeley RB, Ritter MK, Doebley JF, Pe ME, Schmidt RJ (2004) The role of barren stalk1 in the architecture of maize. Nature 432: 630-635

Grattapaglia D, Sederoff RR (1994) Genetic linkage maps of *Eucalyptus grandis* and *Eucalyptus urophylla* using a pseudo-testcross mapping strategy and RAPD markers. Genetics 137: 1121-1137

Haldane JBS, Waddington CH (1931) Inbreeding and linkage. Genetics 16: 357-374

Haley CS, Knott SA (1992) A simple method for mapping quantitative trait loci in line crosses using flanking markers. Heredity 69: 315-324

Huang N, Parco A, Mew T, Magpantay G, McCouch S, Guiderdoni E, Xu J, Subudhi P, Angeles ER, Khush GS (1997) RFLP mapping of isozymes, RAPD and QTLs for grain shape, brown planthopper resistance in a doubled haploid rice population. Mol Breed 3: 105-113

Jansen RC, Stam P (1994) High resolution mapping of quantitative traits into multiple loci via interval mapping. Genetics 136: 1447-1455

Kao C-H (2000) On the differences between maximum likelihood and regression interval mapping in the analysis of quantitative trait loci. Genetics 156: 855-865

Kao C-H, Zeng Z-B (1997) General formulas for obtaining the MLEs and the asymptotic variance-covariance matrix in mapping quantitative trait loci when using the EM algorithm. Biometrics 53: 653-665

Kao C-H, Zeng Z-B (2002) Modeling epistasis of quantitative trait loci using Cockerham's model. Genetics 160: 1243-1261

Kao C-H, Zeng Z-B, Teasdale RD (1999) Multiple interval mapping for quantitative trait loci. Genetics 152: 1203-1216

Lande R, Thompson R (1990) Efficiency of marker-assisted selection in the improvement of quantitative traits. Genetics 124: 743-756

Lander ES, Botstein D (1989) Mapping Mendelian factors underlying quantitative traits using RFLP linkage maps. Genetics 121: 185-199

Lander E, Kruglyak L (1995) Genetic dissection of complex traits: Guidelines for interpreting and reporting linkage results. Nat Genet 11: 241-247

Li B, McKeand S (2004) Forest Genetics and Tree Breeding in the Age of Genomics: Progress and Future. IUFRO Joint Conference of Division 2, Conference Proceedings, 490 p

Li CB, Zhou AL, Sang T (2006) Rice domestication by reducing shattering. Science 311: 1936-1939

Lin M, Lou XY, Chang M, Wu RL (2003) A general statistical framework for mapping quantitative trait loci in non-model systems: Issue for characterizing linkage phases. Genetics 165: 901-913

Lu Q, Cui YH, Wu RL (2004) A multilocus likelihood approach to joint modeling of linkage, parental diplotype and gene order in a full-sib family. BMC Genetics 5: 20

Lynch M, Walsh B (1998) Genetics and Analysis of Quantitative Traits. Sinauer, Sunderland, MA, USA

Ma C-X, Casella G, Wu RL (2002) Functional mapping of quantitative trait loci underlying the character process: A theoretical framework. Genetics 161: 1751-1762

Mackay TFC (1996) The nature of quantitative genetic variation revisited: lessons from *Drosophila* bristles. BioEssays 18: 113-121

Mackay TFC (2001) Quantitative trait loci in *Drosophila*. Nat Rev Genet 2: 11-20

Martin OC, Hospital F (2006) Two and three-locus tests for linkage analysis using recombinant inbred lines. Genetics 173: 451-459

Martinez O, Curnow RN (1992) Estimating the locations and the sizes of the effects of quantitative trait loci using flanking markers. Theor Appl Genet 85: 480-488

Mather K (1943) Polygenic inheritance and natural selection. Biol Rev 18: 32-65

Mayer M (2005) A comparison of regression interval mapping and multiple interval mapping for linked QTL. Heredity 94: 599-605

Meng X-L, Rubin DB (1993) Maximum likelihood via the ECM algorithm: A general framework. Biometrika 80: 267-278

Paterson AH, Lander ES, Hewitt JD, Peterson S, Lincoln SE, Tanksley SD (1988) Resolution of quantitative traits into Mendelian factors by using a complete linkage map of restriction fragment length polymorphisms. Nature 335: 721-726

Paterson AH (2006) Leafing through the genomes of our major crop plants: strategies for capturing unique information. Nat Rev Genet 7: 174-184

Pearson K (1894) Contribution to the mathematical theory of evolution. Philos Trans Roy Soc London A 185: 71-110

Peltonen L, McKusick VA (2001) Dissecting human disease in the postgenomic era. Science 291: 1224-1229

Piepho HP (2000) A mixed model approach to mapping quantitative trait loci in barley on the basis of multiple environment data. Genetics 156: 253-260

Piepho HP (2001) A quick method for computing approximate thresholds for quantitative trait loci detection. Genetics 157: 425-432

Satagopan JM, Yandell BS, Newton MA, and Osborn TC (1996) A Bayesian approach to detect quantitative trait loci using Markov chain Monte Carlo. Genetics 144: 805-816

Shrimpton AE, Robertson A (1988a) The isolation of polygenic factors controlling bristle score in *Drosophila melanogaster*. I: Allocation of third chromosome sterno-pleural bristle effects to chromosome sections. Genetics 118: 437-443

Shrimpton AE, Robertson A (1988b) The isolation of polygenic factors controlling bristle score in *Drosophila melanogaster*. II: Distribution of third chromosome bristle effects within chromosome sections. Genetics 118: 445-459

Sillanpaa MJ, Arjas E (1998) Bayesian mapping of multiple quantitative trait loci from incomplete inbred line cross data. Genetics 148: 1373-1388

Sillanpaa MJ, Arjas E (1999) Bayesian mapping of multiple quantitative trait loci from incomplete outbred offspring data. Genetics 151: 1605-1619

Titterington DM (1997) Mixture distributions. In: Encyclopedia of Statistical Sciences. Wiley, NY, USA, pp 399-407

Via S, Gomulkiewicz R, de Jong G, Scheiner S, Schlichting CD, van Tienderen PH (1995) Adaptive phenotypic plasticity: Consensus and controversy. Trends Ecol Evol 5: 212-217

Wall JD, Pritchard JK (2003) Haplotype blocks and linkage disequilibrium in the human genome. Nat Rev Genet 4: 587-597

Walsh B (2001) Quantitative genetics in the age of genomics. Theor Pop Biol 59: 175-184

Wang H, Nussbaum-Wagler T, Li BL, Zhao Q, Vigouroux Y, Faller M, Bomblies K, Lukens L, Doebley JF (2005) The origin of the naked grains of maize. Nature 436: 714-719

Wilcox PL, Amerson HV, Kuhlman EG, Liu BH, O'Malley DO, Whetten R, Sederoff RR (1996) Detection of a major gene for resistance to fusiform rust disease in loblolly pine by genomic mapping. Proc Natl Acad Sci USA 93: 3859-3864

Wu RL (1997) Genetic control of macro- and microenvironmental sensitivities in Populus. Theor Appl Genet 94: 104-114

Wu RL (1998) The detection of plasticity genes in heterogeneous environments. Evolution 52: 967-977

Wu RL (1999) Mapping quantitative trait loci by genotyping haploid tissues. Genetics 152: 1741-1752

Wu RL, Zeng Z-B (2001) Joint linkage and linkage disequilibrium mapping in natural populations. Genetics 157: 899-909

Wu RL, Lin M (2006) Functional mapping - How to study the genetic architecture of dynamic complex traits. Nat Rev Genet 7: 229-237

Wu RL, Ma CX, Casella G (2002) Joint linkage and linkage disequilibrium mapping of quantitative trait loci in natural populations. Genetics 160: 779-792

Wu RL, Ma CX, Casella G (2007) Statistical Genetics of Quantitative Traits: Linkage, Map and QTL. Springer, New York, USA

Xu S, Atchley WR (1995) A random model approach to interval mapping of quantitative genes. Genetics 141: 1189-1197

Yin TM, DiFazio SP, Gunter LE, Riemenschneider D, Tuskan GA (2004) Large-scale heterospecific segregation distortion in Populus revealed by a dense genetic map. Theor Appl Genet 109: 451-463

Zeng Z-B (1993) Theoretical basis for separation of multiple linked gene effects in mapping quantitative trait loci. Proc Natl Acad Sci USA 90: 10972-10976

Zeng Z-B (1994) Precision mapping of quantitative trait loci. Genetics 136: 1457-1468

Zeng Z-B, Liu J, Stam LF, Kao C-H, Mercer JM, Laurie CC (2000) Genetic architecture of a morphological shape difference between two *Drosophila* species. Genetics 154: 299-310

7 Comparative Mapping

Amy Frary[1], Sami Doganlar[2]* and Milind B. Ratnaparkhe[3]

[1]Department of Biological Sciences, Clapp Lab, Mount Holyoke College, South Hadley, MA 01075, USA

[2]Department of Biology, Izmir Institute of Technology, Gulbahcekoyu Campus, Urla 35430, Izmir, Turkey

[3]210 Waters Hall, Division of Plant Sciences, University of Missouri-Columbia, MO 65211, USA

*Corresponding author: samidoganlar@iyte.edu.tr

1   INTRODUCTION

In the mid 1980s, restriction fragment length polymorphism (RFLP) analysis was first applied to plants for the purposes of creating genetic linkage maps. Using the maps developed for major crop species, the genes controlling qualitative and quantitative traits could be detected and then selected for (via closely linked molecular markers) in breeding programs. Advances in DNA marker technology not only allowed the rapid generation of high-resolution plant genetic maps, but also facilitated detailed comparisons among species. When complementary molecular markers are mapped across related species, it is then possible to align the chromosomes of those species to create comparative linkage maps. In this way, genomic similarities between species are revealed so that genetic information about one species may be extended to others and evolutionary inferences drawn.

The first comparative genetic mapping experiments in plants were performed on members of the Solanaceae and Poaceae families, and studies soon extended to the Brassicaceae. Overall, studies in these and other families have shown that plant genomes have remained surprisingly conserved during evolution. At the chromosomal level, one can find extensive conservation of gene content (synteny) and gene order (colinearity) despite the huge differences in genome size that occur in many plant families. And yet, when one looks at the level of genes, one finds many small scale changes in the genome (such as insertion, deletion, duplication, and translocation) that interrupt colinearity. Herein, we provide an overview of comparative mapping, its methods benefits, limitations, and applications. Because the size of the literature makes it impossible to be comprehensive, we have merely highlighted some of the major findings from the past two decades of work in four angiosperm families: the Poaceae, Solanaceae, Brassicaceae, and Fabaceae (Table 1).

2 METHODOLOGY AND HISTORY

One of the prerequisites for comparative mapping is a segregating population for each species of interest. Another requirement is a common set of markers/probes that can be used to merge existing linkage maps and to ascertain conservation and rearrangement of linkage groups.

Table 1 Some examples of comparative mapping in different crops.

Crop	Comparative analysis	Reference
Maize	Sorghum, Sugarcane	Dufour et al. 1996
Rice	Wheat, Maize	Ahn et al. 1993
Rice	Wheat, Barley	Saghai Maroof et al. 1996
Rice	Wild Rice	Kennard et al. 2002
Rice	Maize	Wilson et al. 1999
Soybean	Mungbean, Common Bean	Concibido et al. 1996
Cucumber	Melon	Park et al. 2004
Rice	Barley	Perovic et al. 2004
Wheat	Wild Wheat	Krugman et al. 1997
Rice	Grasses	Zhang et al. 2001
Wheat	Rice	Guyot et al. 2004
Pearl millet	Rice, Foxmillet	Devos et al. 2000
Tomato	Potato	Tanksley et al. 1992
Eggplant	Tomato	Doganlar et al. 2002a,b
Sorghum	Maize	Whitkus et al. 1992
Barley	Rice	Dunford et al. 2002
Brassica rapa	B. napus, B. oleracea	Osborn et al. 1997
	Arabidopsis, B. napus	Kole et al. 2001, 2002b
	B. napus	Kole et al. 2002a

Conservation of gene sequences during evolution allows molecular markers derived from one species to be used in genetic mapping experiments in closely related species. The markers of choice for many studies, have been RFLP markers: cDNA, EST, gene, or random genomic DNA sequences that reveal restriction-site polymorphisms in the DNA of different individuals in a genomic DNA blot hybridization. An alternative approach uses PCR followed by restriction site analysis (cleaved amplified polymorphic sequence, CAPS markers) or sequencing (single nucleotide polymorphism, SNP markers) to detect polymorphism and to genotype individuals within the mapping population. Regardless of the methodology used, markers that are informative for comparative mapping should be single copy in a reference species, be highly conserved in the majority of target species, and provide good genome coverage in all species. In this way, the degree of conservation of gene repertoire and marker order in the genomes of the target species can be revealed.

However, practical limitations on the size of segregating populations and the number of markers used lower the overall resolution of comparative mapping studies. Moreover, for evolutionary comparisons to be valid, it is crucial that only orthologous loci be compared. Difficulties in ascertaining orthology in plant genomes arise due to ancient and recent gene duplication and polyploidization events. For these reasons, the information gained from comparative genetic mapping experiments is often confined to the level of macrostructure, the gross organization of chromosomes. The genomic region between two adjacent mapped markers can potentially contain a large number of individual genes. Thus, while comparative maps can reveal a great deal about macrosynteny and macrocolinearity, they tell us little about the specific arrangement of genes (their order, orientation, and spacing) on a fine scale.

Rates of sequence divergence among the species of interest must also be low enough to ensure hybridization of the probe to the target locus. Failure of a probe to detect its complementary sequence within a genome can be attributed to the sequence being altogether absent or merely too diverged (DNA sequences must have at least 70% identity to cross-hybridize). Comparative mapping experiments between distantly related species are hampered by a low degree of sequence similarity unless highly conserved ESTs or gene sequences are used as homologous anchor markers. Thus, in the majority of cases, comparative mapping experiments have examined species belonging to the same family.

Early comparative genetic mapping in plants involved members of the Solanaceae family (Bonierbale et al. 1988; Tanksley et al. 1988, 1992). While tomato, potato, and pepper share the same haploid chromosome number, their genome sizes show a two- to four-fold variation. Mapping experiments established that most markers are well conserved among these three species. In addition, near perfect conservation of marker order was found between potato and tomato, with only five chromosomal inversions distinguishing the two genomes. However, the genome of pepper showed significant differences in gene number and order, not surprising given the greater evolutionary distance between *Capsicum* and the two *Solanum* species (Prince et al. 1993).

Comparative mapping of the Poaceae, the most economically important family of plants, followed closely upon the work in the Solanaceae and revealed a degree of relatedness among the major cereal crops much greater than previously expected. Despite a 35-fold variation in genome size, gene repertoire and gene order along chromosomes are so highly conserved that the maps of six cereal genomes (rice, wheat, maize, foxtail millet, sugarcane and sorghum) were readily aligned (Figure 1), allowing a putative ancestral grass genome to be derived (Moore et al. 1995).

For species where linkage relationships have been conserved throughout the genome, consensus maps can be developed. A consensus map is a genetic map of a species developed from all available information from previously existing linkage maps and based on conservation of linkage groups composed of homologous and homoeologous chromosomes. One of the earliest consensus maps was developed for Triticeae species, based on a common set of markers mapped in *Triticum aestivum*, *T. tauschii*, and *Hordeum* species (Van Deynze et al. 1995a, b). Conserved homoeologous segments in the Triticeae tribe, rice, wheat, and oat could then be identified. By merging information from closely related species, consensus maps are useful for cross-referencing genetic information from more distantly related species and understanding the ways in which plant genomes have diverged.

A complementary approach to traditional comparative mapping is to develop comparative physical maps through mapping and sequencing of orthologous BACs. While whole-genome analysis is not feasible using this strategy, the microstructure of limited regions of the genome can be revealed. Moreover, as comprehensive sequence data become available for more plant species, these reference genomes will make the analysis of microsynteny easier.

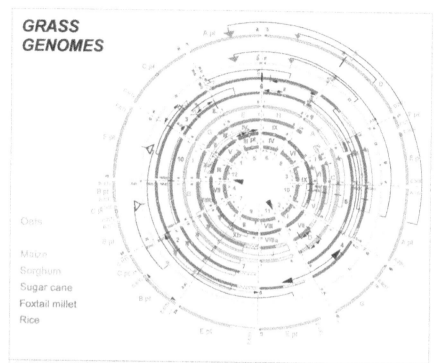

GRASS GENOMES

Oats
Maize
Sorghum
Sugar cane
Foxtail millet
Rice

Fig. I A consensus grass comparative map: Arrows indicate inversions and transpositions necessary to describe present-day chromosomes. A remarkable degree of genome conservation has been established in comparative genetic mapping experiments for the Poaceae family, although genome sizes vary as much as 40-fold between some of the species. This led to the development of the circular grass genome figure–'Grass genomes line up and form a circle'– (Moore et al. 1995).

Some of the earliest work on microsynteny focused on rice, maize, and sorghum genomes. For example, the genomic regions containing the *a1* and *sh2* genes were compared in these three grasses (Chen et al. 1997, 1998) to reveal that while the genes were separated by less than 20 kb in the smaller genomes of rice and sorghum, they were 140 kb apart in maize. Moreover, comparative analysis of the sorghum and rice sequences in the region showed that only exon sequences were highly conserved. Similar investigation of the *adh* region of maize and sorghum identified nine identically ordered putative genes in a region encompassing less than 80 kb in sorghum but nearly three times that (225 kb) in maize (Tikhonov et al. 1999).

3 APPLICATIONS OF COMPARATIVE MAPPING

3.1 Development of Linkage Maps

One can use anchor markers, identified from mapping several related species, in constructing new molecular linkage maps. A researcher can thus select from an existing set of markers to develop a genetic linkage map that is both comprehensive and directly comparable to established maps. In addition, related species can also serve as a convenient source of markers for increasing map density.

3.2 Predicting Gene Location and Function

Comparative maps are useful tools in trait mapping and gene discovery experiments. Depending on the extent of colinearity between the genomes under consideration, the position of loci underlying particular traits in one species can be predicted to lie within orthologous regions of another species. Even the complex traits that are the basis of agricultural productivity can be molecularly dissected using a comparative approach. For example, comparative quantitative trait loci mapping has shown that the quantitative trait loci (QTL) for domestication traits like shattering and seed size are located in corresponding positions in maize, rice and sorghum despite an estimated 65 million years since their divergence. These results imply that these domestication-related genes have a common origin and conserved function (Paterson et al. 1995). Thus, comparative mapping allows knowledge of the genetic and molecular basis of traits to be transferred between different crop species, facilitating map-based selection of traits in breeding programs.

3.3 Gene Cloning

Given a high degree of genome colinearity, comparative genome mapping experiments can serve as an efficient tool for transferring information and resources from well-characterized genomes, such as those of Arabidopsis and rice, to related plants. Thus, one can use information from a species with a small genome to clone genes from species with larger and more complex genomes. The genetic mapping of a particular trait is performed in the species with the large genome, and fine-mapping of the locus can be expedited by using markers from any number of species that are syntenic in the genomic region of interest, thus obviating the need to develop new markers. Map-based cloning can then be initiated using resources from closely related model organisms for which ample genomic information has been obtained. For example, a leaf rust resistance gene was isolated from bread wheat using a diploid/polyploid shuttle mapping strategy (Huang et al. 2003). Of course, such an approach is subject to pitfalls. An attempt to clone a gene conferring stem-rust resistance in barley using markers derived from rice failed despite extensive colinearity between the homoeologous chromosome segments because the gene was simply not present in the corresponding region in barley (Kilian et al. 1997; Brueggeman et al. 2002).

3.4 Analysis of Genome Evolution

Within a species, comparative analysis among chromosomes can reveal important aspects of the genetic history of that species, in particular the extent of genomic duplication. Detailed mapping and sequence analyses have now shown that even simple plant genomes contain tremendous amounts of segmental duplication. On a practical level, this insight into genome complexity permits more informed selection of an appropriate system in which to locate and clone genes of interest. On a broader scale, comparing the structure and organization of related genomes provides information as to how species evolve. Thus, comparative maps can be used to make inferences about how gene rearrangement and duplication, DNA sequence amplification and mobility, and polyploidization have shaped the genetic architecture of plants. In this way, key mechanisms and events in genome evolution can be elucidated. For example, mapping and sequencing data from comparisons with tomato and rice provide evidence that a series of duplication events has taken place in the genome of Arabidopsis within the last 180 million years (Ku et al. 2000; Mayer et al. 2001). Moreover, comparative mapping experiments have revealed that rates of genome evolution can differ significantly between various plant lineages. Thus, while the genomes of *Arabidopsis thaliana* and *Brassica nigra* have experienced some 90 rearrangements since their divergence around 35 million years ago (Mya) (Lagercrantz 1998), in the 60 million years since the radiation of the grasses, evolution has proceeded at a much slower pace so that the genomes of ten cereal species can be readily aligned (Devos and Gale 2000).

Comparative studies have shed light on the evolution of intergenic regions. Studies in maize first showed the dynamic nature of intergenic regions and the role that retrotransposons and other repeated DNA elements have played in genome evolution. These findings provided support for the idea that genome size depends more on the abundance of repetitive elements than on the number of functional genes as more than

50% of a large and complex genome may consist of interspersed repetitive DNAs (Bennetzen et al. 1998).

4 COMPARATIVE MAPPING IN GRASSES (POACEAE)

The grass family consists of around 700 genera and about 10,000 diverse species (Gaut 2002) including the most important food crops rice, wheat, maize, sugarcane, and sorghum. RFLP maps exist for all of these species allowing comparisons of gene content and order to be made. Early genetic mapping experiments assessed the degree of similarity among homoeologous chromosomes of the A, B, and D genomes of wheat (Chao et al. 1989) and this work soon extended to other members of the Triticeae (for example, see Devos et al. 1992, 1993; Namuth et al. 1994). Overall, this work revealed conservation of gene order between wheat and its relatives as well as evidence of duplication and translocation events. In addition, linkage mapping in maize (Helentjaris et al. 1988) and comparisons between maize and sorghum (Whitkus et al. 1992) revealed duplicated regions, supporting the hypothesis that maize is an ancient tetraploid. Thus, comparative analyses have helped to elucidate aspects of evolution within the cereals.

On a broader taxonomic scale, initial studies revealed remarkable genome colinearity in rice, wheat, sorghum, and maize, species belonging to different subfamilies of the Poaceae (Hulbert et al. 1990; Ahn et al. 1993). Despite the fact that the major cereal species diverged over 60 million years ago and possess different numbers of chromosomes, different ploidy levels, and a 35-fold variation in genome size, remarkably few rearrangements have been detected at the level of gross chromosomal structure (Figure 2). In fact, the genomes of ten grass species have been aligned to generate a consensus map based on 30 linkage groups (Devos and Gale 2000).

As evidence for conservation of gene content and order in the Poaceae accumulated, strategies were devised for exploiting this conservation. A set of anchor probes was developed from barley, oat, and rice cDNAs in order to increase the efficiency of molecular mapping (Van Deynze et al. 1998). Moreover, it was suggested that grasses could be considered a single genetic system (Bennetzen and Freeling 1993), with knowledge in one species being sufficient to inform/direct work in the others. In recent years, rice has emerged as the reference species for grass genomics because it is a diploid with a small genome (~440 Mbp) that has been completely sequenced (Yu et al. 2002, 2005). Sorghum's comparatively small genome (~736 Mbp) makes it an obvious choice as a

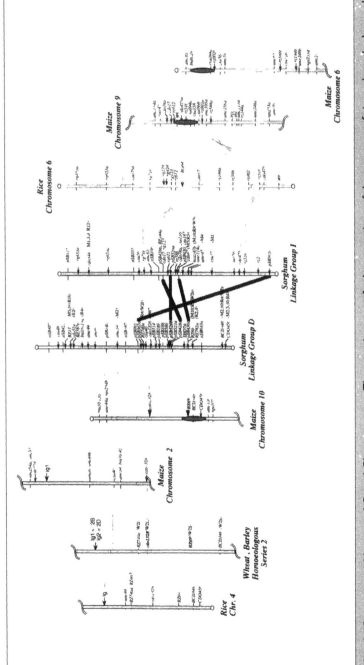

Fig. 2. Comparative Mapping of Cereal Chromosomes. The genetic maps of relevant regions of the genomes of rice, wheat, barley, and maize are aligned to the map of sorghum. Mapping experiment have shown a large degree of colinearity among diverse grasses. Approximate locations of centromeres (ovals within linkage groups), telomeres (circles at ends of linkage groups), and breakpoints of chromosomal rearrangements (double squiggles) distinguishing taxa have been indicated (Paterson et al. 2000).

second model species for the grasses. It is more closely related to both maize and sugarcane than is rice, sharing a common ancestor with maize ~12 Mya and with sugarcane ~5 Mya (Paterson et al. 2005). Thus, efforts are underway to integrate the genetic and physical maps of tropical grasses using the sorghum genome as a framework (Draye et al. 2001; Bowers et al. 2003).

It is hoped that the wealth of molecular genetic resources (high-density molecular maps, physical maps, EST collections, and large insert libraries) that currently exist for the economically important grasses can be more effectively utilized toward crop improvement. However, the fine-scale changes in genome structure that exist among the grasses and which remain undetected by most comparative mapping strategies could hamper efforts to transfer knowledge from model systems to other species (Tarchini et al. 2000; Brueggeman et al. 2002). Thus, recent work has focused on ascertaining to what extent macrocolinearity is suggestive of microcolinearity.

Significant colinearity has been reported at the molecular level. Sequence analysis of selected homologous regions in maize and wheat demonstrated that although inversions, duplications, translocation, and individual gene transfer events are present, coding regions are usually well-conserved (Bennetzen 2000; Wicker et al. 2001). However, the spacing between these conserved genes varies considerably with genome size. Furthermore, in the regions studied, interruptions in microcolinearity are generally due to the insertion of repetitive DNA, especially retroelements. Larger scale and higher resolution in silico comparisons of rice sequences with maize (Salse et al. 2004) and wheat (Sorrells et al. 2003) indicate that the deviations from colinearity can be quite complex. Numerous discontinuities in chromosomal organization reveal that grass genomes are evolving at a more rapid rate than previously supposed. Such findings highlight the drawbacks of using low-resolution comparative maps when making predictions about gene position.

5 THE SOLANACEAE

A diverse plant family of undeniable agricultural and economic importance, the Solanaceae encompasses over 90 genera and some 3,000-4,000 species including a number of vegetable and ornamental crops such as tomato, potato, pepper, eggplant, tobacco, and petunia (Knapp et al. 2004). For over a century, solanaceous species (most of which have a haploid chromosome number of 12) have served as model organisms in a number of areas of plant biology including physiology, development, and genetics. As some of the first plants for which molecular linkage maps were developed (Tanksley and Rick 1980; Bernatzky and Tanksley 1986), these species have also been at the forefront of comparative genomics. Mapping of orthologous loci in potato and tomato revealed that their genomes differed by only five chromosomal arm inversions (Figure 3) (Bonierbale et al. 1988; Tanksley et al. 1992).

A comparative map of eggplant reinforced the idea that paracentric inversion has played an important role in the evolution of the family as 23 of the 28 rearrangements differentiating tomato and eggplant were changes of this type (Doganlar et al. 2002a). While the results also suggested that eggplant and tomato are diverged three- to six-fold more than tomato and potato, widespread colinearity between the two genomes means that tomato genetic resources should be very useful in eggplant studies (Figure 3).

Tomato probes have also been used to generate maps for pepper as gene repertoire in the two species appears to be identical. Nevertheless, substantial rearrangements exist between the two genomes (Prince et al. 1993). By mapping over 1,000 markers in pepper, 18 homoeologous linkage blocks were identified (Livingstone et al. 1999). Detailed comparison of genome structure identified a minimum of 22 breakpoints which were characterized as largely arising from inversions and translocations. Once again, paracentric inversion appears to have been a major force shaping chromosome structure in a solanaceous species. The relative timing of these rearrangements during the radiation of the tomato, potato, and pepper genomes was predicted and a hypothetical ancestral genome proposed.

Comparative trait mapping experiments in the Solanaceae have identified orthologous loci and revealed conservation of gene function in the family. For example, fruit traits (weight, shape and color) map to syntenic regions in eggplant, tomato, potato, and/or pepper (Ben Chaim et al. 2001; Doganlar et al. 2002b). Thus, the current genome sequencing initiative in tomato promises to benefit the other economically important members of the Solanaceae.

The resources developed over two decades of molecular genetic analysis in tomato have also been utilized on a larger scale. By screening tomato EST sequences against the complete genomic sequence of Arabidopsis, a conserved ortholog set (COS) of markers was developed (Fulton et al. 2002). These markers, representing single/low copy genes that have been conserved over the 100-150 million years of evolution separating the two species, represent a valuable tool for facilitating comparative genomics and shedding light on gene evolution in dicots.

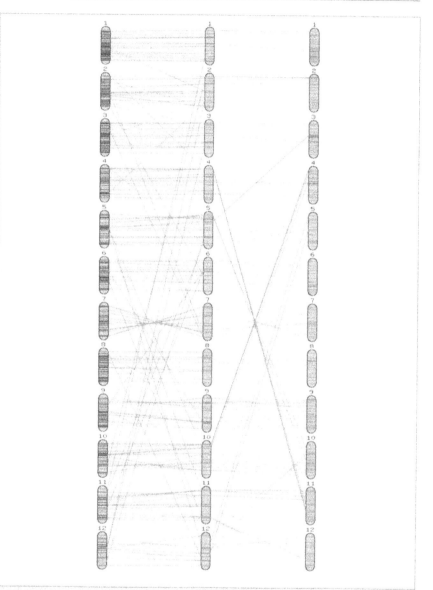

Fig. 3 Comparative map of the tomato (left), potato (middle) and eggplant (right) genomes. Each chromosome is represented by a bar and common markers are connected by lines between the chromosomes. Comparative map of solanaceae chromosomes from three different maps displaying the relationships between the chromosomes. Such analysis shows that five inversions differentiate the tomato and potato genomes while eggplant differs from the others by multiple inversions and translocations (www.sgn.cornell.edu).

6   THE BRASSICACEAE

The Brassicaceae family is widely distributed and comprises more than 3,000 species in approximately 360 genera. The family includes six cultivated species as well as the model plant Arabidopsis. With its small genome (130 Mbp) and short generation time, Arabidopsis has been the focus of intense research in all areas of plant biology and it has the distinction of being the first plant species to be completely sequenced (The Arabidopsis Genome Initiative 2000). Thus, the Brassicaceae provides an ideal forum in which to investigate issues of genome conservation at both macro- and micro-levels. In addition, because base chromosome numbers in the family range from 128 down to 4 with an inferred ancestral state of 8 chromosomes (Koch and Kiefer 2005), comparative genomics can provide insight into the sorts of evolutionary changes that lead to reductions in chromosome number.

Early comparative mapping experiments provided evidence that colinear regions of the *B. nigra*, *B. oleracea*, and *B. rapa* genomes are triplicated, suggesting divergence from a common hexaploid ancestor (Lagercrantz and Lydiate 1996). Moreover, analyses of Arabidopsis and *B. oleracea* and *B. nigra* indicated that, while conserved segments could be identified, chromosomal evolution between the two species has been surprisingly rapid involving numerous rearrangements, insertions, deletions, and chromosome number changes. Thus, initial mapping studies revealed that at least 26 alterations had occurred since the divergence of *B. oleracea* and *A. thaliana* such that colinear segments of the genomes averaged 21.3 cM in length (Kowalski et al. 1994). In contrast, 90 changes separate the more distantly related genomes of *B. nigra* and *A. thaliana* and colinearity extends for only 8 cM on average (Lagercrantz 1998). Despite the extensive reshuffling and duplication of genomic regions within the family, gene sequences are typically highly conserved throughout the Brassicaceae, thus facilitating the development of markers for comparative mapping (Brunel et al. 1999; Acarkan et al. 2000).

The availability of a complete annotated sequence for the *A. thaliana* genome has had a tremendous impact on genomics in the Brassicaceae. Comprehensive sequence analysis verified suspicions that, despite its small size, as much as 60% of the genome of Arabidopsis is duplicated (Blanc et al. 2000). Moreover, the amount of divergence in coding and non-coding regions of the duplicated segments indicates multiple duplication events that took place 100 to 200 million years ago (Vision et al. 2000).

Given the replicated nature of crucifer genomes, distinguishing between orthologous and paralogous genes has been a difficult task. However, recent work between Arabidopsis and *Brassica* species has used the Arabidopsis genome sequence as a means of establishing orthology of the markers used to generate comparative maps. Mapping of orthologs in Arabidopsis and *B. oleracea* revealed that, even though large colinear tracts have been maintained, extensive structural changes have occurred. Results suggest that recent polyploidization followed by diploidization probably contributed to the restructuring of the *B. oleracea* genome (Lukens et al. 2003). Integrated maps of *B. napus* and Arabidopsis also show considerable conservation of gene content and order and support the hypothesis that *Brassica* species have evolved from a hexaploid ancestor (Parkin et al. 2005).

Noteworthy genome colinearity has been found between *A. thaliana* ($n = 5$) and its close relative *Capsella* ($n = 8$) (Boivin et al. 2004). Fourteen large colinear segments define the genomes of these two species; and structural changes in the 10 million years since their radiation appear to have been limited to discrete inversion, fusion, and fission events. Similar mapping analyses in the closest relative of *A. thaliana*, *A. lyrata* ($n = 8$), highlight the minimal changes in chromosome arrangement that have occurred in the 5 million years since the two species separated from a common ancestor (Kuittinen et al. 2004; Yogeeswaran et al. 2005). In addition, integration of the *Capsella*, *A. lyrata*, and *A. thaliana* maps provide insight into the sorts of changes that may have been responsible for chromosome number reduction in the lineage leading to *A. thaliana* and to estimate the rates of chromosomal evolution in the *Arabidopsis* lineage (Koch and Kiefer 2005; Yogeeswaran et al. 2005). Moreover, these estimates have been extended to *Brassica* to suggest much higher rates of mutation than previously reported.

7 THE FABACEAE

An amazingly diverse family of some 20,000 species, the Fabaceae includes a number of important vegetable, forage, and oilseed crops. Thus, linkage maps are available for soybean, lentil, common bean, pea, mungbean, peanut, and alfalfa. Early comparisons of these maps uncovered evidence of significant macrocolinearity between lentil (*Lens*) and pea (*Pisum*) (Weeden et al. 1992), mungbean and cowpea (both *Vigna* species) (Menacio-Hautea et al. 1993), and pea and chickpea (*Cicer*) (Simon and Muehlbauer 1997). A broader scale comparison among *Phaseolus*, *Vigna* and *Glycine* revealed that the genomes of common bean and mungbean were more similar to each other than to soybean (Boutin et al. 1995).

Genome evolution in legumes has been explored through analysis of species with widely divergent genome sizes. The genome of pea is 5 to 10 times larger than that of alfalfa. However, the extent of synteny and colinearity between their maps and the paucity of duplicated regions suggests that multiplication of transposable elements rather than polyploidization is responsible for the large genome of pea (Kalo et al. 2004).

Because molecular genetic analysis in legume species has been complicated by issues of genome size and polyploidy, work has largely focused on two diploid species with moderate-sized genomes: *Medicago truncatula* and *Lotus japonicus*. The genomic structure of these two model species was compared to alfalfa, soybean, pea, mungbean and common bean and a broad conservation of synteny was observed despite large differences in genome size and chromosome number. Not surprisingly, synteny declined with increasing phylogenetic distance. Significant microsynteny of *M. truncatula*, *L. japonicus*, and soybean has also been revealed by sequencing BACs (Cannon et al. 2003; Choi et al. 2004).

Large-scale sequencing projects have been undertaken in the two model legumes and are facilitating more detailed comparisons of genome organization. Analysis of 122 Mb of DNA sequence indicated that more than 75% of genes have been conserved between the two species over ~37 million years of evolution (Young et al. 2005). Thus, these two genomes promise to serve as useful references for understanding genome structure and function in other legumes.

8 COMPARATIVE MAPPING BEYOND FAMILIES

While comparative mapping experiments suggest that closely related plant species possess considerable similarity at the gross chromosomal level, genome structure is not expected to be as well-maintained between families. And yet, rather surprisingly, gene order is conserved in regions of the sorghum, cotton, and *Brassica* genomes, three crop species which span the monocot-dicot divide (Paterson et al. 1996). Gene homology that extends over large evolutionary distances suggests that Arabidopsis may be used as a starting point for the identification of key genes in economically important species. However, such an approach is dependent upon the availability of integrated genetic maps of Arabidopsis and crop plants. Thus efforts have been directed toward mapping Arabidopsis ESTs in a range of dicot species (the EuDicotMap project), including potato, sunflower, sugarbeet, and *Prunus* (almond and cherry) (Dominguez et al. 2003). Three regions of synteny were shared among all five species; this striking degree of conservation may indicate that the regions play an important evolutionary role. While mapping approach can be successfully applied in interfamily investigations of genome organization, the extent of genome colinearity between such distantly related species can be difficult to assess using such a strategy because of resolution issues.

While low density molecular maps simply cannot provide insight into the specific arrangement of genes (their order, orientation, and spacing) on a fine scale, the availability of the complete sequence of the Arabidopsis genome has recently allowed microsynteny and microcolinearity to be investigated across family lines. For example, in an early study, conceptual translations of Arabidopsis DNA sequences were used to identify syntenic regions in soybean (Grant et al. 2000). By comparing amino acid instead of nucleotide sequences, homologous genes and syntenic blocks between these broadly divergent species could be identified. This approach also provided evidence of the duplicated nature of the Arabidopsis genome. When 105 kb (Ku et al. 2000) and 57 kb (Rossberg et al. 2001) regions of tomato were sequenced and compared to Arabidopsis, additional evidence was obtained that a series of duplication events followed by gene loss has taken place in the genome of Arabidopsis. It also appears that, despite some 150 million years of evolution since their divergence, tomato and Arabidopsis carry a similar repertoire of genes dispersed in a network of microsynteny. Comparative analysis of *Medicago truncatula* and Arabidopsis also revealed a network of degenerate microsynteny (Figure 4) and identified the principal mechanisms driving divergence as duplication and rearrangement of chromosome segments followed by loss of information from the duplicated regions (Zhu et al. 2003).

Comparisons of genome microstructure between rice and Arabidopsis provide insight into the sort of evolutionary changes that differentiate monocots and dicots. Sequence analysis of putatively homologous regions of rice chromosome 2 and Arabidopsis chromosome 4 verified the conservation of gene repertoire and order. Nevertheless, duplication events and small scale rearrangements make it difficult to trace evolutionary relationships and confirm orthology (Mayer et al. 2001). By examining the order of homologs identified on the protein level, Liu et al. (2001) discovered short segments (< 30 kb) of colinearity that suggest it is not feasible to predict gene locations across the monocot-dicot divide.

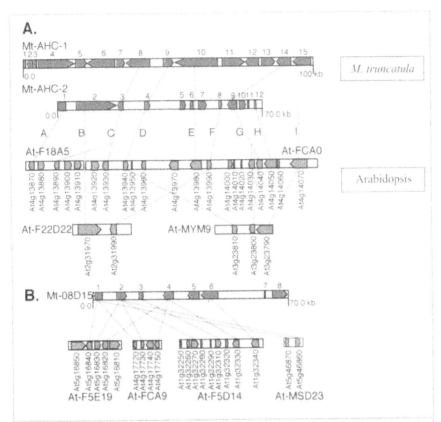

Fig. 4 Microsynteny studies have confirmed that colinearity is also generally observed at the level of genes. Fig. 4 shows the representative picture of microsynteny between sequenced *M. truncatula* BACs and segments of Arabidopsis chromosomes. The orientations of predicted genes are indicated by arrows (Zhu et al. 2003).

Conclusions

Comparative mapping studies have shown that gene sequences are highly conserved in plants. Moreover, in many instances, conserved genes are organized in a similar fashion within plant families and even across large taxonomic distances. Structural rearrangements of chromosomes, including duplication, deletion, translocation, and inversion events can be characterized through studies conducted at the level of macro- and microstructure. In addition, the profound impact that the accumulation of repetitive DNA elements has had on genome size can be readily seen. Moreover, the relative rates of these evolutionary changes can be estimated through comparative analyses. While integrated genetic maps provide an avenue for understanding the structural evolution of genomes, they also help to elucidate the functional importance of genes both within and among plant families. By leveraging knowledge from reference species, the molecular basis of agronomic traits can be more fully explored in plants species whose genomes have been little studied heretofore. In recent years, the power of comparative genomics has been exponentially increased by the availability of whole-genome sequence data. Thus, while it has been two decades since the first comparative mapping studies were published and the tools have changed considerably, the objectives remain the same: to elucidate modes of genome evolution and facilitate crop improvement.

References

Acarkan A, Rossberg M, Koch M, Schmidt R (2000) Comparative genome analysis reveals extensive conservation of genome organization for *Arabidopsis thaliana* and *Capsella rubella*. Plant J 23: 55-62

Ahn S, Anderson JA, Sorrells ME, Tanksley SD (1993) Homoeologous relationships of rice, wheat, and maize chromosomes. Mol Gen Genet 241: 483-490

Ben Chaim A, Paran I, Grube RC, Jahn M, van Wijk R, Peleman J (2001) QTL mapping of fruit-related traits in pepper (*Capsicum annuum*). Theor Appl Genet 102: 1016-1028

Bennetzen JL, Freeling M (1993) Grasses as a single genetic system—genome composition, collinearity and compatibility. Trends Genet 9: 259-261

Bennetzen JL (2000) Comparative sequence analysis of plant nuclear genomes: microcolinearity and its many exceptions. Plant Cell 12: 1021-1029

Bennetzen JL, SanMiguel P, Chen M, Tikhonov A, Francki M, Avramova Z (1998) Grass genomes. Proc Natl Acad Sci USA 95: 1975-1978

Bernatzky R, Tanksley SD (1986) Toward a saturated linkage map in tomato based on isozymes and random cDNA sequences. Genetics 112: 887-898

Blanc G, Barakat A, Guyot R, Cooke R, Delseny M (2000) Extensive duplication and reshuffling in the Arabidopsis genome. Plant Cell 12: 1093-1101

Boivin K, Acarkan A, Mbulu R-S, Clarenz O, Schmidt R (2004) The Arabidopsis genome sequence as a tool for genome analysis in Brassicaceae. A comparison of the Arabidopsis and *Capsella rubella* genomes. Plant Physiol 135: 734-744

Bonierbale MW, Plaisted RL, Tanksley SD (1988) RFLP maps based on a common set of clones reveal modes of chromosomal evolution in tomato and potato. Genetics 120: 1095-1103

Boutin SR, Young ND, Olsen TC, Yu ZH, Shoemaker RC, Vallejos CE (1995) Genome conservation among three legume genera detected with DNA markers. Genome 38: 928-937

Bowers JE, Abbey C, Anderson S, Chang C, Draye X, Hoppe AH, Jessup R, Lemke C, Lennington J, Li Z, Lin Y, Liu S, Luo L, Marler BS, Ming R, Mitchell SE, Qiang D, Reischmann K, Schulze SR, Skinner DN, Wang Y, Kresovich S, Schertz KF, Paterson AH (2003) A high-density genetic recombination map of sequence-tagged sites for Sorghum, as a framework for comparative structural and evolutionary genomics of tropical grains and grasses. Genetics 165: 367-386

Brueggeman R, Rostoks N, Kudrna D, Kilian A, Haan F, Chen J, Druka A, Steffenson B, Kleinhofs A (2002) The barley stem rust-resistance gene *Rpg1* is a novel disease-resistance gene with homology to receptor kinases. Proc Natl Acad Sci USA 99: 9328-9333

Brunel D, Froger N, Pelletier G (1999) Development of amplified consensus genetic markers (ACGM) in *Brassica napus* from *Arabidopsis thaliana* sequences of known biological function. Genome 42: 387-402

Cannon SB, McCombie WR, Sato S, Tabata S, Denny R, Palmer L, Katari M, Young ND, Stacey G (2003) Evolution and microsynteny of the apyrase gene family in three legume genomes. Mol Genet Genom 270: 347-361

Chao S, Sharp PJ, Worland AJ, Warham EJ, Koebner RMD, Gale MD (1989) RFLP-based genetic maps of wheat homoeologous group 7 chromosomes. Theor Appl Genet 78: 495-504

Chen M, SanMiguel P, Oliveira ACD, Woo S-S, Zhang H, Wing RA, Bennetzen JL (1997) Microcolinearity in *sh2*-homologous regions of the maize, rice, and sorghum genomes. Proc Natl Acad Sci USA 94: 3431-3435

Chen M, SanMiguel P, Bennetzen JL (1998) Sequence organization and conservation in *sh2/a1*-homologous regions of sorghum and rice. Genetics 148: 435-443

Choi H-K, Mun J-H, Kim D-J, Zhu H, Baek J-M, Mudge J, Roe B, Ellis N, Doyle J, Kiss GB, Young ND, Cook DR (2004) Estimating genome conservation between crop and model legume species. Proc Natl Acad Sci USA 101: 15289-15294

Concibido V, Young ND, Lange DA, Denny RL, Danesh D, Orf JH (1996) Targeted comparative genome analysis and qualitative mapping of a major partial-resistance gene to the soybean nematode. Theor Appl Genet 93: 234-241

Devos KM, Gale MD (2000) Genome relationships: The grass model in current research. Plant Cell 12: 637-646

Devos KM, Atkinson MD, Chinoy CN, Liu CJ, Gale MD (1992) RFLP-based genetic map of the homoeologous group 3 chromosomes of wheat and rye. Theor Appl Genet 83: 931-939

Devos KM, Atkinson MD, Chinoy CN, Liu CJ, Gale MD (1993) RFLP-based genetic map of the homoeologous group 2 chromosomes of wheat, rye, and barley. Theor Appl Genet 85: 784-792

Devos KM, Pittaway TS, Reynolds A, Gale MD (2000) Comparative mapping reveals a complex relationship between the pearl millet genome and those of foxtail millet and rice. Theor Appl Genet 100: 190-198

Doganlar S, Frary A, Daunay MC, Lester RN, Tanksley SD (2002a) A comparative genetic linkage map of eggplant (*Solanum melongena*) and its implications for genome evolution in the Solanaceae. Genetics 161: 1697-1711

Doganlar S, Frary A, Daunay MC, Lester RN, Tanksley SD (2002b) Conservation of gene function in the Solanaceae as revealed by comparative mapping of domestication traits in eggplant. Genetics 161: 1713-1726

Dominguez I, Graziano E, Gebhardt C, Barakat A, Berry S, Arus P, Delseny M, Barnes S (2003) Plant genome archaelogy: evidence for conserved ancestral chromosome segments in dicotyledonous plant species. Plant Biotech J 1: 91-99

Draye X, Lin Y-R, Qian X, Bowers JE, Burow GB, Morrell PL, Peterson DG, Presting GG, Ren S, Wing RA, Paterson AH (2001) Toward integration of comparative genetic, physical, diversity, and cytomolecular maps for grasses and grains, using the sorghum genome as a foundation. Plant Physiol 125: 1325-1341

Dufour P, Grivet L, D'Hont A, Deu M, Trouche G, Glaszmann JC, Hamon P (1996) Comparative genetic mapping between duplicated segments on maize chromosomes 3 and 8 and homoeologous regions in sorghum and sugarcane. Theor Appl Genet 92: 1024-1030

Dunford RP, Yano M, Kurata N, Sasaki T, Huestis G, Rocheford T, Laurie DA (2002) Comparative mapping of the barley *Ppd-H1* photoperiod response gene region, which lies close to a junction between two rice linkage segments Genetics 161: 825-834

Fulton TM, van der Hoeven R, Eannetta NT, Tanksley SD (2002) Identification, analysis, and utilization of conserved ortholog set markers for comparative genomics in higher plants. Plant Cell 14: 1457-1467

Gaut BS (2002) Evolutionary dynamics of grass genomes. New Phytol 154: 15-28

Grant D, Cregan P, Shoemaker RC (2000) Genome organization in dicots: Genome duplication in Arabidopsis and synteny between soybean and Arabidopsis. Proc Natl Acad Sci USA 97: 4168-4173

Guyot R, Yahiaoui N, Feuillet C, Keller, B (2004) *In silico* comparative analysis reveals a mosaic conservation of genes within a novel colinear region in wheat chromosome 1AS and rice chromosome 5S. Funct Integr Genom 4: 47–58

Helentjaris T, Weber D, Wright S (1988) Identification of the genomic locations of duplicate nucleotide sequences in maize by analysis of restriction fragment length polymorphisms. Genetics 118: 353-363

Huang L, Brooks SA, Li W, Fellers JP, Trick HN, Gill BS (2003) Map-based cloning of a leaf rust resistance gene *Lr21* from the large and polyploid genome of bread wheat. Genetics 164: 655-664

Hulbert, SH, Richter TE, Axtell JD, Bennetzen JL (1990) Genetic mapping and characterization of sorghum and related crops by means of maize DNA probes. Proc Natl Acad Sci USA 87: 4251-4255

Kalo P, Seres A, Taylor SA, Jakab J, Kevei Z, Kereszt A, Endre G, Ellis THN, Kiss GB (2004) Comparative mapping between *Medicago sativa* and *Pisum sativum*. Mol Genet Genom 272: 235-246

Kennard W, Porter R, Grombacher A, Phillips RL (2000) A comparative map of wildrice (*Zizania palustris* L. 2n = 2*x* = 30). Theor Appl Genet 99: 793-799

Kilian A, Chen J, Han F, Steffenson B, Kleinhofs A (1997) Towards map-based cloning of the barley stem rust resistance genes *Rpg1* and *rpg4* using rice as an intergenomic cloning vehicle. Plant Mol Biol 35: 187-195

Knapp S, Bohs L, Nee M, Spooner DM (2004) Solanaceae—a model for linking genomics with biodiversity. Comp Funct Genom 5: 285-291

Koch MA, Kiefer M (2005) Genome evolution among cruciferous plants: a lecture for the comparison of the genetic maps of three diploid species—*Capsella rubella, Arabidopsis lyrata* subsp. *petraea*, and *A. thaliana*. Am J Bot 92: 761-767

Kole C, Quijada P, Michaels SD, Amasino RM, Osborn TC (2001) Evidence for the homology of flowering time genes *VFR2* from *Brassica rapa* and *FLC* from *Arabidopsis thaliana*. Theor Appl Genet 102: 425-430

Kole C, Thormann CE, Karlsson BH, Palta JP, Gaffney P, Yandell B, Osborn TC (2002a) Comparative mapping of loci controlling winter survival and related traits in oilseed *Brassica rapa* and *B. napus*. Mol Breed 9(3): 201-210

Kole C, Williams PH, Rimmer SR, Osborn TC (2002b) Linkage mapping of genes controlling resistance to white rust (*Albugo candida*) in *Brassica rapa* (syn. *campestris*) and comparative mapping to *B. napus* and *Arabidopsis thaliana*. Genome 45: 22-27

Kowalski SP, Lan T-H, Feldmann KA, Paterson AH (1994) Comparative mapping of *Arabidopsis thaliana* and *Brassica oleracea* chromosomes reveals islands of conserved organization. Genetics 138: 499-510

Krugman T, Levy O, Snape JW, Rubin B, Korol AB and Nevo E, (1997) Comparative RFLP Mapping of the chlorotoluron resistance gene (*Su1*) in cultivated wheat (*Triticum aestivum*) and wild wheat (*Triticum dicoccoides*). Theor Appl Genet 94: 46-51

Ku, HM, Vision T, Liu J, Tanksley SD (2000) Comparing sequenced segments of the tomato and Arabidopsis genomes: Large-scale duplication followed by selective gene loss creates a network of synteny. Proc Natl Acad Sci USA 97: 9121-9126

Kuittinen H, de Haan AA, Vogl C, Oikarinen S, Leppala J, Koch M, Mitchell-Olds T, Langley CH, Savolainen O (2004) Comparing the linkage maps of the close relatives *Arabidopsis lyrata* and *A. thaliana*. Genetics 168: 1575-1584

Lagercrantz U, Lydiate DJ (1996) Comparative genome mapping in Brassica. Genetics 144: 1903-1910

Lagercrantz U (1998) Comparative mapping between *Arabidopsis thaliana* and *Brassica nigra* indicates that Brassica genomes have evolved through extensive genome replication accompanied by chromosome fusions and frequent rearrangements. Genetics 150: 1217-1228

Liu H, Sachidanandam R, Stein L (2001) Comparative genomics between rice and Arabidopsis shows scant collinearity in gene order. Genome Res 11: 2020-2026

Livingstone KD, Lackney VK, Blauth JR, van Wijk R, Jahn MK (1999) Genome mapping in *Capsicum* and the evolution of genome structure in the Solanaceae. Genetics 152: 1183-1202

Lukens L, Zou F, Lydiate D, Parkin I, Osborn T (2003) Comparison of a *Brassica oleracea* genetic map with the genome of *Arabidopsis thaliana*. Genetics 164: 359-373

Mayer K, Murphy G, Tarchini R, Wambutt R, Volckaert G, Pohl T, Dusterhoft A, Stiekema W, Entian K-D, Terryn N, Lemcke K, Haase D, Hall CR, van Dodeweerd A-M, Tingey SV, Mewes H-W, Bevan MW, Bancroft I (2001) Conservation of microstructure between a sequenced region of the genome of rice and multiple segments of the genome of *Arabidopsis thaliana*. Genome Res 11: 1167-1174

Menancio-Hautea D, Fatokun CA, Kumar L, Danesh D, Young ND (1993) Comparative genome analysis of mungbean (*Vigna radiata* L. Wilczek) and cowpea (*V. unguiculata* L. Walpers) using RFLP mapping data. Theor Appl Genet 86: 797-810

Moore G, Devos KM, Wang Z, Gale MD (1995) Grasses, line up and form a circle. Curr Biol 5: 737-739

Namuth DM, Lapitan NLV, Gill KS, Gill BS (1994) Comparative RFLP mapping of *Hordeum vulgare* and *Triticum tauschii*. Theor Appl Genet 89: 865-872

Osborn TC, Kole C, Parkin IAP, Sharpe AG, Kuiper M, Lydiate DJ, Trick M (1997) Comparison of flowering time genes in *Brassica rapa*, *B. napus*, and *Arabidopsis thaliana*. Genetics 146: 1123-1129

Park Y, Katzir N, Brotman Y, King J, Bertrand F, Havey M (2004). Comparative mapping of ZYMV resistances in cucumber (*Cucumis sativus* L.) and melon (*Cucumis melo* L.). Theor Appl Genet. 109: 707-712

Parkin IAP, Gulden SM, Sharpe AG, Lukens L, Trick M, Osborn TC, Lydiate DJ (2005) Segmental structure of the *Brassica napus* genome based on comparative analysis with *Arabidopsis thaliana*. Genetics 171: 765-781

Paterson AH, Lin Y-R, Li Z, Schertz KF, Doebley JF, Pinson SRM, Liu S-C, Stansel JW, Irvine JE (1995) Convergent domestication of cereal crops by independent mutations at corresponding genetic loci. Science 269: 1714-1718

Paterson AH, Lan TH, Reischmann KP, Chang C, Lin YR, Liu SC, Burow MD, Kowalski SP, Katsar CS, DelMonte TA, Feldmann KA, Schertz KF, Wendel JF (1996) Toward a unified genetic map of higher plants, transcending the monocot-dicot divergence. Nat Genet 14: 380-382

Paterson AH, Bowers JE, Birow MD, Draye X, Elsik CG, Jiang C-X, Katsar CS, Lan T-H, Lin Y-R, Ming R, Wright RJ (2000) Comparative genomics of plant chromosomes. Plant Cell 12: 1523-1540

Paterson AH, Freeling M, Sasaki T (2005) Grains of knowledge: Genomics of model cereals. Genome Res 15: 1643-1650

Prince JP, Pochard E, Tanksley SD (1993) Construction of a molecular linkage map of pepper and a comparison of synteny with tomato. Genome 36: 404-417

Perovic D, Stein N, Zhang H, Drescher A, Prasad M, Kota R, Kopahnke D, Graner A (2004) An integrated approach for comparative mapping in rice and barley with special reference to the *Rph16* resistance locus. Funct Integr Genomics 4(2): 74-83

Rossberg M, Theres K, Acarkan A, Herrero R, Schmitt T, Schumacher K, Schmitz G, Scmidt R (2001) Comparative sequence analysis reveals extensive microcolinearity in the lateral suppressor regions of the tomato, Arabidopsis, and Capsella genomes. Plant Cell 13: 979-988

Saghai Maroof MA, Yang GP, Biyashev RM, Maughan PJ, Zhang Q (1996) Analysis of the barley and rice genomes by comparative RFLP linkage mapping. Theor Appl Genet 92: 541-551

Salse J, Piegu B, Cooke R, Delseny M (2004) New *in silico* insight into the synteny between rice (*Oryza sativa* L.) and maize (*Zea mays* L.) highlights reshuffling and identifies new duplications in the rice genome. Plant J 38: 396-409

Simon CJ, Muehlbauer FJ (1997) Construction of a chickpea linkage map and its comparison with maps of pea and lentil. J Hered 88: 115-119

Sorrells ME, La Rota M, Bermudez-Kandianis CE, Greene RA, Kantety R, Munkvold JD, Miftahudin, Mahmoud A, Ma X, Gustafson PJ, Qi LL, Echalier B, Gill BS, Matthews DE, Lazo GR, Chao S, Anderson OD, Edwards H, Linkiewicz AM, Dubcovsky J, Akhunov ED, Dvorak J, Zhang D, Nguyen HT, Peng J, Lapitan NLV, Gonzalez-Hernandez JL, Anderson JA, Hossain K, Kalavacharla V, Kianian SF, Choi D-W, Close TJ, Dilbirligi M, Gill KS, Steber C, Walker-Simmons MK, McGuire PE, Qualset CO (2003) Comparative DNA sequence analysis of wheat and rice genomes. Genome Res 13: 1818-1827

Tanksley SD, Rick CM (1980) Isozymic gene linkage map of the tomato: applications in genetics and breeding. Theor Appl Genet 57: 161-170

Tanksley SD, Bernatzky R, Lapitan NL, Prince JP (1988) Conservation of gene repertoire but not gene order in pepper and tomato. Proc Natl Acad Sci USA 85: 6419-6423

Tanksley SD, Ganal MW, Prince JP, de Vicente MC, Bonierbale MW, Broun P, Fulton TM, Giovannoni JJ, Grandillo S, Martin GB, Messeguer R, Miller JC, Miller L, Paterson AH, Pineda O, Roder MS, Wing RA, Wu W, Young ND (1992) High density molecular linkage maps of the tomato and potato genomes. Genetics 132: 1141-1160

Tarchini R, Biddle P, Wineland R, Tingey S, Rafalski A (2000) The complete sequence of 340 kb of DNA around the rice *Adh1-Adh2* region reveals interrupted colinearity with maize chromosome 4. Plant Cell 12: 381-391

The Arabidopsis Genome Initiative (2000) Analysis of the genome sequence of the flowering plant *Arabidopsis thaliana*. Nature 408: 796-815

Tikhonov AP, SanMiguel PJ, Nakajima Y, Gorenstein NM, Bennetzen JL, Avramova Z (1999) Colinearity and its exceptions in orthologous *adh* regions of maize and sorghum. Proc Natl Acad Sci USA 96: 7409-7414

Van Deynze AE, Nelson JC, O'Donoughue JS, Ahn SN, Siripoonwiwat W, Harrington SE, Yglesias ES, Braga DP, McCouch SR, Sorrells ME (1995a) Comparative mapping in grasses. Oat relationships. Mol Gen Genet 249: 349-356

Van Deynze AE, Nelson JC, Yglesias ES, Harrington SE, Braga DP, McCouch SR, Sorrells ME (1995b) Comparative mapping in grasses. Wheat relationships. Mol Gen Genet 248: 744-754

Van Deynze AE, Sorrells ME, Park WD, Ayres NM, Fu H, Cartinhour SW, Paul E, McCouch SR (1998) Anchor probes for comparative mapping of grass genera. Theor Appl Genet 97: 356-369

Vision TJ, Brown DG, Tanksley SD (2000) The origins of genomic duplications in Arabidopsis. Science 290: 2114-2117

Weeden NF, Muehlbauer FJ, Ladizinsky G (1992) Extensive conservation of linkage relationships between pea and lentil genetic maps. J Hered 83: 123-129

Whitkus R, Doebley J, Lee M (1992) Comparative genome mapping of sorghum and maize. Genetics 132: 1119-1130

Wicker T, Stein N, Albar L, Feuillet C, Schlagenhauf E, Keller B (2001) Analysis of a contiguous 211 kb sequence of diploid wheat (*Triticum monococcum* L.) reveals multiple mechanisms of genome evolution. Plant J 26: 307-316

Wilson WA, Harrington SE, Woodman Wl, Lee M, Sorrells ME, McCouch SR (1999) Can we infer the genome structure of progenitor maize through comparative analysis of rice, maize, and the domesticated panicoids? Genetics 153: 453-473

Yogeeswaran K, Frary A, York TL, Amenta A, Lesser AH, Nasrallah JB, Tanksley SD, Nasrallah ME (2005) Comparative genome analyses of Arabidopsis spp.: Inferring chromosomal rearrangement events in the evolutionary history of *A. thaliana*. Genome Res 15: 505-515

Young ND, Cannon SB, Sato S, Kim D, Cook DR, Town CD, Roe BA, Taabata S (2005) Sequencing the genespaces of *Medicago truncatula* and *Lotus japonicus*. Plant Physiol 137: 1174-1181

Yu J, Hu S, Wang J, Wong GK-S, Li S, Liu B, Deng Y, Dai L, Zhou Y, Zhang, X, et al. (2002) A draft sequence of the rice genome (*Oryza sativa* L. ssp. *indica*). Science 296: 79-92

Yu J, Wang J, Lin W, Li SG, Li H, Zhou J, Ni PX, Dong W, Hu SN, Zeng CQ, Zhang JG, Zhang Y, Li RQ, et al. (2005) The genomes of *Oryza sativa*: A history of duplications. PLOS Biol 3: 266-281

Zhang J, Zheng HG, Aarti A, Pantuwan G, Nguyen TT, Tripathy JN, Sarial AK, Robin S, Babu RC, Nguyen BD, Sarkarung S, Blum A, Nguyen HT (2001) Locating genomic regions associated with components of drought resistance in rice: comparative mapping within and across species. Theor Appl Genet 103: 19-29

Zhu H, Kim D-J, Baek J-M, Choi H-K, Ellis LC, Kuester H, McCombie WR, Peng H-M, Cook DR (2003) Syntenic relationships between *Medicago truncatula* and Arabidopsis reveal extensive divergence of genome organization. Plant Physiol 131: 1018-1026

8 Map-based Cloning of Genes and Quantitative Trait Loci

Hong-Bin Zhang

Department of Soil and Crop Sciences, Texas A&M
University, 2474 TAMU, College Station, Texas 77843-2474,
USA; E-mail: hbz7049@tamu.edu

1 INTRODUCTION

A major goal of genome research is to isolate the genes and genome loci controlling quantitative traits, i.e., quantitative trait loci (QTL), of scientific and/or economic importance. For the purposes of gene and QTL isolation, several methods have been developed and used. These include: [1] Library screening, [2] DNA subtraction, [3] gene expression differential display including cDNA differential display, microarray or GeneChip, and serial analysis of gene expression (SAGE), [4] expressed sequence tag (EST) and genome sequencing, [5] T-DNA or transposon gene tagging, [6] polymerase chain reaction (PCR)-based candidate gene cloning, [7] antisense- or RNA interference (RNAi)-based gene expression interference, and [8] map-based or positional cloning (hereafter both are referred to as map-based cloning).

While some of the methods, such as map-based cloning, clone genes or QTLs in a manner from mutated phenotype to gene or QTL sequence, which is so-called forward genetics approach, others, such as T-DNA or transposon gene tagging and antisense- or RNAi-based gene expression interference, clone genes or QTLs in a manner from nucleotide sequence to phenotype, which is so-called reverse genetics approach. Some methods, such as library screening, T-DNA or transposon tagging, antisense- or RNAi-based gene expression interference, and map-based cloning, allow to determine the functions of the target genes or QTLs individually, whereas others, such as gene expression differential display, EST and genome sequencing, and DNA subtraction, can only provide information on the potential functions of the targeted genes in a large scale. Some methods, such as library screening, PCR-based candidate gene cloning and antisense- or RNAi-based gene expression interference, are based on the existing knowledge or prior assumptions of gene sequences or biochemistry, but others, such as T-DNA or transposon gene tagging and map-based cloning, are based on the phenotype that the targeted gene or QTL controls. In the T-DNA or transposon gene tagging method, T-DNA or a transposable element is cloned in a plant-transformation-competent binary vector and transformed into plants. Insertion of the T-DNA or transposon into the host plant genome would, if it inserts into a gene or its regulatory element, create mutation at the locus that could be identified phenotypically (Krysan et al. 1999; Brutnell 2002; Jeong et al. 2002). Since the T-DNA or transposon gene tagging method needs techniques for large-scale transformation in the host plant and is relatively random in gene tagging, implying that some of the genes of interest may escape from the tagging, map-based cloning has emerged as the method of choice for cloning targeted genes or QTLs that are known only by phenotype. This is especially true as high-density DNA marker maps, large-insert bacterial artificial chromosome (BAC)-based integrated physical maps and whole-genome sequence maps have been developed for a number of species of economic and/or scientific importance.

Many traits considered important to agriculture, such as crop yield and quality, and resistance to biotic and abiotic stresses, remain known only by phenotype even though the genes for a number of them have been cloned and characterized using different methods. For many of the traits, mutants already exist naturally or have been induced experimentally by using physical (such as radiation) or chemical (such as ethyl-methane sulfonate) method through plant genetics research and breeding programs. Many of these mutations may not be achievable by the T-DNA or transposon tagging method. These mutants are not only useful sources for plant genetic studies and breeding, but also provide valuable materials for isolation of genes or QTLs controlling the traits. Unlike the mutants induced by the T-DNA or transposon gene tagging method in which the nucleotide sequences of the T-DNA or transposon could be used to isolate the genes or QTLs (Krysan et al. 1999; Brutnell 2002; Jeong et al. 2002), there is no sequence information available for isolation of the naturally occurring and physically or chemically-induced mutants. Therefore, map-based cloning has been the method of choice to isolate the genes or QTLs controlling such mutations.

Map-based cloning has been successfully used to clone many genes and QTLs of biological and/or agricultural importance. Table 1 shows some examples of the genes and QTLs cloned by map-based cloning from crop plant species with different genome sizes. Map-based cloning was first used to clone the gene responsible for cystic fibrosis in human (Rommens et al. 1989). The method was first introduced in plants to clone the *ABI3* gene (Giraudat et al. 1992) and the omega-3 fatty acid desaturase gene (Arondel et al. 1992) in the small-genome (120 Mb/ haploid genome) of plant model species, *Arabidopsis thaliana*. Map-based cloning was demonstrated in crop plant species with larger genomes by successful isolation of the bacterial speck disease resistance gene, *Pto*, in tomato of 950 Mb/haploid genome (Arumuganathan and Earle 1991; Martin et al. 1993b). Almost all of the genes cloned by map-based cloning at an early time are those controlling simple Mendelian traits. As quantitative traits are considered to be controlled by multiple genes, with each having a minor effect and their expression being readily subjected to the variation of environments, cloning of the loci controlling these traits by map-based cloning had once been considered to be difficult, if not impossible. The success in map-based cloning of the QTLs controlling tomato fruit weight (*fw 2.2*) (Frary et al. 2000), and tomato sugar content (*Brix 9-2-5*) (Fridman et al. 2000), *A. thaliana* flowering time (*FRIGIDA*) (Johanson et al. 2000) and rice heading date (*Hd1*) (Yano et al. 2000) shed light on isolation of QTLs by map-based cloning. Since simple Mendelian and quantitative traits differ in genetic behavior, the strategies of cloning the genes controlling these types of traits by map-based cloning may vary. Nevertheless, map-based cloning of genes or QTLs generally follows a similar procedure including targeted gene genetic mapping, physical mapping of the gene-containing region (optional), chromosome landing or walking, high-resolution mapping, and gene identification.

2 DETAILED PROCEDURE OF MAP-BASED CLONING

Figure 1 shows a flow chart of map-based cloning of genes or QTLs. Map-based cloning often consists of four or five steps and each step is described in detail. The order of the steps in the procedure may be changed, depending on availability of tools for a map-based cloning project.

Table I Examples of genes and QTLs cloned by map-based cloning from crop plant species with different genome sizes

Name	Gene or QTL[a]	Functions	Species	Genome size[b] (Mb/haploid)	Reference
Hd1	QTL	Photoperiod response/heading	Rice	430	Yano et al. 2000
Hd6	QTL	Photoperiod response/heading	Rice	430	Takahashi et al. 2001
Rf-1	Gene	Fertility restorer	Rice	430	Komeri et al. 2004
sh4	QTL	Grain shattering	Rice	430	Li et al. 2006
Pto	Gene	Bacterial speck blight resistance	Tomato	950	Martin et al. 1993b
chloronerva	Gene	Iron uptake	Tomato	950	Ling et al. 1999
fw2.2	QTL	Fruit weight	Tomato	950	Frary et al. 2000
Brix 9-2-5	QTL	Sugar content	Tomato	950	Fridman et al. 2000
Mlo	Gene	Fungal broad spectrum resistance	Barley	4,870	Büschges et al. 1997
Rpg1	Gene	Fungal stem rust resistance	Barley	4,870	Brueggeman et al. 2002
Ppd-H1	QTL	Photoperiod response/flowering time	Barley	4,870	Turner et al. 2005
VRN1	Gene	Vernalization/flowering	Triticum monococcum	5,750	Yan et al. 2003
VRN2	Gene	Vernalization/flowering	Triticum monococcum	5,750	Yan et al. 2004
Lr10	Gene	Fungal leaf rust resistance	Wheat	15,960	Feuillet et al. 2003
Lr21	Gene	Fungal leaf rust resistance	Wheat	15,960	Huang et al. 2003
Pm3b	Gene	Fungal powdery mildew resistance	Wheat	15,960	Yahiaoui et al. 2004
NAC	QTL	Protein, zinc and iron contents	Wheat	15,960	Uauy et al. 2006

[a]The 'gene' indicates those controlling Mendelian traits.
[b]The genome size data are from Arumuganathan and Earle (1991).

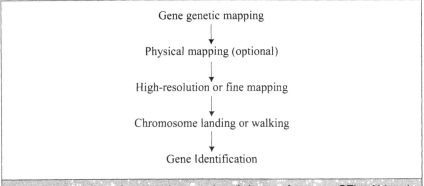

Gene genetic mapping

↓

Physical mapping (optional)

↓

High-resolution or fine mapping

↓

Chromosome landing or walking

↓

Gene Identification

Fig. 1 A flow chart of positional or map-based cloning of genes or QTLs. Although each step of the map-based cloning procedure is presented in order, some of the steps could be carried out in a different order or combination.

2.1 Genetic Mapping or Tagging of the Targeted Genes or QTLs with DNA Markers

Genetic mapping or tagging of genes or QTLs with DNA markers remains the sole approach to date to associating the genes or QTLs known only by phenotype with DNA sequences; therefore, the step is essential to isolate the genes known only by phenotype. To genetically map or tag a gene of interest, a population segregating for the targeted trait is a prerequisite. There are often two cases that are frequently found in research. One is that the gene or QTL of interest has been genetically mapped to a genetic map of a population segregating for the gene and the other case is that the gene does not segregate in the population of any existing genetic maps and is not mapped genetically. In the first case, if the flanking DNA markers are closely linked to, or co-segregate with the gene of interest, they could be directly used to estimate the physical/genetic distance ratio in the interval of the targeted gene locus and conduct chromosome landing or walking (Figure 1). Otherwise, more closely linked DNA markers should be isolated (see below). For the second case, since no DNA marker linked to the gene of interest is available, it is necessary to identify one or more DNA markers linked (even distantly linked) to the gene. The marker(s) will then be mapped, if possible, to an existing genetic map of the species or related species so that a number of DNA markers for the genomic region of interest may be identified from the genetic map and used to map the gene using the population segregating for the targeted gene. Several strategies have been used to isolate additional DNA markers that are more closely linked to the targeted gene previously mapped genetically, or isolate DNA markers linked to the targeted genes for which there are no markers available.

2.1.1 Genome-wide Interval Mapping

This strategy has been used to map the genes, especially QTLs, for which no DNA markers are available (e.g., Alpert et al. 1995). The population segregating for the targeted gene is used for the mapping work. DNA markers are selected from the existing genetic maps of the species in an interval of 10-30 cM between neighboring markers, and used to construct a framework map of the population. The targeted genes or QTLs are mapped to the framework map while it is developed. Once the gene or QTL is mapped, additional DNA markers could be selected from the genomic interval of the existing genetic map containing the gene locus or QTL and used to search for DNA markers closer to the gene. Considering that quantitative traits are controlled by multiple genes and being readily subjected to variation of environments, repeated experiments in multiple environments, for instance, in different years and/or at multiple locations, should be conducted to map quantitative traits. For this purpose, recombinant inbred lines (RILs) are often developed from the mapping population if quantitative traits are the target of research. Moreover, because genetic mapping is polymorphism- and recombination frequency-dependent, selection of a proper population that has sufficiently high genetic recombination frequency and polymorphism for the DNA markers would greatly affect the success of the mapping experiment. Alternatively, DNA markers that are highly polymorphic in the population should be selected and used.

Bulked Segregant Analysis (BSA)

This strategy has been used to tag and develop DNA markers for the targeted gene that is not mapped genetically (Figure 2A), or develop additional markers for the targeted gene genetically mapped (Figure 2B). To demonstrate the strategy, here the gene controlling a disease resistance is used as an example, with the letter 'R' for resistant allele and 'r' for susceptible allele of the gene. A population consisting of 100 or more plants segregating for the resistant trait is developed and phenotyped (see high-resolution mapping below). To tag the gene for which no DNA markers are available, the extreme types of the segregants (resistant vs. susceptible) are selected from the population based on the trait phenotype, usually 10 - 20 plants from each type. DNA is isolated from each of the selected plants and equally bulked into two

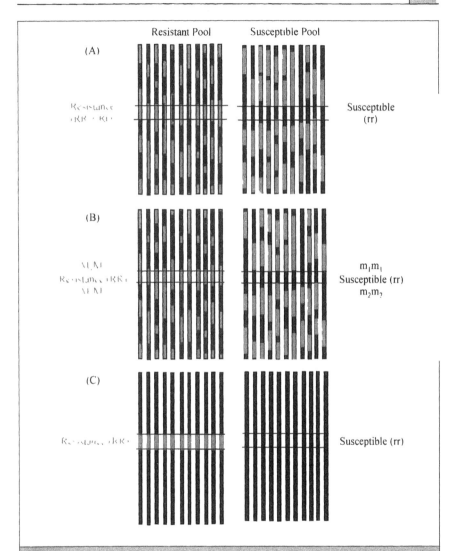

Resistant Pool   Susceptible Pool

(A)

Resistance   Susceptible
(RR + Rr)   (rr)

(B)

M₁M₁
Resistance (RR)   Susceptible (rr)
M₂M₂   m₂m₂

(C)

Resistance (RR)   Susceptible (rr)

Fig. 2 Methods widely used to tag a gene of interest with DNA markers or generate new DNA markers for a genomic region containing the gene. The DNA pools to be used for these purposes are constructed from the extreme types of the trait segregating population only based on plant phenotype such as disease resistant versus susceptible (A) or based on both plant phenotype and the genotypes of flanking DNA markers (B), or from a pair of NILs for the gene of interest (C). The capital letter 'R' is for the disease resistant allele whereas the small letter 'r' is for the disease susceptible allele. The capital letter 'M' is for the DNA markers flanking the targeted gene and having a genotype of the resistant parent and the small letter 'm' is for the DNA markers flanking the targeted gene and having a genotype of the susceptible parent.

DNA pools according to their phenotypes, for instance, resistant segregant pool and susceptible segregant pool (Michelmore et al. 1991; Zhang et al. 2000). As the plants are selected only based on phenotype, the two pools are essentially the same in chromatin composition except for the targeted trait locus region (Figure 1A). If the resistance is dominant over the susceptible, the resistant pool is expected to consist of two types of genotypes, homozygous resistant (*RR*) and heterozygous resistant (*Rr*), whereas the susceptible pool only consists of homozygous susceptible genotype (*rr*). The two pools differ by the fact that the resistant pool has the loci of both alleles *R* and *r*, while the susceptible pool only has the *r* allele locus. The discrepancy will allow isolation of DNA markers potentially closely linked to the targeted gene when they are screened with DNA markers.

However, to isolate additional markers for a gene previously mapped, the selected two extreme types of segregants often are also genotyped with the DNA markers flanking the targeted gene. Segregants are selected based on both plant phenotype and genotypes of the flanking markers – only the segregants homozygous for the corresponding parents in flanking markers are used to construct the DNA pools (Giovannoni et al. 1991; Wing et al. 1994). As shown in Figure 2B, the resistant pool only consists of the resistant homozygous segregants (*RR*) and the susceptible pool only consists of susceptible homozygous segregants (*rr*), as indicated by their flanking markers. Since the two pools differ at the gene locus, with one containing the *R* allele locus and the other containing the *r* allele locus, the pools constructed based on both targeted trait phenotype and flanking marker genotype have a better opportunity of isolating DNA marker for the targeted gene than those constructed solely based on targeted trait phenotype.

Near-Isogenic Line (NIL) Analysis

NILs, pairs of lines differing at a single locus or genomic region, are desirable resources for isolation of DNA markers for both mapped and unmapped genes (Figure 2C; Martin et al. 1991). NILs are widely available in plant breeding programs. As the NILs are homozygous, one or few plants from each line are needed for pool construction. NIL pairs provide a similar probability to the bulked segregant pools constructed based on both targeted trait phenotype and flanking marker genotypes.

The pair of the segregant or NIL pools is then screened with DNA markers. A number of types of DNA markers have been developed in the past decades (Peters et al. 2003; see Chapter 2 of this volume for an excellent review). These include restriction fragment length polymorphism (RFLP), random amplified polymorphic DNA (RAPD), sequence-tagged site (STS), cleaved amplified polymorphic sequence (CAPS), simple sequence repeat (SSR) or microsatellite, amplified fragment length polymorphism (AFLP), and single nucleotide polymorphism (SNP). The RFLP, STS, CAPS, SSR and SNP are more likely locus-specific, thus being well portable between populations, whereas RAPD and AFLP are highly likely multiple-loci markers, thus often being poorly portable between populations. This suggests that a fragment (band) of RFLP, STS, CAPS, SSR or SNP mapped to a locus in one population will highly likely be mapped to the same locus in another population, whereas the fragments of an AFLP or RAPD that could be mapped may vary from population to population. Theoretically, although all types of the DNA markers could be used to screen the segregant or NIL pools for targeted gene tagging and mapping, RAPD or AFLP has been more widely used to genetically tag and map a gene of interest due to their multiple-loci nature (Giovannoni et al. 1991; Martin et al. 1991; Wing et al. 1994; Alpert and Tanksley 1996; Büschges et al. 1997; Zhang et al. 2000). Nevertheless, it should be pointed out that the RAPD or AFLP fragments identified to be polymorphic between the pools are more likely derived from repeated elements, which is especially true for the species with large genomes. This may make it difficult to use the markers for subsequent local physical mapping and chromosome walking. In comparison, the single- or low-copy markers, such as RFLP, STS, CAPS, SSR or SNP, would have no or minimal problems in this regard.

Once DNA markers that are potentially closely linked to the targeted gene or QTL are identified, a local genetic map of the targeted trait is developed by the standard molecular genetic mapping method using the entire mapping population of the targeted trait. For the single- or low-copy markers, the mapping experiments are straightforward. However, for the multiple-loci markers, attention needs to be taken to ensure that the same fragment or band as the one identified by the pool analysis be mapped. For quantitative traits, multiple-environmental and repeated experiments are needed to accurately phenotype each lines or plants of the mapping population. This is because the accurate mapping of the targeted trait is crucial to the success of map-based cloning (see below).

2.2 Physical Mapping of the Targeted Region

Physical mapping of the targeted region is to physically measure the distance in kilobase pairs (kb) between the targeted gene and its flanking markers so that the feasibility of approaching to the gene from a flanking marker by chromosome walking using a large-insert DNA library could be estimated. Several methods have been used to physically map a genomic region of interest: pulsed-field gel electrophoresis (PFGE) (Wing et al. 1994; Alpert and Tanksley 1996), chromosome walking using large-insert clones (Zhang et al. 1994), and chromosome in situ hybridization. Although each method has advantages and disadvantages, the PFGE method is more desired for this purpose.

Figure 3 shows how to physically map a genomic region of interest using the PFGE method. Megabase DNA embedded in low-melting-point agarose plugs is isolated (Zhang et al. 1995; Wu et al. 2004b; Ren et al. 2005; He et al. 2007), completely digested with several rare cutting restriction enzymes, respectively (Martin et al. 1993a; Wing et al. 1994; Alpert and Tanksley 1996; He et al. 2007), fractionated by PFGE, and Southern-blotted onto nylon membrane. The Southern blot is first hybridized with one of the flanking DNA markers of the targeted gene, and exposed to X-ray film. The probe is removed and the blot is then re-hybridized with the other flanking marker of the gene. The size of the smallest fragment (band) that both probes hybridize is the maximally physical distance between the two flanking markers. Since the physical distance estimated by the PFGE method represents a maximal distance between the markers, the more restriction enzymes are used, the more accurate estimation will be obtained (Martin et al. 1993a; Wing et al. 1994; Alpert and Tanksley 1996).

Since the flanking markers, if closely linked to the targeted gene, are frequently used directly as probes to screen a large-insert DNA library for chromosome walking, the physical mapping step is often considered to be optional. Nevertheless, it is helpful and sometimes essential for effective map-based cloning of genes and QTLs. This is because physical mapping of the targeted trait locus can provide at least two types of information that are useful for map-based cloning. One is the ratio of physical distance (in kb) to genetic distance (in centiMorgans or cM). Although in some regions of a genome physical distance well corresponds to genetic distance, in other regions the correspondence could be very poor. This implies that the close genetic distance between DNA markers and the targeted gene does not always mean closer physical distance. The situation may be opposite, i.e., a closer genetic distance likely implies a distant physical distance, given the fact that genetic distance is genetic recombination-dependent. For the regions in which recombination is suppressed or 'cold' , the genetic distances are lower, but the ratios of physical distance to genetic distance are increased (Lin et al. 1999; Chen et al. 2002; Li et al. 2007). On the other hand, for the regions in which recombination is promoted or 'hot', the

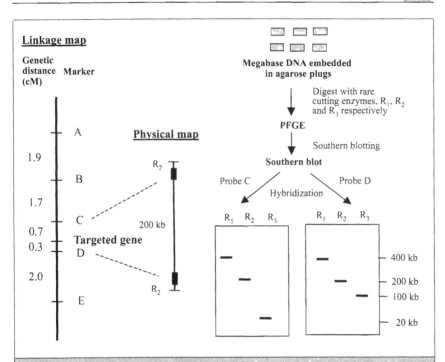

Fig. 3 A flow chart of physical mapping of a genomic region of interest by pulsed-field gel Southern blot hybridization. The figure shows that the DNA markers C and D that are 1.0 cM apart are co-localized to a 200-kb R_2 fragment. Therefore, the physical distance between the two markers is ≤ 200 kb, suggesting that the physical/genetic distance ratio in the targeted gene region is ≤ 200 kb/cM. Note that since the targeted gene is localized to a 200-kb R_2 fragment, the fragment can be directly cloned into a vector with R_2 as a cloning site, as described by He et al. (2007) (also see 4.5 Targeted Gene Source Genotypes and Source Libraries).

genetic distance may become larger, but the physical/genetic distance ratios are likely decreased. Physical mapping of the targeted locus will provide the information about the physical/genetic distance ratio in the region containing the targeted gene, thus allowing estimation of the feasibility of approaching the gene from its flanking DNA markers by chromosome walking.

The other information that could be derived from the physical mapping experiment is the restriction site map of the targeted locus. Such knowledge is essential to direct cloning of the targeted locus using the BAC cloning technique (Fu and Dooner 2000; He et al. 2007). The megabase-size DNA embedded in low-melting-point agarose plugs is digested with the enzyme used to physically map the targeted locus (e.g., R_2 in Figure 3) and fractionated by PFGE under the PFGE condition used for the physical mapping. The DNA fragments potentially containing the targeted locus, justified by size, are excised from the pulsed-field gel, eluted and cloned in a BAC cloning vector (Fu and Dooner 2000; He et al. 2007). The clones containing the targeted gene could be identified from the locus-enriched library by using the flanking markers as probes. This targeted locus BAC cloning technique is especially useful for map-based cloning of the genes or QTLs from the genotypes for which no large-insert BAC libraries are available (see below).

2.3 High-Resolution or Fine-Mapping of Targeted Gene Locus

High-resolution or fine-mapping of the genomic region containing the targeted gene or QTL is to genetically map the gene locus at a fine-scale and delimit it to an interval that is physically manageable, such as genome sequencing and genetic transformation. It is a crucial step toward map-based cloning of the gene or QTL. Fine-mapping is based on the principle that the effect of a gene or QTL can not extend across the crossover points nearby the gene locus (Figure 4). This implies that only the plants that carry the recombination segments containing the targeted gene or QTL has the expected phenotype of the gene or QTL. According to this principle, a gene or QTL of interest could be delimited to an interval containing a single targeted gene or a DNA fragment that is manageable in size between two genetic recombination crossover points. Therefore, the success of high-resolution mapping depends on not only the availability of DNA markers for, but also the recombination frequency in the targeted region. High-resolution mapping is, in fact, to identify and order the genetic recombination crossover points between the targeted gene and its flanking markers (e.g., Martin et al. 1993a; Alpert and Tanksley 1996; Yano et al. 2000; Yan et al. 2003, 2004; Li et al. 2006).

As a general strategy of high-resolution mapping, a large mapping population segregating for the locus of the targeted gene or QTL is necessary to identify and order the recombinants between the targeted gene and its flanking DNA markers. Several types of mapping populations have been used to fine-map a targeted gene or QTL, depending on the genetic behavior of the trait controlled by the gene. For simple Mendelian trait fine-mapping, F_2 progenies have been used (Zhang et al. 1994, 2000; Ling et al. 1999; Brueggeman et al. 2002; Huang et al. 2003; Yan et al. 2003, 3004; Yaniaoui et al. 2004; Li et al. 2006),

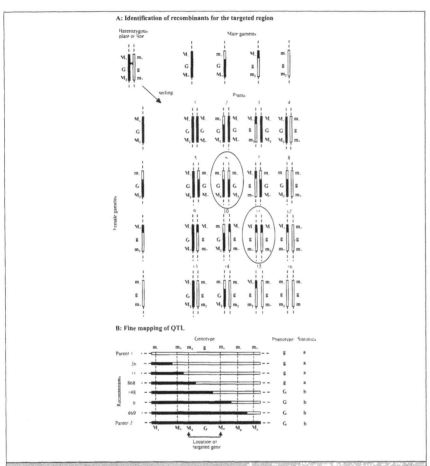

Fig. 4 Fine-mapping of genes or QTLs: Identification of recombinants for the region containing the targeted gene (A) and fine-mapping of a QTL (B). The heterozygous plant or line could be F₁ of the two parents of the mapping population or a pair of NILs, or a heterogeneous inbred family (HIF) that segregates for the region containing the targeted gene. The capital letter 'G' is for the dominant allele of the targeted gene and the small letter 'g' is for the recessive allele of the targeted gene. The capital letter 'M' is for the DNA markers flanking the targeted gene and having a genotype of the dominant gene parent (parent 2) and the small letter 'm' is for the DNA markers flanking the targeted gene and having a genotype of the recessive gene parent (parent 1).

The letters 'a' and 'b' in Panel B indicate the significance level of difference between different recombinant lines in the targeted gene region, with same letters indicating that there is no statistically significant difference in the trait of interest between the lines, but different letters indicating that there is significant difference at a level of $P \leq 5\%$ for the trait of interest between the lines.

whereas for complex trait QTL fine-mapping, backcross (BC) progenies at middle generation (e.g., $BC_3 - BC_5$) (Yano et al. 2000; Takahashi et al. 2001) or F_2 populations of near-isogenic lines (NILs) derived by selfing the cross between the two lines of an NIL pair for the targeted locus (Alpert and Tanksley 1996; Fridman et al. 2000; Uauy et al. 2006). Such an NIL population could be also developed from heterogeneous inbred families (HIFs) derived by selfing the heterogeneous recombinant inbred lines (RILs at generations F_5 to F_6) for the targeted region (Tuinstra et al. 1997) (Figure 4A). RILs are developed from the F_2 plants of a cross by continuously selfing single-seed descents and HIFs for a targeted region could be generated by selfing the heterozygous RILs identified with flanking DNA markers for the region. The purposes of developing the BC or NIL populations for QTL fine-mapping are to create a relatively consistent genetic background for all recombinants in the targeted region so that the targeted quantitative trait could be phenotyped accurately and in a repeated experiment across multiple environments. How large the fine-mapping population should be depends mainly on the recombination frequency of the targeted region, seeming not on the genome size of the species (for explanation, see below). If the region is high in recombination frequency, indicating the lower ratio of physical distance to genetic distance, fewer plants or lines are needed for the fine-mapping of the region. Otherwise, a larger mapping population is essential. In previous map-based cloning projects, the populations consisting of 500 to 10,000 plants or lines have been used (Jander et al. 2002). This size of mapping population implies that if the average physical/genetic distance ratio for a genome is 1,000 kb/cM, the targeted gene could be mapped to an interval of 400 [1,000 kb/(500 x 1%) = 200 kb between the gene and its flanking marker on either side] down to 20 kb [1,000 kb/(10,000 x 1%) = 10 kb] if there is one recombinant between the targeted gene and its flanking markers on either side of the gene.

Once a large mapping population has been developed, genotyping and phenotyping its large number of plants or lines become a challenge due to a huge amount of work. Figure 4 summarizes the generalized principle and procedure of fine-mapping a gene or QTL (e.g., G/g in Figure 4). Several strategies have been used in high-resolution mapping. For simple Mendelian trait fine-mapping and if the trait is readily phenotyped, a large F_2 population is planted. Plants with recessive trait are selected and genotyped using flanking DNA markers (Maritin et al. 1993a; Wing et al. 1994; Budiman et al. 2004). This is because recessive plants are homologous at the trait locus and their genotypes are in consistency with their phenotypes. Consequently, the phenotyping data could be directly used for recombination analysis. As QTL fine-mapping is labor-intensive and it is costly to phenotype a large population accurately, phenotyping the population is often conducted after genotyping. The idea is that the entire population is first screened with the flanking DNA markers (e.g., M_1 and M_2 in Figure 4) to identify the recombinants in the targeted region, and the recombinants are then phenotyped (Alpert and Tanksley 1996). This is because only the recombinants in the targeted region could provide information useful for high-resolution mapping of the targeted gene. The recombinants for a targeted region are identified by examining the genotypes of the flanking DNA markers (Figure 4A). Since a very limited number of plants of the mapping population, depending on the distance between the flanking markers and recombination frequency in the targeted region, are likely to be recombinants, the number of plants to be phenotyped will be significantly reduced.

To facilitate screening of the mapping population for recombinants in the targeted region, PCR-based, co-dominant markers are more desirable than other types of markers (Figure 4A) and a plant pooling strategy has been used (Tanksley et al. 1995; Budiman et al. 2004). As described above, the PCR-based, co-dominant markers include SSRs, SNPs and CAPs. These DNA markers are highly polymorphic, which is needed to genotype the mapping population, and readily manageable in high-throughput as they need less amount of DNA in assay, amenable to multiplex (for SSRs), more informative (than dominant markers), and cost-efficient. To screen the mapping population using the plant pooling strategy, tissues are collected from each plant or line of the mapping population and pooled in equal amounts at five or more plants per pool. DNA is isolated from the pooled tissues and used for population genotyping using the flanking DNA markers. Once the pools containing one or more recombinant plants are identified, DNA is isolated from each plant tissue of the pools and genotyped using the flanking DNA markers. The latter work identifies individual recombinants of the pools and further verifies the pool analysis results.

Phenotyping the mapping population or regional recombinants is straightforward for simple Mendelian traits; however, experiments with biological replicates and multiple environments, such as multiple years and/or multiple locations, are necessary to accurately phenotype the recombinants in quantitative traits (Alpert and Tanksley 1996). Unlike the simple traits that can be phenotyped by simply scoring 'presence' or 'absence', quantitative traits are much more complex and must be phenotyped by quantitative measurement. Therefore, statistical tools are needed to analyze the quantitative trait data for QTL fine-mapping

(Figure 4B). On the other hand, biological replicated experimental designs are essential to phenotype quantitative traits to fit statistical analysis. It is the reason that F_2 mapping populations have been used for fine mapping of simple Mendelian traits, whereas RILs, BC lines or NILs are used to fine-map the QTL of quantitative traits.

To order the DNA markers within the targeted region delimited by the flanking markers used to screen the mapping population for recombinants, the recombinants must be further analyzed with the markers. They are ordered in the targeted region according to the number of recombinants between each DNA marker and the targeted gene (Figure 4B) (Martin et al. 1993a; Alpert and Tanksley 1996; Fridman et al. 2000; Takahashi et al. 2001; Yan et al. 2003, 2004; Li et al. 2006; Uauy et al. 2006). The more recombinants between a marker and the targeted gene, the farther the marker is from the targeted gene. Therefore, all markers in the targeted region could be ordered, relative to the targeted gene, as long as there are a sufficient number of recombinants available for the region. The marker having no recombinant with the targeted gene is considered to be co-segregating with the gene. As described above, if the analyzed original population size is 10,000 plants, the plant genome average physical/genetic distance ratio is 1,000 kb/cM, and there is one recombinant between the targeted gene and closest flanking marker on either side, the targeted gene can be located to a physical interval of 20 kb.

2.4 Chromosome Walking or Landing

Chromosome walking is to approach the targeted gene from a flanking DNA marker using a large-insert DNA library. This step could be pursued while or after fine-mapping of the gene locus is conducted. Figure 5A shows a generalized procedure of chromosome walking. The flanking DNA marker (e.g., M_4) is used as a probe to screen a large-insert BAC or yeast artificial chromosome (YAC) library. Since large-insert BAC libraries are low in chimeric clones, stable in the host cells and readily manipulated (Wu et al. 2004b; He et al. 2007), they have emerged as the choice of large-insert DNA libraries for chromosome walking. Positive BACs are selected, analyzed and assembled into a BAC contig according to the restriction pattern similarities of the clones (Wu et al. 2005). The outmost ends of the contig are isolated by plasmid rescue (Zhang et al. 1994), inverse-PCR (IPCR) (Zhang et al. 1994) and/or BAC-end sequencing (Li et al. 2007), and genetically mapped using the recombinants identified in the fine-mapping or the entire mapping population. The genetic distances of the BAC-ends from the targeted

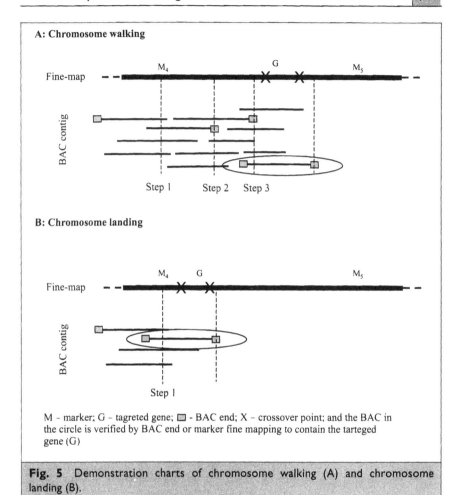

Fig. 5 Demonstration charts of chromosome walking (A) and chromosome landing (B).

gene are determined. This is the first step of a chromosome walk. As needed, the BAC-end closer to the targeted gene, justified by the number of recombinants between the BAC-end and targeted gene, is used as a probe to continue the chromosome walk as done for the first step until a BAC containing the targeted gene is isolated, as indicated by genetically mapping its ends to either side of the gene. In the case that the flanking marker is physically close to the targeted gene, the BACs containing the targeted locus could be identified at the first step of the chromosome walk. This is called chromosome landing (Tanksley et al. 1995).

The success of a chromosome walk is significantly affected by the genome architecture and physical/genetic distance ratio of the targeted genomic region, and the insert size of the large-insert DNA library (Zhang et al. 1996; Ren et al. 2005; for more detail, see below). In general, the larger the genome of targeted species, the larger portion of repeated sequences the genome tends to contain. This is because the numbers of genes are relatively consistent among species whereas the genome sizes of species could vary up to tens of thousands of folds. The differences of the genome sizes among species are mainly attributed to the content of repeated sequences. Repeated sequences could not only dilute the gene density of a genome, but also could block chromosome walking. For instance, if the ends of BACs of the first walk (Figure 5A) contain repeated elements, it is difficult, if not impossible, to continue the subsequent walks as when they are used as probes to screen a large-insert BAC library, the BACs originated from different genomic regions could be isolated. It is time-consuming and might be difficult to identify the BACs originated from the targeted genomic region. Therefore, the larger the genome of targeted species, the smaller the probability of successfully completing a walk of a genomic region and consequently, the smaller the probability of cloning a targeted gene or QTL by map-based cloning. Nevertheless, this does not mean that it is impossible to isolate genes or QTLs from species with large genomes by map-based cloning. It has been observed that the distribution of repeated sequences and the physical/genetic distance ratio along a genome vary too largely. (Lin et al. 1999; Li et al. 2007). For some regions of a large genome such as the gene-rich regions, the gene density and physical/genetic distance ratio may be the same as the small genomes (e.g., Wing et al. 1994; Zhang et al. 1994; Alpert and Tanksley 1996). Therefore, it has been proven that it is possible to isolate genes and QTLs from the species with large genomes such as barley (Büscheges et al. 1997; Brueggeman et al. 2002) and wheat (Huang et al. 2003; Yan et al. 2003, 2004; Uauy et al. 2006) that have a genome size of 4,760 Mb/haploid genome and 15,960 Mb/haploid genome, respectively (Table 1).

Moreover, the development of whole-genome integrated physical/genetic maps (e.g., Marra et al. 1999; Chang et al. 2001; Tao et al. 2001; Chen et al. 2002; Wu et al. 2004a; Li et al. 2007) and sequence maps (The Arabidopsis Genome Initiative 2000; International Rice Genome Sequencing Project 2005) have significantly accelerated map-based cloning of genes and QTLs (Zhang and Wing 1997; Jander et al. 2002). One of the rationales is that there is no chromosome walking necessary to access a targeted gene or QTL from a flanking DNA marker using integrated physical/genetic maps and/or whole-genome sequence maps (as described by Ren et al. 2005) genome sequence maps are a type of genome physical maps). This is because both integrated physical/genetic maps and whole-genome sequence maps are marked with both genetic distance in cM and physical distance in kb. The ratios of genetic/physical distances are known for different genomic regions. Based on the physical/genetic distance ratio for a targeted genomic region, the position of the targeted gene and the clones containing the gene could be predicted and identified directly from the integrated physical/genetic map or genome sequence map, without any need for chromosome walking. Therefore, Zhang and Wing (1997) defined this process as 'gene golfing' as it is similar to playing a game of golf. It is also because no chromosome walking is required to access a targeted gene from its flanking DNA marker using integrated physical/genetic maps and sequence maps that the influence of the repeated sequences on map-based cloning would be greatly reduced (also see below).

Additionally, comparative genetics and genomics studies in the past decades have shown that the gene content and order are highly conserved among related species, especially among the grass species, including Triticeae (wheat, barley, rye), maize, sorghum, rice, foxtail millet, sugar cane, and oats (Moore et al. 1995; Chen et al. 1997; Gale and Devos 1998). The colinearity between species with large and small genomes could provide useful information for chromosome walking in large-genome species using the species having smaller genomes, such as walking a Triticeae chromosome using the completed rice genome sequence (International Rice Genome Sequencing Project 2005). Examples in this regard may be the chromosome walking toward the barley *RPG1* stem rust disease resistance gene using the colinearity of barley-rice (Kilian et al. 1997; Han et al. 1999) and the wheat vernalization genes *VRN1* and *VRN2* using the colinearities between wheat and rice and between wheat and sorghum (Yan et al. 2003, 2004). Nevertheless, the microsynteny between the species may be violated (Fu and Dooner 2002; Brunner et al. 2003) and/or the targeted gene or QTL may be absent in the reference genome (Brueggeman 2002; Brunner et al. 2003). Therefore, caution must be made to walk a large genome chromosome using a smaller genome using the genome colinearity knowledge.

2.5   Gene Identification

This step is essential to identify the targeted gene or QTL by map-based cloning. Once a BAC or BAC contig is isolated and confirmed to contain the targeted gene, as delimited by recombinants flanking the gene locus through chromosome walking and high-resolution mapping, identification and verification of the targeted gene from the BAC or contig become the major task of map-based cloning. While different procedures have been used to identify the targeted gene from a BAC containing it, the straightforward approach seems to first completely sequence the BAC or contig, given the fact that it has been routine to sequence hundreds of kilobase pairs of DNA using the recent sequencing technology and it is highly likely that the BAC or contig contains more than one genes. Sequencing of the targeted gene is also necessary to characterize the gene. The sequenced BAC or contig is annotated to identify the genes potentially contained in the BAC or contig. Then, each of the putative genes contained in the BAC or contig is subjected to the following analyses.

BLAST Search

This is the most economical and fastest method. Comparative analysis of each candidate gene contained in the BAC or contig against those structurally and functionally characterized in GenBank will provide useful information about the gene of the targeted trait (Yano et al. 2000; Huang et al. 2003; Yan et al. 2004). For instance, approximately 3/4 of the plant disease resistance genes cloned to date encode proteins characterized by the conserved nucleotide-binding site (NBS) and leucine-rich repeat (LRR) domains (Takken and Joosten 2000). If the targeted gene of map-based cloning is responsible for resistance to a pathogen and the predicted protein of a candidate gene also contains NBS and LRR domains, the candidate gene is more likely the targeted gene.

Comparative Sequence Analysis

Since the mutation of the targeted gene is responsible for the phenotypic variation of the targeted trait, the wild and mutated alleles of the gene must somehow differ in coding sequences, regulatory sequences, or both. Comparative sequence analysis of the targeted genes isolated from phenotypically contrasting lines of the species will reveal the nature of the mutation, thus providing useful information about the targeted genes (Li et al. 2006). According to the sequences of the annotated genes contained in the BAC or contig, primer pairs are designed for each gene and used for PCR to amplify the corresponding genes from the phenotypically contrasting lines such as resistant versus susceptible lines for resistant gene cloning. The PCR products are then used as templates to sequence the corresponding gene from the lines. If difference is consistently detected for a gene at the nucleotide sequence and its predicted amino sequence between the wild and mutated lines studied, the gene is likely to be the targeted gene.

Gene Expression Analysis

Many of genes express tissue-specifically and/or are environment-inducible. These features provide another characteristic of assisting at identification of the targeted gene from the candidate genes contained in the BAC or contig. For instance, the wheat vernalization gene *VRN2* tends to express at the apices, the critical point for transition from vegetative growth to reproductive growth and responses to low temperature (Yan et al. 2004). The rice grain shattering gene *sh4* prefers to express at the junction between flower and pedicel (Li et al. 2006). Quantitative real-time reverse transcription PCR (RT-PCR) and northern blot analysis have been widely used to analyze the expression of the candidate genes in particular tissues and/or with particular treatments (Yan et al. 2004; Li et al. 2006). The RT-PCR is a relatively new technique to quantify gene expression, in which mRNA is first reverse transcribed into cDNA and the cDNA is then used as templates of conventional PCR.

BLAST search, comparative sequencing and gene expression analysis provide useful information on which of the candidate genes contained in the BAC or contig is more likely to be the targeted gene; however, a genetic complementation experiment is necessary to verify the prediction of the above results on the targeted gene. Several methods have been used for the genetic complementation experiment.

Genomic DNA Transformation

Genomic DNA constructs are made for each candidate gene that is likely the targeted gene in a plant-transformation-competent binary vector (e.g., pCLD04541) from the BAC or contig containing the targeted gene according to the sequence of the BACs (Ling et al. 1999; Frary et al. 2000; Li et al. 2006). To facilitate the expression of the transgene in the host plant, the construct must include both the coding region and the regulatory sequences of the genes. The DNA constructs are then transformed into a mutant genotype of the targeted trait, which is often the mutant parent of the mapping population, by *Agrobacterium*-mediated transformation or bombardment method. The conversion of the host plant mutant type into the wild type will be indicative of that the DNA construct contains the targeted gene.

cDNA Transformation

An alternative method is to transform cDNA clones into plants for genetic complementation experiment (e.g., Martin et al. 1993b). This method is especially used when no sequence information is available for the BAC or contig. The BAC or contig containing the targeted gene is employed as a probe to screen a cDNA library potentially containing the targeted gene, the positive cDNA clones are isolated, sequenced and subjected to fine-mapping, BLAST, comparative sequence analysis and expression study as described above. The most likely candidate cDNA clone(s) is/are subcloned into a plant-transformation-competent binary vector (e.g., pBl121) with a promoter (e.g., the cauliflower mosaic virus 35S promoter) to drive the inserted gene to express. There are two possible orientations of cloning a cDNA insert into a binary vector. One orientation is the start codon of the cDNA gene is just in the downstream of or proximal to the promoter, which is called sense construct. The other orientation is that the start codon of the cDNA gene distal to the driving promoter, which is called anti-sense construct. This could be determined by PCR analysis using the cDNA constructs as templates and the primer pair, one from the vector and the other from insert cDNA. The sense construct is transformed into the mutant genotype plants either by the *Agrobacterium*-mediated and/or by the bombardment method. As the transformation of genomic DNA construct above, if the cDNA construct converts the mutant phenotype into the wild phenotype, the cDNA must contain the targeted gene. The anti-sense construct is transformed into the wild genotype of the targeted trait. If the expression of the wild type gene is repressed significantly, the cDNA construct should contain the targeted gene (also see RNAi).

RNA Interference (RNAi)

RNA interference or RNA-mediated interference (RNAi) was discovered by Fire et al. (1998). It is a mechanism for RNA-guided regulation of gene expression that is commonly observed in plants and animals. RNAi involves a process of generation of double-stranded RNA (dsRNA) and small inhibitory RNA (siRNA) that interferes with the expression of the gene complementary to the dsRNA. dsRNA is trimmed in cells by an RNAase ribonuclease named as dicer into short dsRNA called siRNA of 20-25 nucleotides long. An siRNA can be processed into single-stranded sense and anti-sense RNAs. The single-stranded sense RNA is degraded in cells whereas the single-stranded anti-sense RNA interacts with several proteins to form the ribonuclease-containing multi-protein complex (RISC) that binds to and degrade the sense mRNA that are complementary to the anti-sense RNA, thus the translation of the targeted mRNA being inhibited. This phenomenon has been recently used in the genetic complementation experiments for gene identification in map-based cloning of genes and QTLs (e.g., Yan et al. 2004; Uauy et al. 2006). A segment unique to each candidate gene contained in the BAC or contig is cloned in forward and reverse orientations into a plant-transformation-competent, RNAi-specific binary plasmid vector such as pMDC161 to facilitate dsRNA formation and transformed into the wild genotype of the targeted traits. If the performance of the targeted trait of the transgenic plants is significantly repressed, the candidate gene must be the targeted gene.

For all three transformation experiments described above, it is necessary to verify the phenotypes of the primary transgenic plants (T_0) by analyzing their T_1 progeny. The T_0 plant, if the transgene is dominant, will immediately complement the mutant type and convert it into wild type in phenotype; however, it is hemizygous and the phenotyping result is dependent on a single primary transformant plant. Therefore, it is desirable to further genetically analyze its T_1 progeny to verify the genotype and phenotype of the T_0 plant. This is especially true for the targeted gene controlling quantitative trait because the progeny test is essential to accurately phenotype a quantitative trait.

3 EXAMPLES OF MAP-BASED CLONING IN PLANTS

Many genes and several QTLs have been cloned to date by map-based cloning (Table 1). To further demonstrate the procedure of map-based cloning, two examples, one for map-based cloning of genes controlling simple Mendelian traits and the other for map-based cloning of QTLs, are presented here as paradigms.

3.1 Map-based Cloning of Genes Controlling Mendelian Traits

Cloning of the gene conferring resistance to the bacterial speck disease of tomato caused by *Pseudomonas syringae* pv. tomato (Martin et al. 1993b) is used as an example. The disease resistance is controlled by a single dominant gene, *Pto*, representing the first Mendelian trait gene cloned by map-based cloning in crop plants.

The *Pto* gene was introgressed from a wild tomato species, *Lycopersicon pimpinellifolium*, into the cultivated tomato, *L. esculentum* cv. Rio Grande by six generations of backcrosses to the cultivated tomato, followed by selfing for one generation. This resulted in a pair of NILs named Rio Gande and Rio Grande-PtoR. To genetically map the gene with DNA markers, Martin et al. (1991) isolated the DNAs from the NILs, Rio Gande and Rio Grande-PtoR, and comparatively analyzed the DNAs by PCR using 140 RAPD primers. While the majority of the RAPD products were identical between the two NILs, seven of the 140 primers produced the fragments that were present in one NIL but absent in the other NIL. Using the polymorphic RAPD fragments generated by four of the seven primers and two small populations (total 47 F_2 plants) segregating for the gene, they mapped the *Pto* gene to chromosome 5 and developed a local genetic map for the gene locus. To more accurately map the gene and isolate more DNA markers, they mapped the gene with the RAPD markers to a high-density tomato genetic map using a population of 80 plants though it did not segregate for the gene (Tanksley et al. 1992). As a result, the *Pto* gene was mapped to an interval of 16.5 cM between CD64 and TG96 on chromosome 5.

Since the *Pto* gene was approximately 6.8 cM away from TG96 and 9.7 cM from CD64 and it was, on average, approximately 750 Mb/cM for the tomato genome (Tanksley et al. 1992), it was difficult to access the gene locus by chromosome walking from either of the two flanking markers using a large-insert tomato YAC library (YAC was the only system available for large-insert library construction at that time). Therefore, it was necessary to identify more DNA markers to map or order them in the *Pto* region. For the purpose, Martin et al. (1993a) identified a total of 28 DNA markers for the region from the high-density genetic map of tomato (Tanksley et al. 1992) and by further analyzing the Rio Grande/Rio Grande-PtoR NILs with random tomato cDNA and genomic DNA clones, and RAPD. These markers were then analyzed for RFLP with six restriction enzymes and 18 of them were shown to be polymorphic at least with one restriction enzyme. To order the 18 polymorphic DNA markers by crossover points occurring in the region, a population consisting of 1,200 F_2 plants was developed from a cross between the Rio Grande and Rio Grande-PtoR NILs. Since the disease resistance is dominant and to avoid subsequent progeny test on a large number of F_2 plants, only the homozygous recessive plants (*pto/pto*) were selected for genotyping with the 18 DNA markers. A total of 251 homozygous recessive plants were selected and analyzed by RFLP with the 18 DNA markers. Consequently, 85 homologous recessive, recombinant F_2 plants were identified. Mapping these markers using the recombinant plants resulted in a high-resolution linkage map for the *Pto* region, with the marker TG538 co-segregating with *Pto*, and TG475 being 0.4 cM and TG504 12.0 cM from *Pto*, flanking on either side of the gene.

To estimate the feasibility of approaching the *Pto* locus from the DNA markers by chromosome walking using the available tomato YAC library with an average insert size of 145 kb constructed from the tomato cultivars VFNT cherry and Rio Grande-PtoR (Martin et al. 1992), the maximal physical/genetic distance ratio was determined by pulsed-field gel electrophoresis (PFGE). The megabase-size DNA were isolated from

Rio Grande and Rio Grande-PtoR, digested with eight rare-cutting enzymes, subjected to PFGE, blotted and probed with TG538, TG475 and TG504, respectively. The result showed that the TG538 and TG475 hybridized to a common 435-kb *Sal*I fragment and a common 450-kb *Sfi*I fragment in Rio Grande-PtoR, suggesting the maximal physical distance between these two markers to be 435 kb, and it was feasible to approach the *Pto* locus from the markers using the tomato YAC library.

The chromosome walking toward the *Pto* locus was conducted by screening the tomato YAC library (Martin et al. 1992) using the DNA marker TG538 co-segregating with the *Pto* gene as a probe. An YAC clone, PTY538-1, was identified and estimated to be about 400 kb in size. Both end sequences of the YAC, named PTY538-1L (for left arm end) and PTY538-1R (for right arm end), were isolated by IPCR and genetically mapped to the high-resolution linkage map of the *Pto* region using the above Rio Grande x Rio Grande-PtoR F_2 population. The PTY538-1L was mapped to 1.8 cM apart from *Pto*, and PTY538-1R co-segregated with *Pto*. To confirm whether the YAC PTY538-1 contained the *Pto* locus, additional 1,300 F_2 plants from the Rio Grande x Rio Grande-PtoR population were screened using TG538 and PTY538-1R. Recombinant analysis between the two markers and the *Pto* trait indicated that the YAC PTY538 spans the *Pto* locus. To find the *Pto* candidate gene on the YAC clone, DNA was isolated from the PTY538-1 clone and hybridized to a cDNA library constructed from the leaves of Rio Grande-PtoR infected with avirulent *P. syringae* pv. *tomato* (Martin et al. 1993b). Approximately 200 clones were identified and 30 of them were investigated by further using the 85 recombinant F_2 plants identified as above for the high-resolution mapping of the *Pto* region. A cDNA clone, CD127, was found to co-segregate with *Pto*, suggesting that it was likely the *Pto* candidate gene.

Nevertheless, when a Southern hybridization of the genomic DNAs of Rio Grande and Rio Grande-PtoR and YAC PTY538-1 was conducted with CD127, many fragments resulted, suggesting the cDNA clone could either contain exons spanning a large genomic region or represent a multigene family. To distinguish between these two possibilities, the CD127 was used as a probe to rescreen the leaf cDNA library and 14 other cross-hybridizing clones were found that represented six different classes of related genes. Further, since the YAC PTY538-1 detected all (but one) of the genomic fragments detected on the Rio Grande-PtoR Southern blot, it was concluded that PTY538-1 contained a multigene family.

To determine if the homologs of the cDNA CD127 contained the *Pto* gene, genetic complementation experiments were conducted by genetic transformation with the cDNA clones. Two cDNA clones representing the two size classes of the clones, CD127 (1.2-kb insert) and CD186 (2.4-kb insert), were subcloned into the plant-transformation-competent binary vector pBI121 in sense orientation, resulting in the constructs pPTC5 (CD127) and pPTC8 (CD186). The two constructs were transformed into a susceptible tomato cultivar, Moneymaker. Two pPTC8 resistant plants were obtained, whereas no pPTC5 resistant plants were found, suggesting that CD186 was the better candidate for the *Pto* gene than CD127. To further verify this conclusion, one of the two primary (R_0) pPTC8 transgenic plants, expected to have a genotype of *Pto/pto*, was crossed to the susceptible cultivar control, Rio Grande (*pto/pto*). Among the 22 plants, 9 were found to contain the CD186 sequence and resistant to the bacterial speck pathogen, and the remaining 13 plants that did not contain the CD186 sequence were susceptible to the bacterial speck. The segregation of the progeny fitted to a 1:1 single gene segregation ratio and confirmed that the CD186 is the transcript of the *Pto* gene. The insert of the clone CD186 was then sequenced and a 321-amino acid open reading frame (ORF) was found. BLAST analysis of the ORF of the CD186 revealed that the *Pto* gene is similar to a serine-threonine protein kinase.

3.2 Map-based Cloning of QTLs

The procedure of map-based cloning of QTLs is essentially the same as that of map-based cloning of simple Mendelian genes. The major difference of the former from the latter is phenotyping of the targeted trait due to their difference in genetic behavior. While simple Mendelian traits can be phenotyped by simple counting, quantitative traits must be phenotyped by measurement and are readily influenced by environments. Therefore, replicated experiments are desired to phenotype the mapping populations and recombinants identified for the region of the targeted QTL. To demonstrate the map-based cloning of QTLs, the cloning of the QTL for tomato fruit weight, *fw 2.2*, has provided an excellent paradigm (Frary et al. 2000).

To map fruit weight genetically in tomato, an introgression line F_2 derived from *L. esculentum* (tomato) x *L. pennellii* and a BC_1 population derived from *L. esculentum* x *L. pimpinellifolium* segregating for fruit weight were used (Alpert et al. 1995). The QTL *fw 2.2* was mapped near DNA markers TG91 and TG167 on chromosome 2 in both populations. It accounted for 30 - 40% of the total phenotypic variance, indicating that it is a major QTL. The mapping study also showed that the QTL is partially dominant in the species.

To fine-map the QTL, two experiments were conducted first (Alpert and Tanksley 1996). One was the development of an NIL F_2 population consisting of 3,472 plants from a cross of NILs *L. esculentum* cv. M82-1-8 x IL2-5 (*L. pennellii* introgression line) and the other was identification of additional DNA markers for the *fw* 2.2 region. Three new DNA markers, TG686, TG687 and *HSF24*, were obtained for the region. The first two were obtained by analysis of a pair of NILs for the QTL using RAPD primers as described by Martin et al. (1991) and the third one was obtained from the tomato genetic map (Tanksley et al. 1992). The NIL F_2 population was screened with the two DNA markers, CD66 and TG361, flanking the QTL to identify the recombinants between them. Fifty-five recombinants were identified and used to fine-map the QTL. Six DNA markers that were mapped to the *fw* 2.2 region were used to screen a tomato YAC library (Martin et al. 1992) and six YACs were isolated. To determine the relationship of the clones, the arm ends of the YACs were isolated by IPCR and fine-mapped with the 55 recombinants in the region. Using the data, a physical map spanning the *fw* 2.2 region was constructed. To establish the relationship between the markers, YACs and fruit weight, a five-replicate experiment was conducted in field at two locations (Davis, California and Ithaca, New York). The five replications of individual recombinants, along the two controls (NIL 393-2 containing the QTL locus derived from *L. pennellii* and cultivar M82-1-8 containing the QTL locus from cultivated tomato), were evaluated for fruit weight by weighing 10 representative fruits from each plant and subjected to statistical analysis to determine whether the differences of fruit weights between the recombinants were due to environmental effects or genotype differences. These experiments placed the *fw* 2.2 locus to an interval of approximately 0.01 cM flanked by TG91 and *HSF24*, and included it within two YACs: YAC264 (210 kb) and YAC355 (300 kb).

The physical/genetic distance ratio in the *fw* 2.2 region was estimated by pulsed-field gel Southern blot analysis. Megabase-size DNA was isolated, digested with five rare-cutting enzymes, subjected to pulsed-field gel electrophoresis, blotted onto membrane and probed consequentially with the DNA markers in the region. Since TG91 and TG167 flanking the QTL were hybridized to the same fragments of a few enzymes with the smallest *Mlu*I fragment being $\leq$ 150 kb, the physical distance between the two markers was 150 kb or smaller.

To identify the candidate gene for the *fw* 2.2 QTL, a cDNA library was constructed from the small-fruited genotype, *L. pennellii* LA716, and screened using the YAC264 and YAC355 DNA as probes (Frary et al. 2000). Four unique cDNA clones were identified and mapped to the high-resolution genetic map of the region constructed using the NIL F_2 population consisting of 3,472 plants. The four cDNAs were then used as probes to screen a plant-transformation-competent genomic DNA binary cosmid library constructed from the small-fruited *L. pennellii*. Four cosmid clones, each corresponding to one of the four cDNA clones, were transformed into two large-fruited tomato cultivars. Since the primary transgenic (T_0) plants are hemizygous and the *fw* 2.2 QTL is partially dominant, the T_0 plants were self-pollinated to obtain segregating T_1 progeny. The fruit weights of the progeny were evaluated and statistically analyzed. Two transgenic plants, both of which contained the same cosmid clone cos50, showed that their fruit weights were significantly reduced relative to the large-fruited host cultivars, suggesting that the *fw* 2.2 QTL was contained in the cos50. Sequence analysis of the cos50 revealed that the cosmid clone had two ORFs, cDNA 44 that was used to isolate cos50, and *ORFX* that was not detected in the cDNA library. Further analysis of the two ORFs using the recombinants for the *fw* 2.2 region in the previously fine-mapping indicated that *ORFX* is the *fw* 2.2 QTL gene. It is expressed early in floral development, controls carpel cell number, and has a sequence suggesting structural similarity to the human oncogene c-H-*ras* p21.

4   NOTES OF MAP-BASED CLONING

The approach of map-based cloning has been so far successfully used to clone a number of genes and QTLs in plants (Table 1). These projects have provided a wealth of valuable information and experience for map-based cloning of genes and QTLs in plants.

4.1   Phenotyping Accuracy

Accurate phenotyping of the plants that are used in the mapping, especially fine-mapping, is crucial to the success of map-based cloning. This is because mis-scoring or low penetrance in one plant could mislead chromosome walking from flanking DNA markers to the targeted gene, resulting in the failure of map-based cloning. Therefore, for map-based cloning of genes controlling Mendelian traits, it is often necessary to repeatedly score each plant of the population such as at different developmental stages. For instance of fine-mapping of the *Pto* gene (Martin et al. 1993a), the NIL F_2 plants used were screened for homozygous recessive plants (*pto/pto*) using a pleiotropic or tightly linked gene, *Fen*. In the first screening, 18% of the *pto/pto* plants were mis-scored. Therefore, the second screening was conducted using an improved method and repeated twice to verify the phenotypes of the plants. For fine-mapping of QTLs, such as the tomato fruit weight QTL, *fw 2.2* (Alpert and Tanksley 1996), a repeated experiment with multiple environments is essential to accurately phenotype each line of the mapping population. The replication of the experiment is also necessary to analyze the phenotypic data using statistical tools. Furthermore, use of mapping populations that have identical or similar genetic background but the targeted locus, such as NIL F_2 populations (e.g., Martin et al. 1993a; Alpert and Tanksley 1996; Fridman et al. 2000; Yano et al. 2000; Takahashi et al. 2001; Uauy et al. 2006), will significantly enhance the phenotyping accuracy of plants.

4.2 Variation of Genetic Recombination Rate along a Chromosome

Map-based cloning had been once considered to be an effective approach to map-based cloning of genes and QTLs only in the species with small or moderate genome sizes, such as Arabidopsis (Arondel et al. 1992; Giraudat et al. 1992; Johanson et al. 2000), rice (Yano et al. 2000; Takahashi et al. 2001; Li et al. 2006) and tomato (Martin et al. 1993b; Ling et al. 1999; Frary et al. 2000). Recent success of map-based cloning in the species with large genomes has demonstrated that it is feasible to isolate genes and QTLs in large-genome species by map-based cloning. Examples in this regard include map-based cloning of the broad spectrum disease resistance gene *Mlo* (Büschges et al. 1997) and flowering gene *Ppd-H1* (Turner et al. 2005) in barley (4,870 Mb/haploid genome), and the leaf rust resistance genes *Lr21* (Huang et al. 2003) and *Lr10* (Feuillet et al. 2003), the vernalization genes *VRN1* and *VRN2* (Yan et al. 2003, 2004), the powdery mildew resistance gene *Pm3b* (Yahiaoui et al. 2004) and the *NAC* gene (Uauy et al. 2006) in wheat (15,960 Mb/haploid genome) and related species (5,750 Mb/haploid genome). This is because gene density and genetic recombination frequency or physical/genetic distance ratio vary by many fold along a chromosome of a genome (e.g., Lin et al. 1999; Li et al. 2007). For instance, tomato (950 Mb/haploid genome) has an average physical/genetic distance ratio of 750 kb/cM (Tanksley et al. 1992); however, it is 86 kb/cM for the *jointless* gene region (Zhang et al. 1994), 1,087 kb/cM for the *Pto* region (Martin et al. 1993a), ≤ 950 Kb/cM for the *fw 2.2* region (Alpert and Tanksely 1996), 4 Mb/cM for the *Tm-2a* region (Ganal et al. 1989) and 25 Mb/cM for the *jointless-2* region (Budiman et al. 2004). The variation of the physical/genetic distance ratio is > 290-fold. Wheat has an average physical/genetic distance ratio of 4.4 Mb/cM (Faris and Gill 2002), but it is 340 kb/cM for the *Lr21* region (Huang et al. 2003). This number is approximately 13-fold smaller than the mean ratio of the wheat genome and 2.2-fold smaller than the mean ratio of the tomato genome. The variation of gene density and physical/genetic distance ratio in a genome provides an opportunity of cloning genes and QTLs residing at the gene-rich and/or low physical/genetic distance ratio regions from large-genome species, but casts a challenge of cloning genes and QTLs residing at the gene-poor and/or high physical/genetic distance ratio regions from small-genome species.

4.3 Use of Plant-Transformation-Competent BIBAC and TAC Libraries

Large-insert DNA libraries have been proven essential for map-based cloning of genes and QTLs (e.g., Martin et al. 1993b; Frary et al. 2000; Yan et al. 2003, 2004; Uauy et al. 2006). There are two types of large-insert DNA libraries available: YAC and large-insert bacterial clones (LBC) including BAC, bacteriophage P1-derived artificial chromosome (PAC), plant-transformation-competent binary BAC (BIBAC), conventional large-insert plasmid-based bacterial clone (PBC) and transformation-competent artificial chromosome (TAC) (Wu et al. 2004b; Ren et al. 2005; He et al. 2007). In comparison with YAC, BAC, PAC and PBC libraries, plant-transformation-competent BIBAC (Hamilton et al. 1996, 1999; Hamilton, 1997) and TAC (Tao and Zhang 1998; Liu et al. 1999, 2002; He et al. 2003) libraries have additional advantages for map-based cloning of genes and QTLs. This is because in addition to their utility as large-insert clones as YAC, BAC, PAC and PBC, BIBAC and TAC can be directly transformed into plants via *Agrobacterium* for gene identification by genetic complementation, without need of subcloning into a binary vector for genetic transformation. This is especially of significance for identification of genes and QTLs that span a large (e.g., > 50 kb) genomic region, physically cluster, or reside in group. In contrast, the sequences containing such genes and QTLs are difficult to be subcloned into a binary vector for genetic complementation due to their large span in the genome, if they are contained in a YAC, BAC, PAC or PBC. Nevertheless, for identification of independent genes and QTLs that span a small genomic region, subcloning of the BIBAC or TAC containing the targeted gene may be necessary to determine which of the genes contained in the BIBAC or TAC is the targeted gene. Therefore, to streamline the procedure of map-based cloning, BIBAC and TAC libraries have been developed for a number of plant species (*http://hbz7.tamu.edu*).

4.4 Chromosome Walking versus Whole-Genome Physical Mapping and Sequencing

Chromosome walking seems straightforward to access a targeted gene from its flanking marker using a large-insert DNA library. However, it is time-consuming (Patocchi et al. 1999; Yan et al. 2003, 2004; Komeri et al. 2004; Yahiaoui et al. 2004) and sometimes difficult, if not impossible, to approach to a targeted gene by chromosome walking due to the abundance of repeated sequences in the region containing the targeted gene (Ganal et al. 1989; Budiman et al. 2004). This is especially true for species with large, complex genomes (Tanksley et al. 1995). Therefore, many of the genes and QTLs cloned by map-based cloning were actually accessed by chromosome landing (e.g., Martin et al. 1993b; Zhang et al. 1994; Alpert and Tanksley 1996; Büschges et al. 1997; Ling et al. 1999; Frary et al. 2000; Fridman et al. 2000; Yano et al. 2000; Takahashi et al. 2001; Huang et al. 2003; Turner et al. 2005; Li et al. 2006; Uauy et al. 2006). Nevertheless, for chromosome landing it is essential to develop a sufficient number of DNA markers that are physically closely linked to and fine-map the targeted gene region. So, the development of DNA markers physically closely linked to the gene and capacity of fine-mapping of the gene region are crucial to the success of chromosome landing and thus that of map-based mapping. Although the development of PCR-based DNA marker technology and availability of high-density genetic maps for many crop species have made it easier to isolate DNA markers for and fine-mapping of a targeted gene region, whole-genome BAC/BIBAC-based physical maps and whole-genome sequence maps integrated with genetic maps will revolutionize the procedure of map-based cloning. As described by Zhang and Wing (1997) and Jander et al. (2002), with physical and/or sequence maps, a unlimited number of DNA markers can be developed for any genomic region for fine-mapping and the targeted gene can be accessed by in silico chromosome walking (Jander et al. 2002) or 'gene golfing' (Zhang and Wing 1997). Therefore, the process of map-based cloning can be accelerated by several folds.

4.5 Targeted Gene Source Genotypes and Source Libraries

It has been demonstrated that large-insert genomic and/or cDNA libraries constructed from the genotypes containing the targeted genes are needed for map-based cloning of a gene or QTL (Martin et al. 1993b; Zhang et al. 1994; Alpert and Tanksley 1996; Büschges et al. 1997; Ling et al. 1999; Patocchi et al. 1999; Frary et al. 2000; Huang et al. 2003; Komeri et al. 2003; Yan et al. 2003, 2004; Turner et al. 2005; Li et al. 2006; Uauy et al. 2006). For different purposes of genomics and genetics research, including map-based cloning, large-insert BAC and BIBAC libraries and cDNA libraries have been developed for a number of genotypes of all major crops and model plant species (*http:// hbz7.tamu.edu;   http://www.genome.clemson.edu/groups/bac/;   http:// bacpac.chori.org/*). These BAC/BIBAC and cDNA libraries have promoted map-based cloning of genes and QTLs in different crop species, but many genes or QTLs of interest are not contained in the existing libraries. Although BAC or BIBAC libraries could be readily constructed from the targeted gene source genotypes, it is time-consuming and costly to construct a whole-genome BAC or BIBAC library. To overcome this conflict, Fu and Dooner (2000) developed a BAC vector containing a rare-cutting enzyme site for large DNA fragment cloning. Therefore using such vectors, the large DNA fragment containing the targeted gene could be directly cloned from the gene source genotype as described by He et al. (2007) without the need of constructing a large-insert BAC or BIBAC library for the whole genome of the gene source genotype.

5 CONCLUDING REMARKS

Isolation of a gene or QTL by map-based cloning, in general, is a tedious, time-consuming process. Except for the model plant species, such as *Arabidopsis* for which whole-genome sequence is available, that is readily transformable and has a short life cycle, it usually takes several years for many crop species to isolate a gene or QTL by map-based cloning (e.g., see Martin et al. 1991, 1993a, b; Alpert et al. 1995; Alpert and Tanksley 1996; Frary et al. 2000). Development of high-density genetic maps, whole-genome integrated physical/genetic maps, whole-genome sequence maps, a wealth of readily used DNA markers, and high-throughput DNA marker techniques will significantly shorten the process of map-based cloning, especially targeted gene genetic mapping and fine-mapping, and make it much easier to approach to the gene from its flanking markers and identify the candidate genes for gene identification by genetic complementation. Whole-genome, BAC and/or BIBAC-based physical maps have been developed for *A. thaliana* (Marra et al. 1999; Chang et al. 2001), *indica* rice (Tao et al. 2001), *japonica* rice (Chen et al. 2002; Li et al. 2007), soybean (Wu et al. 2004a) and maize (Nelson et al. 2005), and are under construction for several other plant species, including cotton, sorghum, tomato and wheat. Whole-genome sequences have been generated for *A. thaliana* (The Arabidopsis Genome

Initiative 2000), indica rice (Yu et al. 2002), japonica rice (Goff et al. 2002; International Rice Genome Sequencing Project 2005), and poplar (Tuskan et al. 2006) and the whole genomes of maize, soybean, tomato, *Medicago*, and *Lotus* are under sequencing. There is no doubt that these whole-genome physical maps and sequences will make map-based cloning in crop plants routine and thus, boost map-based cloning of genes and QTLs important to agricultural and biological sciences.

References

Alpert KB, Grandillo S, Tanksley SD (1995) *fw* 2.2: a major QTL controlling fruit weight is common to both red- and green-fruited tomato species. Theor Appl Genet 91: 994-1000

Alpert KB, Tanksley SD (1996) High-resolution mapping and isolation of a yeast artificial chromosome contig containing *fw*2.2: A major fruit weight quantitative trait locus in tomato. Proc Natl Acad Sci USA 93: 15503-15507

Arondel V, Lemieux B, Hwang I, Gibson S, Goodman HN, Somerville CR (1992) Map-based cloning of gene controlling omega-3 fatty acid desaturation in *Arabidopsis*. Science 258: 1353-1355

Arumuganathan K, Earle ED (1991) Nuclear DNA content of some important plant species. Plant Mol Biol Rep 9: 208-218

Brueggeman R, Rostoks N, Kudrna D, Kilian A, Han F, Chen J, Druka A, Steffenson B, Kleinhofs A (2002) The barley stem rust-resistance gene *Rpg1* is a novel disease-resistance gene with homology to receptor kinases. Proc Natl Acad Sci USA 99: 9328-9333

Brunner S, Heller B, Feuillet C (2003) A large rearrangement involving genes and low-copy DNA interrupts the microcolinearity between rice and barley at the *Rph7* locus. Genetics 164: 673-683

Brutnell TP (2002) Transposon tagging in maize. Funct Integr Genom 2: 4-2

Budiman MA, Chang S-B, Lee S, Yang TJ, Zhang H-B, de Jong H, Wing RA (2004) Localization of *jointless-2* gene in the centromeric region of tomato chromosome 12 based on high resolution genetic and physical mapping. Theor Appl Genet 108: 190-196

Büschges R, Hollricher K, Panstruga R, Simons G, Wolter M, Frijter A, van Daelen R, van der Lee T, Diergaarde P, Groenendijk J, Töpsch S, Vos P, Salamini F, Schilze-Lefert P (1997) The barley *Mlo* gene: A novel control element of plant pathogen resistance. Cell 88: 695-705

Chang Y-L, Tao Q, Scheuring C, Ding K, Meksem K, Zhang H-B (2001) An integrated map of *Arabidopsis thaliana* for functional analysis of its genome sequence. Genetics 159: 1231-1242

Chen M, Presting G, Barbazuk WB, Goicoechea JL, Blackmon B, Fang G, Kim H, Frisch D, Yu Y, Sun S, Higingbottom S, Phimphilai J, Phimphilai D, Thurmond S, Gaudette B, Li P, Liu J, Hatfield J, Main D, Farrar K, Henderson C, Barnett L, Costa R, Williams B, Walser S, Atkins M, Hall C,

Budiman MA, Tomkins JP, Luo M, Bancroft I, Salse J, Regad F, Mohapatra T, Singh NK, Tyagi AK, Soderlund C, Dean RA, Wing RA (2002) An integrated physical and genetic map of the rice genome. Plant Cell 14: 537-545

Chen M, SanMiguel P, de Oliveira AC, Woo S-S, Zhang H-B, Wing RA, Bennetzen JL (1997) Microcolinearity in *sh2*-homologous regions of the maize, rice and sorghum genomes. Proc Natl Acad Sci USA 94: 3431-3435

Faris JD, Gill BS (2002) Genomic targeting and high-resolution mapping of the domestication gene Q in wheat. Genome 45: 706-718

Feuillet C, Travella S, Stein N, Albar L, Nublat A, Keller B (2003) Map-based isolation of the leaf rust disease resistance gene *Lr10* from the hexaploid wheat (*Triticum aestivum* L.) genome. Proc Natl Acad Sci USA 100: 15253-15258

Fire A, Xu SQ, Montgomery MK, Kostas SA, Driver SE, Mello CC (1998) Potent and specific genetic interference by double-stranded RNA in *Caenorhabditis elegans*. Nature 391: 806-811

Frary A, Nesbitt TC, Frary A, Grandillo S, van der Knaap E, Cong B, Liu J, Meller J, Elber R, Alpert KB, Tanksley SD (2000) *fw2.2*: A quantitative trait locus key to the evolution of tomato fruit size. Science 289: 85-88

Fridman E, Pleban T, Zamir D (2000) A recombination hotspot delimits a wild-species quantitative trait locus for tomato sugar content to 484 bp within an invertase gene. Proc Natl Acad Sci USA 97: 4718-4723

Fu H, Dooner HK (2000) A gene-enriched BAC library for cloning large allele-specific fragments from maize: Isolation of a 240-kb contig of the *bronze* region. Genome Res 10: 866-873

Fu H, Dooner HK (2002) Intraspecific violation of genetic colinearity and its implications in maize. Proc Natl Acad Sci USA 99: 9573-9578

Gale MD, Devos KM (1998) Comparative genetics in the grasses. Proc Natl Acad Sci USA 95: 1971-1974

Ganal MW, Young ND, Tanksley SD (1989) Pulsed field gel electrophoresis and physical mapping of large DNA fragments in the *Tm-2a* region of chromosome 9 in tomato. Mol Gen Genet 215: 395-400

Giovannoni JJ, Wing RA, Ganal MW, Tanksely SD (1991) Isolation of molecular markers from specific chromosomal intervals using DNA pools from existing mapping populations. Nucl Acids Res 19: 6553-6568

Giraudat J, Hauge BM, Valon C, Smalle J, Parcy F, Goodman HM (1992) Isolation of the *Arabidopsis ABI3* gene by positional cloning. Plant Cell 4: 1251-1261

Goff SA, Ricke D, Lan T-H, Presting G, Wang R, Dunn M, Glazebrook J, Sessions A, Oeller P, Varma H, Hadley D, Hutchison D, Martin C, Katagiri F, Lange BM, Moughamer T, Xia Y, Budworth P, Zhong J, Miguel T, Paszkowski U, Zhang S, Colbert M, Sun W-L, Chen L, Cooper B, Park S, Wood TC, Mao L, Quail P, Wing R, Dean R, Yu Y, Zharkikh A, Shen R, Sahasrabudhe S, Thomas A, Cannings R, Gutin A, Pruss D, Reid J, Tavtigian S, Mitchell J, Eldredge G, Scholl T, Miller RM, Bhatnagar S, Adey N, Rubano T, Tusneem N, Robinson R, Feldhaus J, Macalma T, Oliphant A, Briggs S (2002) A draft sequence of the rice genome (*Oryza sativa* L. ssp. *japonica*). Science 296: 92-100

Hamilton CM (1997) A binary-BAC system for plant transformation with high-molecular-weight DNA. Gene 200: 107-116

Hamilton CM, Frary A, Lewis C, Tanksley SD (1996) Stable transfer of intact high molecular weight DNA into plant chromosomes. Proc Natl Acad Sci USA 93: 9975-9979

Hamilton CM, Frary A, Xu Y, Tanksley SD, Zhang H-B (1999) Construction of tomato genomic DNA libraries in a binary-BAC (BIBAC) vector. Plant J 18: 223-229

Han F, Kilian A, Chen JP, Kudrna D, Steffenson B, Yamamoto K, Matsumoto T, Sasaki T, Kleinhofs A (1999) Sequence analysis of a rice BAC covering the syntenous barley *Rpg1* region. Genome 42: 1071-1076

He L, Du C, Li Y, Scheuring C, Zhang H-B (2007) Large-insert bacterial clone libraries and their applications. In: Liu Z (ed) Aquaculture Genome Technologies. Blackwell Publ, Ames, Iowa, USA, pp 215-244

He R-F, Wang Y, Shi Z, Ren X, Zhu L, Weng Q, He G-C (2003) Construction of a genomic library of wild rice and *Agrobacterium*-mediated transformation of large-insert DNA linked to BPH resistance locus. Gene 321: 113-121

Huang L, Brooks SA, Li W, Fellers JP, Trick HN, Gill BS (2003) Map-based cloning of leaf rust resistance gene *Lr21* from the large and polyploid genome of bread wheat. Genetics 164: 655-664

International Rice Genome Sequencing Project (2005) The map-based sequence of the rice genome. Nature 436: 793-800

Jander G, Norris SR, Rounsley SD, Bush DF, Levin IM, Last RL (2002) Arabidopsis map-based cloning in the post-genome era. Plant Physiol 129: 440-450

Jeong D-H, An S, Kang H-G, Moon S, Han J-J, Park S, Lee HS, An K, An G (2002) T-DNA insertional mutagenesis for activation tagging in rice. Plant Physiol 130: 1636-1644

Johanson U, West J, Lister C, Michaels S, Amasino R, Dean C (2000) Molecular analysis of *FRIGIDA*, a major determinant of natural variation in *Arabidopsis* flowering time. Science 290: 344-347

Kilian A, Chen J, Han F, Steffenson B, Kleinhofs A (1997) Towards map-based cloning of the barley stem rust resistance genes *Rpg1* and *rpg4* using rice as an intergenomic cloning vehicle. Plant Mol Biol 35: 187-195

Komori T, Ohta S, Murai N, Takakura Y, Kuraya Y, Suzuki S, Hiei Y, Hidemasa Imaseki H, Nitta N (2004) Map-based cloning of a fertility restorer gene, *Rf-1*, in rice (*Oryza sativa* L.). Plant J 37: 315-325

Krysan PJ, Young JC, Sussman MR (1999) T-DNA as an insertional mutagen in Arabidopsis. Plant Cell 11: 2283-2290

Li C, Zhou A, Sang T (2006) Rice domestication by reduced shattering. Science 311: 1936-1939

Li Y, Uhm T, Ren C, Wu C, Santos TS, Lee M-K, Yan B, Santos F, Zhang A,

Scheuring C, Sanchez A, Millena AC, Nguyen HT, Kou H, Liu D, Zhang H-B (2007) A plant-transformation-competent BIBAC/BAC-based map of rice for functional analysis and genetic engineering of its genomic sequence. Genome 50: 278-288

Lin X, Kaul S, Rounsley S, Shea TP, Benito M-l, Town CD, Fujii CY, Mason T, Bowman CL, Barnstead M, Feldblyum TV, Buell CR, Ketchum KA, Lee J, Ronning CM, Koo HL, Moffat KS, Cronin LA, Shen M, Pai G, Van Aken S, Umayam L, Tallon LJ, Gill JE, Adams MD, Carrera AJ, Creasy TH, Goodman HM, Somerville CR, Copenhaver GP, Preuss D, Nierman WC, White O, Eisen JA, Salzberg SL, Fraser CM, Venter JC (1999) Sequence and analysis of chromosome 2 of the plant *Arabidopsis thaliana*. Nature 402: 761-768

Ling H-Q, Koch G, Bumlein Ganal MW (1999) Map-based cloning of *chloronerva*, a gene involved in iron uptake of higher plants encoding nicotianamine synthase. Proc Natl Acad Sci USA 96: 70-98-7103

Liu Y-G, Shirano Y, Fukaki H, Yanai Y, Tasaka M, Tabata S, Shibata D (1999) Complementation of plant mutants with large genomic DNA fragments by a transformation-competent artificial chromosome vector accelerates positional cloning. Proc Natl Acad Sci USA 96: 6535-6540

Marra M, Kucaba TA, Sekhon M, Hiller L, Martienssen R, Chinwalla A, Crockett J, Fedele J, Grover H, Gund C, McCombie WR, McDonald K, McPherson J, Mudd N, Parnell L, Schein J, Seim R, Shelby P, Waterston R, Wilson R (1999) A map for sequence analysis of the *Arabidopsis thaliana* genome. Nat Genet 22: 265-270

Martin GB, Williams JGK, Tanksley SD (1991) Rapid identification of markers linked to a *Pseudomonas* resistence gene in tomato using random primers and near-isogenic lines.Proc Natl Acad Sci USA 88: 2336-2340

Martin GB, Williams JGK, Tanksley SD (1992) Construction of a yeast artificial chromosome library of tomato and identification of cloned fragments linked to two disease resistance loci. Mol Gen Genet 223: 25-32

Martin GB, de Vicente MC, Tanksley SD (1993a) High-resolution linkage analysis and physical characterization of the *Pto* bacterial resistance locus in tomato. Mol Plant-Micr Interact 6: 26-34

Martin GB, Brommonschenkel SH, Chunwongse J, Frary A, Ganal MW, Spivey R, Wu T, Earle ED, Tanksley SD (1993b). Map-based cloning of a protein kinase gene conferring disease resistance in tomato. Science 262: 1432-1436

Michelmore RW, Paran l, Kesseli RV (1991) Identification of markers linked to disease resistance genes by bulked segregant analysis: a rapid method to detect markers in specific genomic regions using segregating populations. Proc Natl Acad Sci USA 88: 9828-9832

Moore G, Devos KM, Wang Z, Gale MD (1995) Cereal genome evolution. Grasses, line up and form a circle. Curr Biol 5: 737-739

Nelson WM, Bharti AK, Butler E, Wei F, Fuks G, Kim H, Wing RA, Messing J, Soderlund C (2005) Whole-genome validation of high-information-content fingerprinting. Plant Physiol 139: 27-38

Patocchi A, Vinatzer BA, Gianfranceschi S, Zhang H-B, Sansavini S, Gessler C

(1999) Construction of a 550-kb BAC contig spanning the genomic region containing the apple scab resistance gene *Vf*. Mol Gen Genet 262: 884-891

Peters JL, Cnudde F, Gerats T (2003) Forward genetics and map-based cloning approaches. Trends Plant Sci 8: 484-491

Ren C, Xu ZY, Sun S, Lee M-K, Wu C, Zhang H-B (2005) Genomic libraries for physical mapping. In: Meksem K and Kahl G (eds) The Handbook of Plant Genome Mapping: Genetic and Physical Mapping. Wiley-VCH Verlag GmbH, Weinheim, Germany, pp 173-213

Rommens JM, Iannuzzi MC, Kerem B, Drumm ML, Melmer G, Dean M, Rozmahel R, Cole JL, Kennedy D, Hidaka N, Zsiga M, Buchwald M, Riordan JR, Tsui LC, Collins FS (1989) Identification of the cystic fibrosis gene: chromosome walking and jumping. Science 245: 1059-1065

Takahashi Y, Shomura A, Sasaki T, Yano M (2001) *Hd6*, a rice quantitative trait locus involved in photoperiod sensitivity, encodes the α subunit of protein kinase CK2. Proc Natl Acad Sci USA 98: 7922-7927

Takken FLW, Joosten MHAJ (2000) Plant resistance genes: Their structure, function and evolution. Eur J Plant Pathol, 106: 699-713

Tanksley SD, Ganal MW, Prince JP, de Vicente MC, Bonierbale MW, Broun P, Fulton TM, Giovannoni JJ, Grandillo S, Martin GB, Messeguer R, Miller L, Paterson AH, Pineda O, Roder MS, Wing RA, Wu W, Young ND (1992) High density molecular linkage maps of the tomato and potato genomes. Genetics 132: 1141-1160

Tanksley SD, Ganal MW, Martin GB (1995) Chromosome landing: a paradigm for map-based cloning in plants with large genomes. Trends Genet 11: 63-68

Tao Q, Chang Y-L, Wang J, Chen H, Schuering C, Islam-Faridi MN, Wang B, Stelly DM, Zhang H-B (2001) BAC-based physical map of the rice genome constructed by restriction fingerprint analysis. Genetics 158: 1711-1724

Tao Q-Z, Zhang H-B (1998) Cloning and stable maintenance of DNA fragments over 300 kb in *Escherichia coli* with conventional plasmid-based vectors. Nucl Acids Res 26: 4901-4909

The Arabidopsis Genome Initiative (2000) Analysis of the genome sequence of the flowering plant *Arabidopsis thaliana*. Nature 408: 796-815

Tuinstra MR, Ejeta G, Goldsbrough PB (1997) Heterogeneous inbred family (HIF) analysis: a method for developing near-isogenic lines that differ at quantitative trait loci. Theor Appl Genet 95: 1005-1011

Turner A, Beales J, Faure S, Dunford RP, Laurie DA (2005) The pseudo-response regulator Ppd-H1 provides adaptation to photoperiod in barley. Science 310: 1031-1034

Tuskan GA, DiFazio S, Jansson S, Bohlmann J, Grigoriev I, Hellsten U, Putnam N, Ralph S, Rombauts S, Salamov A, Schein J, Sterck L, Aerts A, Bhalerao RR, Bhalerao RP, Blaudez D, Boerjan W, Brun A, Brunner A, Busov V, Campbell M, Carlson J, Chalot M, Chapman J, Chen G-L, Cooper D, Coutinho PM, Couturier J, Covert S, Cronk Q, Cunningham R, Davis J, Degroeve S, Déjardin A, dePamphilis C, Detter J, Dirks B, Dubchak I, Duplessis S, Ehlting J, Ellis B, Gendler K, Goodstein D, Gribskov M,

Grimwood J, Groover A, Gunter L, Hamberger B, Heinze B, Helariutta Y, Henrissat B, Holligan D, Holt R, Huang W, Islam-Faridi N, Jones S, Jones-Rhoades M, Jorgensen R, Joshi C, Kangasjärvi J, Karlsson J, Kelleher C, Kirkpatrick R, Kirst M, Kohler A, Kalluri U, Larimer F, Leebens-Mack J, Leplé J-C, Locascio P, Lou Y, Lucas S, Martin F, Montanini B, Napoli C, Nelson DR, Nelson C, Nieminen N, Nilsson O, Pereda V, Peter G, Philippe R, Pilate G, Poliakov A, Razumovskaya J, Richardson P, Rinaldi C, Ritland K, Rouzé P, Ryaboy D, Schmutz J, Schrader J, Segerman B, Shin H, Siddiqui A, Sterky F, Terry A, Tsai C-J, Uberbacher E, Unneberg P, Vahala J, Wall K, Wessler S, Yang G, Yin T, Douglas C, Marra M, Sandberg G, van de Peer Y, Rokhsar D (2006) The poplar genome was duplicated 60 to 65 million years ago, marking the emergence of this tree family, but overall has evolved more slowly than that of *Arabidopsis*. Science 313: 1596-1604

Uauy C, Distelfeld A, Fahima T, Blechl A, Dubcovsky J (2006) A NAC gene regulating senescence improves grain protein, zinc, and iron content in wheat. Science 314: 1298-1301

Wing RA, H-B Zhang, Tanksley SD (1994) Map-based cloning in crop plants: Tomato as a model system: I. Genetic and physical mapping of *jointless*. Mol Gen Genet 242: 681-688

Wu C, Sun S, Nimmakayala P, Santos FA, Springman R, Meksem K, Ding K, Lightfoot D, Zhang H-B (2004a) A BAC and BIBAC-based physical map of the soybean genome. Genome Res 14: 319-326

Wu C, Xu Z, Zhang H-B (2004b) DNA Libraries. In: Meyers RA (ed) Encyclopedia of Molecular Cell Biology and Molecular Medicine. edn 2, Vol 3. Wiley-VCH Verlag GmbH, Weinheim, Germany, pp 385-425

Wu C, Sun S, Lee M-K, Xu ZY, Ren C, Zhang H-B (2005) Whole genome physical mapping: An overview on methods for DNA fingerprinting. In: Meksem K and Kahl G (eds) The Handbook of Plant Genome Mapping: Genetic and Physical Mapping. Wiley-VCH Verlag GmbH, Weinheim, Germany, pp 257-284

Yahiaoui N, Srichumpa P, Dudler R, Keller B (2004) Genome analysis at different ploidy levels allows cloning of the powdery mildew resistance gene *Pm3b* from hexaploid wheat. Plant J 37: 528-538

Yan L, Loukoianov A, Tranquilli G, Helguera M, Fahima T, Dubcovsky J (2003) Positional cloning of the wheat vernalization gene *VRN1*. Proc Natl Acad Sci USA 100: 6263-6268

Yan L, Loukoianov A, Blechl A, Tranquilli G, Ramakrishna W, SanMiguel P, Bennetzen JL, Echenique V, Dubcovsky J (2004) The wheat *VRN2* gene is a flowering repressor down-regulated by vernalization. Science 303: 1640-1644

Yano M, Yuichi Katayose Y, Ashikari M, Yamanouchi U, Monna L, Fuse T, Baba T, Yamamoto K, Umehara Y, Nagamura Y, Sasaki T (2000) *Hd1*, a major photoperiod sensitivity quantitative trait locus in rice, is closely related to the *Arabidopsis* flowering time gene *CONSTANS*. Plant Cell 12: 2473-2484

Yu J, Hu S, Wang J, Wong GK-S, Li S, Liu B, Deng Y, Dai L, Zhou Y, Zhang X,

Cao M, Liu J, Sun J, Tang J, Chen Y, Huang X, Lin W, Ye C, Tong W, Cong L, Geng J, Han Y, Li L, Li W, Hu G, Huang X, Li W, Li J, Liu Z, Li L, Liu J, Qi Q, Liu J, Li L, Li T, Wang X, Lu H, Wu T, Zhu M, Ni P, Han H, Dong W, Ren X, Feng X, Cui P, Li X, Wang H, Xu X, Zhai W, Xu Z, Zhang J, He S, Zhang J, Xu J, Zhang K, Zheng X, Dong J, Zeng W, Tao L, Ye J, Tan J, Ren X, Chen X, He J, Liu D, Tian W, Tian C, Xia H, Bao Q, Li G, Gao H, Cao T, Wang J, Zhao W, Li P, Chen W, Wang X, Zhang Y, Hu J, Wang J, Liu S, Yang J, Zhang G, Xiong Y, Li Z, Mao L, Zhou C, Zhu Z, Chen R, Hao B, Zheng W, Chen S, Guo W, Li G, Liu S, Tao M, Wang J, Zhu L, Yuan L, Yang H (2002) A draft sequence of the rice genome (*Oryza sativa* L. ssp. *indica*). Science 296: 79-92

Zhang H-B, Martin GB, Tanksley SD, Wing RA (1994) Map-based cloning in crop plants: Tomato as a model system II. Isolation and characterization of a set of overlapping YACS encompassing the *jointless* locus. Mol Gen Genet 244: 613-621

Zhang H-B, Zhao X-P, Ding X-L, Paterson AH, Wing RA (1995) Preparation of megabase-size DNA from plant nuclei. Plant J 7: 175-184

Zhang H-B, Woo S-S, Wing RA (1996) BAC, YAC and Cosmid Library Construction. In: Foster G, Twell D (eds) Plant Gene Isolation: Principles and Practice. John Wiley & Sons, England pp 75-99

Zhang H-B, Wing RA (1997) Physical mapping of the rice genome with BACS. Plant Mol Biol 35: 115-127

Zhang H-B, Budman MA, Wing RA (2000) Gentic mapping of *jointless*-2 to tomato chromosome 12 using RFLP and RAPD markers. Theor Appl Genet 100: 1183-1189

9 | Bioinformatics: Fundamentals and Applications in Plant Genetics, Mapping and Breeding

David Edwards[1]* and Jacqueline Batley[2]

[1]Australian Centre for Plant Functional Genomics, Institute for Molecular Biosciences and School of Land, Crop and Food Sciences, University of Queensland, Brisbane, QLD 4072, Australia

[2]Australian Centre for Plant Functional Genomics, Legume Research and School of Land, Crop and Food Sciences, University of Queensland, Brisbane, QLD 4072, Australia

*Corresponding author: Dave.Edwards@acpfg.com.au

1 WHAT IS BIOINFORMATICS?

Bioinformatics is a relatively new field of research that has evolved from the requirement to process and apply the vast quantities of data being produced through the application of high throughput biotechnology. Bioinformatics is not yet mature and the definition of bioinformatics varies depending on the background of the person using the term. Applied bioinformatics may be loosely defined as 'the structuring of biological information to enable logical interrogation', and as such is inherent to virtually all crop research.

Applied plant bioinformatics is an integral part of crop research, and as such, the objectives of plant bioinformatics reflect the objectives of crop research, the improvement of crop production, increased yield and quality, and a reduction in the environmental footprint. Bioinformatics plays an essential role, from up-stream research in the sequencing and characterization of crop and model plant genomes, through to trait analysis and optimization of breeding strategies. This is demonstrated by the range of software packages, databases and web resources available (Table 1).

One of the challenges of applied bioinformatics is to take the information being produced by advances in gene and genome sequencing and use it for crop improvement. The gap between the genome and the phenome, between the genomic DNA sequence of a crop and what crop breeders and farmers measure in the field, remains to be closed. However, bridging technologies such as microarray gene expression analysis (transcriptomics), high-throughput protein and biochemical analysis (proteomics and metabolomics) supported by bioinformatics is gradually linking the genome and the phenome (Edwards and Batley 2004).

2 UNDERSTANDING THE GENOME

Bioinformatics evolved from the requirement to process, characterize and apply the information being produced by advanced high-throughput DNA sequencing technology. Since the development of the Sanger sequencing method, the production of DNA sequence data has continued to grow exponentially. At the same time, faster DNA sequence search methods have been combined with increasingly large computer systems to process this data.

Initial sequencing projects focussed on single-pass sequencing of expressed genes, to produce expressed sequence tags (ESTs). Genes are specifically expressed in tissues in the form of messenger RNA (mRNA). This mRNA is extracted and enzymatically reverse-transcribed to produce complementary DNA (cDNA), which is then cloned into plasmid vectors. The cloned cDNAs are then sequenced from one direction to produce the sequence tag. EST sequencing is a cost-effective method for the rapid discovery of gene sequences that may be associated with development or environmental responses in the tissues from which the mRNA was extracted (Adams et al. 1991). However, there are limitations to this method. Highly expressed genes are over-represented in the mRNA sample, and as the number of EST sequences produced from a cDNA library increases, fewer novel genes are identified. Thus, while EST sequencing is a valuable means to rapidly identify genes moderately or highly expressed in a tissue, the method rarely yields genes that are expressed at lower levels, including many genes that encode regulatory proteins that are only produced in very small quantities.

Table 1 Selected web resources for plant bioinformatics.

Programme	Website	Reference
ArrayExpress	www.ebi.ac.uk/arrayexpress/	Brazma et al. 2003; Parkinson et al. 2005
AtEnsembl	http://atensembl.arabidopsis.info	Hubbard et al. 2005; Love et al. 2006
BarleyBase/PLEXdb	www.barleybase.org/	Shen et al. 2005
BASC bioinformatics	http://bioinformatics.pbcbasc.latrobe.edu.au/basc/cgi-bin/index.cgi	Love et al. 2006; Erwin et al. 2007
BGI-RIS Rice Information System	http://rice.genomics.org.cn	Zhao et al. 2004
EMBL	www.ebi.ac.uk/embl/	Kanz et al. 2005
Gene Ontology Annotation (GOA)	www.ebi.ac.uk/GOA/	Camon et al. 2005
Graingenes	http://wheat.pw.usda.gov/GG2/index.shtml	Matthews et al. 2003; Carollo et al. 2004
Gramene	www.gramene.org/	Ware et al. 2002a,b
HarvEST	http://harvest.ucr.edu/	
ICIS	http://www.icis.cgiar.org:8080/	McLaren et al. 2005
KEGG	http://www.genome.jp/kegg/	Kanehisa et al. 2004
Legume Information System (LIS)	www.comparative-legumes.org/	Gonzales et al. 2005
Maize Genetics and Genomics Database (MaizeGDB)	www.maizegdb.org/	Lawrence et al. 2004
MetaCyc, AraCyc	http://metacyc.org/	Mueller et al. 2003; Krieger et al. 2004a, b; Zhang et al. 2005

Contd.

Contd.

MIPS Plant Genome Information Resources (PlantsDB)	*http://mips.gsf.de/*	Mewes et al. 2004
NASCArrays	*http://affymetrix.arabidopsis.info/*	Craigon et al. 2004
National Center for Biotechnology Information (NCBI)	*www.ncbi.nih.gov/*	
NCBI Gene expression Omnibus (GEO)	*www.ncbi.nlm.nih.gov/geo/*	Barrett et al. 2005
PlantGDB	*www.plantgdb.org/*	Dong et al. 2004; Dong et al. 2005
SNPServer	*http://hornbill.cspp.latrobe.edu.au/ snpdiscovery.html*	Savage et al. 2005
SSR Taxonomy Tree	*http://bioinformatics.pbcbasc.latrobe.edu.au/ cgi-bin/ssr_taxonomy_browser.cgi*	Jewell et al. 2006
SwissProt	*www.ebi.ac.uk/swissprot/*	Boeckmann et al. 2003
The Arabidopsis Information Resource (TAIR)	*www.arabidopsis.org/*	Huala et al. 2001; Garcia-Hernandez et al. 2002; Rhee et al. 2003; Weems et al. 2004; Reiser and Rhee 2005
The Institute for Genomic Research (TIGR)	*www.tigr.org/*	Lee et al. 2005
UniProt Knowledgebase UniProtKB	*http://www.uniprot.org*	Bairoch et al. 2005

An alternative to EST sequencing is genome sequencing, aiming to sequence either the whole genome or portions of the genome. This method removes the bias associated with gene expression level. However, genomes and particularly crop plant genomes tend to be very large. Genome sequencing has only become feasible with the development of capillary-based high-throughput sequencing technology, and the recent development of pyrosequencing technology suggests that genome sequencing will become increasingly common (Margulies et al. 2005). The first plant genome to be sequenced was that of the model plant *Arabidopsis thaliana* (The Arabidopsis Genome Initiative 2000). More recently, the sequence of larger crop genomes has been attempted, with the completion of the genome sequence for the model monocot and major crop, rice (Li et al. 2002; Sasaki et al. 2002). Ongoing sequencing projects for crop and non-crop plant species include *Brassica, Medicago,* lotus, tomato, potato, poplar, soybean, Capsella, papaya, Eucalyptus, grape, *Mimulus guttatus, Triphysaria versicolor,* banana, sorghum, maize and wheat (Jackson et al. 2006). The increasing availability of plant genome information will greatly enhance our understanding of these plants and assist in the improvement of a wide variety of crops. In addition, the genome sequence for numerous plant pathogens, including *Agrobacterium tumefaciens, Burkholderia cenocepacia, Clavibacter michiganensis, Erwinia carotovora, Erwinia chrysanthemi, Leifsonia xyli, Onion yellows phytoplasma, Pseudomonas syringae, Ralstonia solanacearum, Spiroplasma kunkelii, Xanthomonas axonopodis, Xanthomonas campestris* and *Xylella fastidiosa* are being elucidated, and this information will lead to improved methods for reducing the impact of these and related pathogens on crop production.

DNA sequence data is crucial to our understanding of crop growth and development, as the sequence of the genes, or allelic variants of the sequenced genes, are responsible for almost all of the heritable differences between crop varieties and ecotypes. This information, often referred to as the genetic blueprint, is the foundation for all additional information from the genome to the phenome.

One challenge of genome sequencing lies in the fact that only a small portion of the genome encodes genes, and that these genes are surrounded by regulatory elements that are often difficult to characterise, as well as repetitive or 'junk' DNA. Several statistical methods have been applied to identify candidate genes within genomic sequence, and these are being refined as additional gene and genome sequence becomes available, providing evidence of gene expression and conservation of functional parts of the genome.

3 GENE AND GENOME SEQUENCE DATABASES

DNA sequence data often forms the core of bioinformatics systems and several systems have been developed for the maintenance, annotation and interrogation of DNA sequence information. The largest of the DNA sequence repositories is the International Nucleotide Sequence Database Collaboration (INSDC), made up of the DNA Data Bank of Japan (DDBJ) at The National Institute of Genetics in Mishima, Japan (Ohyanagi et al. 2006), GenBank at the National Center of Biotechnology Information in Bethesda, USA (Benson et al. 2006), and the European Molecular Biology Laboratory (EMBL) Nucleotide Sequence Database, maintained at the European Bioinformatics Institute (EBI) in the UK (Cochrane et al. 2006). The Institute for Genomic Research (TIGR) based at Rockville, Maryland, USA (Lee et al. 2005) maintain various data types including genomic sequences and annotation, while MIPS, the Munich Institute for Protein Sequences (MIPS), Germany, hosts another comprehensive database resource and maintains genome databases for *Arabidopsis thaliana* (MatDB), *Oryza sativa* (Rice, MosDB), *Medicago truncatula* (UrMeLDB), *Lotus japonicus*, *Zea mays* and *Solanum lycopersicum* (Tomato) (Mewes et al. 2004).

As well as the major DNA sequence databases, several smaller, more specialized databases have been developed, supporting either specific sequence data analysis or species. Arabidopsis, as the first plant to be fully sequenced and the model species for studies of dicotyledonous crops, is supported by a comprehensive array of databases that have evolved with improvements in bioinformatics systems and the growing quantities and formats of data. The Arabidopsis Information Resource (TAIR) is a collaborative project between the Carnegie institution of Washington; Department of Plant Biology, Stanford, California and the National Centre for Genome Resources (NCGR), Sante Fe, New Mexico, and provides an extensive web-based resource for *Arabidopsis thaliana* (Huala et al. 2001; Rhee et al. 2003; Weems et al. 2004; Reiser and Rhee 2005). Data includes gene, genetic mapping, protein sequence, gene expression and community data within a relational database. Several tools are available for viewing and analyzing this data including SeqViewer (for the visualization of the genome sequence and associated annotations). The Nottingham Arabidopsis Stock Centre (NASC) has developed an Arabidopsis genome viewer (Love et al. 2006) based on the EnsEMBL browser system (Birney et al. 2004; Hubbard et al. 2005) providing a genome based view of integrated Arabidopsis data. This resource provides a broad range of EnsEMBL features including gene and protein information, links to Affymetrix gene expression data, pointers to germplasm, and extensive data download capabilities.

Brassica vegetable and oilseed crops are closely related to the model plant Arabidopsis, and information from Arabidopsis may readily be used to understand Brassica. The Brassica BASC system provides an integrated resource for querying across a wide range of genetic, genomic and phenotypic data for Brassica and Arabidopsis (Love et al. 2006; Erwin et al. 2007). This system provides access to genetic mapping, functional annotation, comparative genomics and gene expression information with comparison to Arabidopsis. The BASC resource is based on five distinct modules, with three of these providing access to EST, MarkerQTL and microarray gene expression data. Two further modules include an Arabidopsis EnsEMBL genome viewer (Love et al. 2006) and the CMap comparative genetic map viewer (Ware et al. 2002b).

The Legume Information System (LIS) (Gonzales et al. 2005) has been developed to support legume crop improvement and enables the application of genomic information from the model legume *Medicago truncatula* (barrel medic) for the improvement of major crops such as soybean. LIS includes genetic and physical maps as well as annotated ESTs for several legume species enabling the comparison of QTLs, genes and genome sequences.

The HarvEST system is an EST database and viewing software initially developed for cereals and now supporting several species including barley, *Brachypodium*, citrus, Coffea, cowpea, rice, soybean and wheat (*http:/harvest.ucr.edu/*). HarvEST supports gene expression microarray content design, gene function annotation and interfacing with physical and genetic maps. One feature of this system is that it can be run on a stand alone laptop with a user friendly interface without the need for internet connectivity or a significant amount of computing power.

The Gramene Genome Browser (Ware et al. 2002a; Jaiswal et al. 2006) hosts an implementation of the EnsEMBL Genome Browser (Birney et al. 2004; Hubbard et al. 2005) containing collated information on the rice genome and rice genes. This information includes predicted function as assigned by GO annotations (Clark et al. 2005; Lomax 2005) and Pfam domain similarity (Bateman et al. 2004), SNP polymorphisms and rice phenotypes (as quantitative trait loci; QTL). Comparative genome information is provided by the alignment of plant ESTs with the best rice ortholog, and maize synteny information based on the maize physical map. The front web page acts as portal to internal pages in Gramene as well as external links to other databases. Internal links within Gramene include Mapviewer, an implementation of CMap (Ware et al. 2002b), Markers, Genes, QTL and an Ontology browser.

The Beijing Genome Institute (BGI) has established and recently updated a Rice Information System (BGI-RIS) that integrates information and resources for the comparative analysis of rice genomes (Zhao et al. 2004). BGI-RIS combines the genomic data of *Oryza sativa* L. ssp. *indica* (produced by BGI), with *Oryza sativa* L. ssp. *japonica*, along with annotation, genetic markers, expressed genes, repetitive elements and genomic polymorphisms, using graphical interfaces.

MaizeGDB (Lawrence et al. 2004) is a relatively new database system that combines information from the original MaizeDB and ZmDB (Dong et al. 2003) repositories with sequence data from PlantGDB (Dong et al. 2004, 2005). The system maintains information on maize genomic and gene sequences, genetic markers, literature references, as well as contact information for the maize research community. In contrast to many systems, MaizeGDB also hosts manually curated information from primary literature, bulletin boards and many links related to maize research.

4 GENE EXPRESSION INFORMATION

The growing quantity of gene and genome sequence data enables the examination of where and when genes are expressed. This information is valuable as the regulation of gene expression during development or in response to environmental cues provides evidence of the function of genes. Several methods have been developed to measure gene expression, and these methods have increasingly become high-throughput, moving from the labourious one gene at a time hybridization based methods, such as northern blotting and RT-PCR, to massively parallel methods, such as microarray gene expression (Brazma et al. 2003; Craigon et al. 2004; Barrett et al. 2005; Parkinson et al. 2005; Shen et al. 2005; Tang et al. 2005), serial analysis of gene expression (SAGE) (Velculescu et al. 1995; Matsumura et al. 1999; Gibbings et al. 2003; Lee and Lee 2003; Fizames et al. 2004; Robinson et al. 2004b; Poroyko et al. 2005) and massively parallel signature sequencing (MPSS) (Brenner et al. 2000a, b; Meyers et al. 2004a, b; Lu et al. 2005).

Microarray based measurement of gene expression is based on the fixing of DNA features onto a solid support, followed by hybridization with labelled expressed genes representing a pool of mRNA. The DNA features may be in the form of short stretches (20-70 base pairs) of single stranded DNA, each representing individual genes, printed onto glass slides or synthesized in situ on other supporting media such as silicon wafers. Alternatively, the features may represent longer stretches of gene sequences derived from amplified cDNAs. In both cases, the measurement of gene expression relies on hybridization between the labelled message and the DNA on the slide. The amount of hybridizing label is quantified and translated into a gene expression value. As many steps are required to derive gene expression data using microarrays, great care must be taken in interpreting these values. Microarray data is frequently, but not directly comparable between different microarray systems and standardization of experimental procedures is essential for valid interpretation.

The SAGE and MPSS methods for gene expression quantification are based on sequencing rather than hybridization. These methods overcome some of the limitations and bias associated with hybridization based methods of gene expression measurement. Sequence based methods measure the number of times a specific gene is identified within a pool of expressed genes in any specific tissue. Both SAGE and MPSS read short representative fragments of expressed gene sequences. For highly expressed genes, the representative sequence fragment will be found abundantly in the pool, while genes that are expressed at lower levels will be represented by less abundant sequence fragments. Sequence based methods are generally more quantitative with a greater potential dynamic range than hybridization methods.

With the above considerations, microarrays still remain one of the most cost effective methods for the production of gene expression data. A great deal of bioinformatics research is dedicated to the interpretation and analysis of microarray data, partly to manage the quantity of information being produced, and also to develop robust statistical methods for their analysis.

5 GENE EXPRESSION DATABASES

Due to the inherent variability of microarray experiments, standards have been developed to assist in the comparison of various data produced by different groups using different microarray platforms. The MIAME standards describe the 'minimum information about a microarray experiment' that is needed to enable the interpretation of the results of the experiment unambiguously and potentially to reproduce the experiment (Brazma et al. 2001). The MIAME checklist is a condensed description of MIAME principles, designed to help authors, reviewers and editors of scientific journals to meet MIAME requirements and to make microarray data available to the community in a useful way. While standards describing microarray experiments are generally agreed, the format for the storage, interrogation and analysis of gene expression data may be quite varied. Some of the more advanced databases for crop gene expression data incude BarleyBase and the more recent PLEXdb (Plant Expression Database) systems for analysing gene expression in plants and plant pathogens (Shen et al. 2005). PLEXdb provides a web interface integrating microarray data from several plant species enabling the comparative analysis of gene expression and providing an insight into the expression of important agronomic genes. In addition, the Nottingham Arabidopsis Stock Centre (NASC) hosts tools for the interrogation and analysis of large quantities of Arabidopsis gene expression information produced using the Affymetrix array platform (Craigon 2004). This information is integrated with the Arabidopsis EnsEMBL genome browser (Love et al. 2006).

6 THE PROTEOME AND METABOLOME

The ability to convert the expressed genes in the form of mRNA into DNA through the process of reverse transcription, followed by the amplification of this DNA by the polymerase chain reaction (PCR) has greatly assisted the methods for gene expression analysis. The lack of such amplification method for proteins and metabolites has likewise hindered the study and analysis of these important steps in the translation of the genome to the phenome. While there is a direct translation between the gene sequence and the protein sequence through the codon translation code, the function and biochemical role of a protein may not be readily derived from the sequence of amino acids alone.

With the development of metabolomic technologies and the increased interest in molecular pharming, there has been an expansion in the application of bioinformatics tools for the analysis of this information and its integration with genetic and genomic data. KEGG (Kyoto Encyclopedia of Genes and Genomes) is a key bioinformatics resource for the integration of biochemical pathways with genetic and genomic information (Kanehisa et al. 2004). The KEGG databases enable the visualization of complex cellular pathways such as metabolism, signal transduction and the cell cycle, frequently with graphical pathway diagrams. In addition to KEGG, the AraCyc database of biochemical pathways provides a graphical overview of metabolic processes for the model plant Arabidopsis (Mueller et al. 2003; Krieger et al. 2004; Zhang et al. 2005).

Large-scale metabolite studies require their own specific systems for data analysis and integration. While this field of research is still in its relative infancy, several tools have already been developed and the area is likely to expand greatly in the coming years (Fiehn 2002; Sumner et al. 2003).

7 THE PHENOME

The final link in the bioinformatics from the genome to the phenome is the storage, analysis and integration of phenotypic information. This information is generally collated by breeders and agronomists, who have traditionally applied phenotypic information for crop improvement. The International Crop Information System (ICIS) is arguably the most advanced database of this type and enables breeders to load and store information pertaining to genealogy and phenotype (McLaren et al. 2005). ICIS was initially developed by the International Rice Research Institute (IRRI) bioinformatics and biometrics unit and is widely used for rice and wheat and is being extended to include information on legumes. Data are predominantly provided by the International Centre for Agricultural Research in the Dry Areas (ICARDA), International Crop Research Institute for the Semi-Arid Tropics (ICRISAT), as well as the United States Department of Agriculture (USDA). Databases may be queried across multiple datasets to identify varieties which have agronomic qualities such as disease resistance, yield, and response to stress.

TASSEL (Trait Analysis by aSSociation, Evolution and Linkage) software applies trait information and evaluates linkage disequilibrium, nucleotide diversity, and trait associations. It works with many different types of diversity data and connects to the Panzea and Gramene data repositories.

The GERMINATE database is a modular system and includes a Data Integration Module to store information on the nature of the plant samples. A Passport Module stores Multi Crop Passport Descriptors (MCPDs), standards developed jointly by the Food and Agriculture Organization (FAO) and the International Plant Genetic Resource Institute (IPGRI). Passport data include geographical information about accession collection sites, using a data format, which is consistent with Geographic Information System (GIS) programs. The Datasets Module stores genotypic (marker), phenotypic and trait data for accessions in the database, while the ICIS-GMS Module stores and manages pedigree and list information, and has been adapted from the Genealogy Management System (ICIS GMS) module of the ICIS database.

8 FROM GENETICS AND GENOMICS TO PHENOMICS

With the volume of genetic and genomic information being produced by the technologies described above, there is an increasing ability to utilize this information for the improvement of crops. The linking of the heritable material, the DNA sequence of genes and genomes that is passed from one generation to another, with the traits that are associated with the possession of this DNA, is one of the greatest challenges facing biological scientists today. Bioinformatics plays a key role in linking the genome information to the observed phenotype or phenome, bringing together and integrating the information to enable logical interrogation and knowledge discovery by researchers.

The genetic improvement of crops may be through either genetic transformation methods or traditional breeding and selection. Genetic transformation requires an understanding of the gene associated with, and responsible for a trait, while traditional breeding requires only an understanding of the inheritance of the trait. Genomic information may be used to assist traditional breeding by the use of molecular genetic markers. These markers may be considered as flag posts on the genome, allowing researchers and breeders to examine genetic locations and their inheritance during breeding. Markers are differences between individuals at specified locations on the genome, and at their most basic level, represent a difference in DNA sequence between individuals. There are several forms of molecular genetic markers, simple sequence repeats (SSRs), also known as microsatellites, and single nucleotide polymorphisms (SNPs) are two of the most common advanced marker systems in use.

9 MOLECULAR GENETIC MARKERS: SSRs AND SNPs

Recent advances in molecular biology have led to a range of molecular genetic marker techniques that can be used for the screening, characterization and evaluation of diversity in genetic material, and be applied to accelerate and redesign the breeding process. For example, in marker-assisted breeding, markers associated with important characters may replace phenotypic screening. DNA profiling may also be applied for cultivar discrimination and seed purity analysis, genetic mapping and for the genetic diversity analysis of germplasm collections. Molecular marker technology provides a direct means to detect the polymorphisms that arise from changes in the DNA sequence, and can be applied to predict lines most suitable for the generation of new cultivars containing genes and traits of interest. Diversity analysis can identify lines that are genetically dissimilar and can be used to generate a diverse cross or identify a source of novel alleles for genes of agronomic importance.

The development of high-throughput methods for the detection of SNPs and SSRs has led to a revolution in their use as molecular markers.

SNPs and SSRs are frequently the markers of choice in genetic analysis and are used routinely as markers in agricultural breeding programs. They also have many uses in animal and human genetics, such as for the detection of alleles associated with genetic diseases and the identification of individuals. SNPs and SSRs are invaluable as a tool for genome mapping, offering the potential for generating very high-density genetic maps, which can be used to develop haplotyping systems for genes or regions of interest.

SNPs may be considered the ultimate molecular marker since they represent the finest point of genetic variation. In their simplest form, an SNP is a single nucleotide base difference between two DNA sequences. SNPs can be categorized according to nucleotide substitution as either transitions (C/T or G/A) or transversions (C/G, A/T, C/A, or T/G). SNPs are also the most frequent genetic polymorphism, and in any, but the most closely related varieties, the potential number of such markers is enormous (Rafalski 2002). These variations in sequence can have a major impact on how the organism responds to disease and environmental stress (toxins, chemicals, etc). The low mutation rate of SNPs makes them excellent markers for studying complex genetic traits and as a tool for the understanding of genome evolution (Syvanen 2001). SNPs, at any particular site could in principle involve four nucleotides, but in practice they are generally only bi-allelic. However, this disadvantage is offset by their abundance. Information on the frequency, nature and distribution of SNPs in plant genomes is limited. However, the development and application of SNPs in higher plants is increasing, including some crop and tree species, and consequently this is an attractive marker to plant breeders and geneticists. With the increasing availability of public sequence data and the rapid discovery of SNPs in crop species, the development and application of SNP markers will continue to accelerate.

SSRs are short stretches of DNA sequence occurring as tandem repeats of mono-, di-, tri-, tetra- and penta- nucleotides (Engel et al. 1996). However, dinucleotide repeats are the most abundant SSRs found in both plant and animal genomes (Gupta et al. 1996). SSRs have been found in every types of organisms tested so far, including mammals (Di Rienzo et al. 1994), birds (Ellegren 1992; Moran 1993), fish (Rico et al. 1993), insects (Batley et al. 1998), plants (Plaschke et al. 1995) and fungi (Owen et al. 1998). Within eukaryotic genomes, SSRs appear to be more or less uniformly and randomly distributed (Moore et al. 1991) and one would expect to encounter an SSR every 10 kb of sequence (Tautz 1989). However, they are under-represented in coding regions (Cardle et al. 2000) and demonstrate an increased abundance up-stream of expressed genes (Mortimer et al. 2005). The uniqueness and value of SSR markers arises from their multi-allelic and somatically stable nature (Powell et al. 1996) and their co-dominant Mendelian transmission (Litt and Luty 1989; Morgante and Olivieri 1993). These characteristics make them ideal genetic markers for population genetic studies, genome mapping (Dietrich et al. 1992), germplasm analysis (Teulat et al. 2000), disease diagnostics (Bocker et al. 1999), evolutionary studies (Schlötterer et al. 1991; Buchanan et al. 1996; Cruzan 1998), paternity and kinship analyses (Kichler et al. 1999) and clone, strain or varietal identification (Powell et al. 1996).

The polymorphism of SSRs is revealed by PCR-amplification from total genomic DNA using two unique primers, composed of short stretches of nucleotides that flank, and hence define the SSR locus. The polymorphism detected is due to the changes in length of the repeat motif and therefore, the size of the amplified product. Although the same SSR repeat motif may be present in many places within the genome, the stretch of DNA either side of the repeat motif generally represents a unique sequence. It is, therefore, possible to design and select primers complementary to this flanking region that specifically amplify the SSR locus. Amplification products obtained from different individuals are resolved by gel or capillary electrophoresis to reveal this polymorphism, and a resolution of differences of one base pair may be observed (Bruford and Wayne 1993; Ferguson et al. 1995) (Figure 1).

Several tools have been developed for the discovery of molecular genetic markers from DNA sequence data. These tools provide a rich source of markers that can then be applied to genetic trait mapping and marker-assisted selection. SSR Primer is a web-based tool that enables the real time discovery of SSRs within submitted DNA sequences, with the concomitant design of PCR primers for SSR amplification (Robinson et al. 2004a). Alternatively, users may browse an SSR Taxonomy Tree (Jewell et al. 2006) to identify pre-determined SSR amplification primers for any species represented within the GenBank database. This system currently hosts 14 million SSR primer pairs for organisms from simple viruses and bacteria to plants and animals.

The SNP discovery software autoSNP (Barker et al. 2003; Batley et al. 2003) identifies SNPs and insertion/deletion (indel) polymorphisms from bulk sequence data using two measures of confidence; redundancy, defined as the number of times a polymorphism occurs at a locus in a sequence alignment; and co-segregation of SNPs to define a haplotype. The software is written in perl for linux or UNIX and works with sequence assembly software such as CAP3 (Huang and Madan 1999), D2/CAP3 (Burke et al. 1999) or TGICL (Pertea et al. 2003). SNPServer

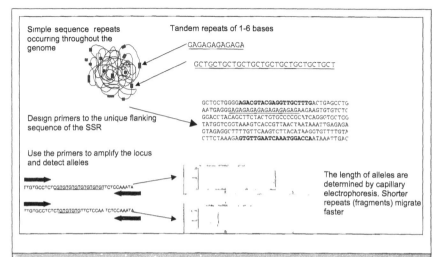

Fig. I   Identification and analysis of simple sequence repeats (SSRs) as molecular genetic markers.

(Savage et al. 2005) is a web-based tool for the real-time discovery of SNPs from DNA sequence data based on the stand alone autoSNP. Following submission of a sequence of interest, SNPServer uses BLAST (Altschul et al. 1990), to identify similar sequences, CAP3 (Huang and Madan 1999) to cluster and assemble these sequences, and then the SNP discovery software autoSNP to detect SNPs and insertion/deletion (indel) polymorphisms. Where the sequence trace files are available, the SNP discovery tool PolyPhred (Bhangale et al. 2006; Stephens et al. 2006) can make use of the base pair quality scores to further differentiate between true SNP polymorphisms and random sequence error. Each of these tools increases the availability of these molecular markers for genetic studies and allows researchers to target molecular markers within specific candidate genes.

10   MOLECULAR MARKER DATABASES

The increased throughput for the discovery and application of molecular genetic markers has led to the requirement for databases hosting the results of molecular marker analysis. These may be integrated within other database systems such as the Brassica BASC MarkerQTL module (Erwin et al. 2007), MaizeGDB (Dong et al. 2003), the Gramene Genome Browser (Ware et al. 2002a; Jaiswal et al. 2006), the Legume Information System (LIS) (Gonzales et al. 2005) and The Arabidopsis Information Resource (TAIR) (Huala et al. 2001; Rhee et al. 2003; Weems et al. 2004; Reiser and Rhee 2005).

One of the most advanced plant molecular genetic databases is GrainGenes (Matthews et al. 2003; Carollo et al. 2005). Molecular genetic information is maintained in this database for barley, wheat, rye, oat and related wild species. GrainGenes integrates genetic data for *Triticeae* and *Avena* that have been produced over several decades, and links this information with the Gramene genome browser (Ware et al. 2002a; Jaiswal et al. 2006). Comprehensive information includes genetic markers, map locations, alleles, key references and disease symptoms.

One of the principal uses of molecular genetic markers is the production of genetic maps and the mapping of heritable traits. While mapping data may be described as lists, graphical representations are more readily understood. CMap, developed by the GMOD consortium (Ware et al. 2002b) is becoming the standard software for genetic map visualization and comparison. CMap enables the comparison of genetic maps that have corresponding mapped loci. This correspondence may be direct, where the locus was identified using the same marker for each map. Alternatively, the locus correspondence may be indirect, for example where there is sequence identity between sequence-based genetic markers and a genome sequence. In either case, comparative genetic mapping using CMap is valuable for the validation of traits that map to the same position in different populations and also for the linkage between crop genetic maps and sequenced model genomes, enabling the identification of candidate genes in the sequenced model species that may be related to as yet unidentified genes within the crop species.

11 MOLECULAR MARKER APPLICATIONS

The application of molecular markers to advance plant breeding is now well established. Modern breeding applies molecular markers for trait identification and marker-assisted breeding and selection. Molecular markers can be used to reduce linkage drag in backcrossing and select for traits that are difficult to score using phenotypic assays. Molecular markers are complementary tools to traditional selection, they can help in increasing the knowledge of the inheritance of selected traits and their genetic association, which in turn may modify breeding strategies or objectives.

The identification of abundant, highly variable loci, detectable by a wide variety of high throughput techniques, has led to a requirement for the rapid analysis of genetic diversity and mapping data. The evaluation of genetic variation ranges from the detection of allelic associations within and between loci, through sub-populations, populations, cultivars, species and inter-species comparisons.

12   POPULATION GENETIC SOFTWARE PACKAGES

A variety of analyses can be conducted on molecular data to estimate parameters concerning population structure. Population estimates of allele and genotype frequencies are normally first tested for correspondence with Hardy-Weinberg equilibrium (HWE). A number of parameters are then estimated, such as the intra- and inter-population genetic diversity (heterozygosity and proportion of polymorphic loci), the extent of interaction between the populations, level of inbreeding and the degree of population sub-structuring. These analyses can be performed using specifically designed computer software programs, a brief description of the main ones is given below. A comprehensive list of all types of population genetic software is also provided on the Rockefeller University Linkage Software List (*http://linkage.rockefeller. edu/ soft/*).

- **AMOVA** (Analysis of molecular variance): This program is only available for Microsoft windows platforms. The program can handle co-dominant data and includes computation of variance components, F-statistics, co-ancestry co-efficients and permutation tests for population statistics. This software is available from: *http://anthropologie.unige.ch/ftp/comp/win/amova/*
- **Arlequin** (Excoffier et al. 2005) is able to handle large samples of molecular data (RFLPs, DNA sequences, SSRs) while retaining the capacity of analyzing conventional genetic data (standard multi-locus data or mere allele frequency data), and boasts a range of features, including intra-population data analysis (diversity indices, distance estimates, maximum-likelihood estimates of allele and haplotype frequencies, tests of selective neutrality) and inter-population tests of genetic structure. The web page (*http://anthropologie.unige.ch/arlequin/*) hosts a Frequently Asked Questions (FAQ) page and bug reports.
- **GENEPOP** (Raymond and Rousset 1995) performs three major tasks, it computes exact tests for HWE, for population differentiation and for genotypic disequilibrium among pairs of loci, it computes estimates of classical population parameters such as Fst, and it converts the GENEPOP file to formats used by other programs. This software is available from: *ftp.cefe.cnrs-mop.fr, http:/ /wbiomed.curtin.edu.au/genepop/*
- **MicroSat** incorporates estimates of genetic distance, each marked by a specific model of evolutionary change. It computes estimates of linearity and diversity measures for each locus. The resultant distance matrices can be incorporated into PHYLIP as input for

cluster analysis and phylogenetic reconstruction. This software is available from: *http://lotka.stanford.edu/microsat*

- **CONVERT** is a program that facilitates transfer of genotypic data amongst commonly used population genetic software packages. CONVERT reads input files of co-dominant marker data, and can convert these to the input formats of GDA, GENEPOP, ARLEQUIN, POPGENE, MICROSAT, PHYLIP, and STRUCTURE. This software is available from: *http://www.agriculture.purdue. edu/ fnr/html/faculty/Rhodes/Students%20and%20Staff/glaubitz/ software.htm*

13   GENETIC DIVERSITY

A common objective of many genetic studies is to describe the amount of genetic variation present within a population. Genetic variation is an important feature of populations, both for short-term fitness of individuals and the long-term survival of the populations, allowing adaptation to changing environmental conditions to occur. An understanding of genetic diversity is also important for establishing breeding strategies for crop improvement.

Genetic diversity can be estimated in a number of ways. One measure is simply the number of polymorphic loci, where a locus is usually considered to be polymorphic if the most common variant has a frequency less than or equal to 0.95. Another, more informative estimate, is average heterozygosity, which is calculated by obtaining the frequency of heterozygotes at each locus and then averaging these frequencies across all loci. A more accurate measure, which takes into account variation between loci is Nei's Gene Diversity (Nei 1987). This is equivalent to the expected heterozygosity. However, genetic change in populations is usually described by changes in allele frequencies. Allele frequencies are preferred over genotypic frequencies because allele frequencies remain relatively stable over time, in contrast to genotypic frequencies, which are more randomized at each new generation.

14   HARDY-WEINBERG EQUILIBRIUM (HWE)

The Hardy-Weinberg principle remains a basic law in population genetics. It was formulated independently, in 1908, by the British mathematician Hardy and the German doctor Weinberg (Hardy 1908; Weinberg 1908). The Hardy-Weinberg law states that in an infinitely large, random mating population, there is a fixed relationship between gene and genotype frequencies such that in the absence of selection, mutation and migration, these frequencies remain constant from generation to generation. This state can be achieved after only a single generation of random crossing. Populations in which the observed genotype frequencies conform to those expected are said to be in 'Hardy-Weinberg Equilibrium' and are sometimes referred to as panmictic (randomly crossing) populations. It is most often the case that more than one of the conditions required for HWE, given above, are not met.

Observed genotype frequencies are usually tested for concordance with the Hardy-Weinberg rule using a chi-square test. Alternative hypotheses are either 'heterozygote deficit' or 'heterozygote excess'. This test is able to detect both small and large deviations from HWE. Deviations from HWE indicate that there is selection operating when genotypes have different probabilities of being included in the sample, crossing is not random or the sample is not from a single population.

15 LINKAGE DISEQUILIBRIUM

Linkage disequilibrium has been noted in many genetic studies of many species (Grice et al. 1996; Goldstein and Weale 2001; Narvel et al. 2001) and is described as the non-random association between alleles at different genetic loci. Linkage is the correlated inheritance and linkage disequilibrium is a correlation between alleles in a population or cross, usually brought about by different alleles being in close chromosomal proximity. Also known as gametic disequilibrium, it is a possible cause of the skewed segregation of molecular markers in a genetic mapping population. Linkage disequilibrium may be caused by a recent mutation, selection or population structure and is frequently applied to identify associations between genes, or between genetic loci and traits.

16 POPULATION SUBDIVISION

Most commonly, species are made up of a number of local populations, which may differentiate from one another depending on the amount of isolation and gene flow between them. It is important to be able to measure the degree to which populations are differentiated and subdivided. Subdivision of a population into smaller units results in a reduction in heterozygosity.

Population subdivision can be measured using Wright's F-statistics with the following fixation indices, F_{IS} and F_{ST} F_{IS}, the inbreeding coefficient, is a measure of the reduction in individual heterozygosity, due to deviations from random crossing in subpopulations relative to regional groupings. A higher value indicates more inbreeding and more homozygotes, a lower value indicates more random mating in the population and therefore, more heterozygotes. F_{ST} provides a measure of population differentiation. It is the reduction of heterozygosity in subpopulations, due to non-random crossing, relative to the total population size. F_{ST} can be used to infer genetic insights about population structure. F_{ST} alone does not give an indication of causal influences, it is still a commonly used estimator of differentiation. Wright suggested quantitative guidelines for the interpretation of F_{ST} values (Table 2) (Wright 1978).

Table 2 Wright's guidelines for the interpretation of F_{ST} values.

F_{ST} value	Level of differentiation
0-0.05	Little genetic differentiation
0.05-0.15	Moderate genetic differentiation
0.15-0.25	Great genetic differentiation
> 0.25	Very great genetic differentiation

There are a number of estimators of F_{ST} based on allele size frequencies, θ is often chosen because it is the most widely used and least biased statistic. Weir and Cockerham's θ (Weir and Cockerham 1984) is analogous to F_{ST} but is independent of sample size, number of alleles at a locus and number of subpopulations sampled.

In applications of population genetics, where the structure of populations is unknown, it is useful to classify individuals in a sample into populations (Pritchard et al. 2000). Using Bayesian assignment programs, estimates of allele frequencies in each population, at a series of unlinked loci are used to compute the likelihood that a given genotype originated in each population (Cornuet et al. 1999). It is assumed that there is an equal likelihood of samples originating from between 1-7 source populations (where k is the number of source populations represented in the sample). Individuals in the sample are assigned to populations, providing a posterior distribution of k, indicating the actual number of populations and degree of substructuring (Dawson and Belkhir 2000).

17 PHYLOGENETIC ANALYSIS

For analysis of germplasm collections and the relationship between accessions, phylogenetic reconstruction programs can be applied to estimate the evolutionary history of alleles between the different cultivars, and is powerful for determining patterns of association. For this analysis, inferences are made about the evolutionary relationships among the cultivars from which the data were obtained, resulting in a phylogenetic tree. All phylogenetic methods have an optimality criterion to evaluate how well alternative trees fit the actual data.

17.1 Phylogenetic Analysis Software Packages

In order to analyse the phylogenetic relationship between cultivars, a number of software packages are freely available for downloading from the web. A list of these and their websites is provided below:

- **PHYLIP** (PHYLogeny Inference Package) is one of the earliest and remains one of the most useful computer packages produced for phylogenetic inference. It consists of over 30 programs which can analyze sequence data, distance matrix data, gene frequencies, discrete or continuous character data and programs for plotting phylogenetic trees. PHYLIP's website contains a wide variety of useful information about phylogenetic reconstruction and other programs related to phylogenetic inference. This software is freely available from *http://evolution.genetics.washington.edu/phylip*

- **PAUP** conducts phylogenetic analysis under the evolutionary assumptions of maximum parsimony. Data are treated as discrete characters with the assumption that the derived phylogenetic tree of the shortest (most parsimonious) length reflects the true phylogeny. The application of maximum parsimony to population data consisting of allele frequencies is not robust. This algorithm is designed for discerning changes in molecular data which reflect a shared, derived pattern of evolution. The new versions of PAUP also perform distance-based and maximum likelihood analysis and these are more versatile and can be used for a wider variety of data types. The program is not freely available, however the cost is only a few hundred dollars (Swofford 1993).

- **TreeView** (Page 1996) is a very convenient program for viewing phylogenetic trees or making publication quality figures from standard output from programs such as PAUP or PHYLIP. The website (*http://taxonomy.zoology.gla.ac.uk/rod/treeview*) also contains software support for the computer program and provides bug fixing tips and related information.

18 APPLICATION OF GENOTYPIC DATA TO GENETIC MAPPING ANALYSIS

Insight into the organization of the plant genome can be obtained by assembling a genetic linkage map using molecular markers, such as SSRs and SNPs. Genetic mapping is the process by which genomic locations (loci) that vary through genetic polymorphism are positioned relative to one another, and places molecular genetic markers on linkage groups based on their segregation in a population. Molecular genetic markers may also allow studies of the conservation of gene order (synteny) and rearrangement between related species. Molecular markers have many advantages over phenotypic markers, they are heritable, easy to score and are not affected by the environment. Genetic linkage maps are prepared by analysing the segregating populations derived from crosses of genetically diverse parents, and estimating the recombination frequency (RF) among genetic loci. The recombination frequency between two loci is defined as the ratio of the number of recombinant chromosomal types generated by a genetic cross to the total number of chromosomal types. The recombination event is related to the physical process of crossover at meiosis. An RF of 0.5 or 50% is diagnostic of loci that are on separate chromosomes or distant on the same chromosome – they are said to be unlinked. RF values less than 0.5 indicate progressively closer linkage. Simple calculations of map order can be performed for small numbers of loci. However, the development of genome-wide genetic maps requires automation of the process using software such MAPMAKER and JOINMAP (Lander et al. 1987; Stam 1993). While the sections below provide an introduction to the software available for genetic mapping analysis, this subject is discussed in detail in Chapter 11.

19 QTL ANALYSIS

When a trait exhibits co-segregation with a genetic locus (marker), the trait is considered to be linked to that marker. Marker-assisted selection is based on the linkage of the molecular marker to a trait of interest. The linked marker may be characterized during the segregation of the trait in a population and applied for the map-based cloning of the underlying gene. Many traits that are important in crop improvement exhibit continuous variation. It has been established that the quantitative pattern of inheritance of these traits arises from the segregation of the alleles of multiple genes where the effects of the gene are often modified by environmental factors. The systematic mapping of quantitative trait loci (QTL) contributing to a continuously variable trait was not feasible before the use of molecular markers, as the inheritance of an entire genome could not be studied using phenotypic markers alone.

A QTL may initially be identified using single marker regression analysis. Whilst not a method of detection in its own right, it provides an indication of marker and trait association at particular regions of the genome. If a QTL is identified, then a significant association between the trait and genotypic data would be expected. This method utilizes an analysis of variance (ANOVA) between the trait and genotypic data and returns the P-values for association of the markers with the traits. QTLs may also be identified and confirmed using interval mapping (IM), composite interval mapping (CIM), multiple interval mapping (MIM) and Bayesian interval mapping (BIM). IM and CIM will usually suffice for the majority of QTL analysis requirements. IM is a standard for QTL analysis and CIM has the property of multiple regression of markers. QTL Cartographer *(http://statgen.ncsu.edu/qtlcart/index.php)* can perform a number of QTL analyses all based on the interval mapping (IM) procedure.

For CIM, permutation analyses can be performed. There are usually 1,000 permutations and this takes a significant amount of computational time. Depending on the number of chromosomes/LG this could take anywhere between 1-2 days to perform. Performing the permutations creates a few files that then sets the base line or new LOD 2.5, subsequent CIM analysis takes about 20 min. A usual approach to QTL analysis is to perform the single marker regression analysis initially, then IM followed by CIM with permutations. The data from the three methods can then be collated and analyzed thoroughly to ensure linkage of markers with the trait and provide sufficient detailed data for confidence in the conclusions drawn.

20   GENETIC MAPPING AND QTL SOFTWARE PACKAGES

In order to perform linkage mapping and QTL analysis, a number of software packages are either available freely to download or to purchase. An exhaustive list of all types of genetic mapping and QTL software is provided on the Rockefeller University Linkage Software List *(http://linkage.rockefeller.edu/soft/)*. These include:

- **MapMaker 3.0** (Lander et al. 1987): This program is only available for UNIX or MSDOS platforms. The program handles co-dominant data and includes computation of variance components, F-statistics, co-ancestry co-efficients and permutation tests for population statistics. The program is freely available for downloading from *http://www.broad.mit.edu/ftp/distribution/software /mapmaker3/*
- **QTLCartographer:** This is a suite of program to map quantitative traits from inbred lines using a map of molecular markers. The program are available via an anonymous ftp server *(http:// statgen.ncsu.edu/qtlcart/index.php)*. Windows QTL Cartographer is a user-friendly version of QTL Cartographer. It has a Graphical User Interface (GUI) and runs under Microsoft Windows
- **JoinMap:** JoinMap 4.0 is advanced Microsoft windows based software for the calculation of genetic linkage maps, dealing with a wide variety of mapping populations (BC_1, F_2, recombinant inbred lines (RILs) (any generation), haploids, doubled haploids, and a full-sib family of cross pollinating species. The program is not freely available and is available for purchase from *http:// www.kyazma.nl/index.php/mc.JoinMap*

21   INTEGRATED BIOINFORMATICS FOR CROP BREEDING

The bioinformatics systems described above enable the structuring and interrogation of biological data for the discovery of candidate genes and genetic markers for a wide range of agronomic traits. As more genome sequence and gene expression information become available for crops, these will be applied for the identification of novel agronomic genes and alleles for crop improvement programs. In addition, the genetic characterization of plant populations provides an insight into the population structure and evolution of crop species, information that may be applied for the identification of diverse alleles and parents for wide crosses. Bioinformatics enables the comparison of species and varieties at the genetic and genomic level, enabling the transfer of information from one crop or model species to crop improvement applications in related species. The development of standards and structured vocabularies for biological experiments such as gene expression analysis, gene functions and observed heritable traits, assists in this comparison. As more genetic and genomic information is applied for crop breeding, there will be an increasing requirement for breeding simulation software, to determine the most precise and rapid way to incorporate traits from several lines or varieties to produce enhanced germplasm expressing the optimal complement of agronomic traits for a specific environment. The application of technology has led to significant increases in crop yield over previous decades, and the application of genomic and post-genomic information for crop improvement, facilitated by advances in bioinformatics, will assist in the continued enhancement of germplasm for future generations.

References

Adams MD, Kelley JM, Gocayne JD, Dubnick M, Polymeropoulos MH, Xiao H, Merril CR, Wu A, Olde B, Moreno RF, Kerlavage AR, Mccombie WR, Venter JC (1991) Complementary DNA sequencing: expressed sequence tags and human genome project. Science 252: 1651-1656

Altschul SF, Gish W, Miller W, Myers EW, Lipman DJ (1990) Basic local alignment search tool. J Mol Biol 215: 403-410

Bairoch A, Apweiler R, Wu CH, Barker WC, Boeckmann B, Ferro S, Gasteiger E, Huang HZ, Lopez R, Magrane M, Martin MJ, Natale DA, O'Donovan C, Redaschi N, Yeh LSL (2005) The universal protein resource (UniProt). Nucl Acids Res 33: 154-159

Bateman A, Coin L, Durbin R, Finn RD, Hollich V, Griffiths-Jones S, Khanna A, Marshall M, Moxon S, Sonnhammer ELL, Studholme DJ, Yeats C, Eddy SR (2004) The Pfam protein families database. Nucl Acids Res 32: D138-D141

Batley J, Barker G, O'Sullivan H, Edwards KJ, Edwards D (2003) Mining for single nucleotide polymorphisms and insertions/deletions in maize expressed sequence tag data. Plant Physiol 132: 84-91

Batley J, Edwards KJ, Wiltshire CW, Glen D, Karp A (1998) The isolation and characterisation of microsatellite loci in the willow beetles, *Phyllodecta vulgatissima* (L.) and *P. vitellinae* (L.). Mol Ecol 7: 1434-1436

Barker G, Batley J, O'Sullivan H, Edwards KJ, Edwards D (2003) Redundancy based detection of sequence polymorphisms in expressed sequence tag data using autoSNP. Bioinformatics 19: 421-422

Barrett T, Suzek TO, Troup DB, Wilhite SE, Ngau W-C, Ledoux P, Rudnev D, Lash AE, Fujibuchi W, Edgar R (2005) NCBI GEO: mining millions of expression profiles - database and tools. Nucl Acids Res 33(1): D562-D566

Benson DA, Karsch-Mizrachi I, Lipman DJ, Ostell J, Wheeler DL (2006) Genbank. Nucl Acids Res 34: D16-D20

Bhangale TR, Stephens M, Nickerson DA (2006) Automating resequencing - based detection of insertion-deletion polymorphisms. Nat Genet 38(12): 1457-1462

Birney E, Andrews TD, Bevan P, Caccamo M, Chen Y, Clarke L, Coates G, Cuff J, Curwen V, Cutts T, Down T, Eyras E, Fernandez-Suarez XM, Gane P, Gibbins B, Gilbert J, Hammond M, Hotz HR, Iyer V, Jekosch K, Kahari A,

Kasprzyk A, Keefe D, Keenan S, Lehvaslaiho H, McVicker G, Melsopp C, Meidl P, Mongin E, Pettett R, Potter S, Proctor G, Rae M, Searle S, Slater G, Smedley D, Smith J, Spooner W, Stabenau A, Stalker J, Storey R, Ureta-Vidal A, Woodwark KC, Cameron G, Durbin R, Cox A, Hubbard T, Clamp M (2004) An overview of Ensembl. Genome Res 14: 925-928

Bocker T, Ruschoff J, Fishel R (1999) Molecular diagnostics of cancer predisposition: hereditary non-polyposis colorectal carcinoma and mismatch repair defects. Biochimica et Biophysica Acta - Reviews on Cancer 1423: 1-10

Boeckmann B, Bairoch A, Apweiler R, Blatter M-C, Estreicher A, Gasteiger E, Martin MJ, Michoud K, O'Donovan C, Phan I, Pilbout S, Schneider M (2003) The SWISS-PROT protein knowledgebase and its supplement TrEMBL in 2003. Nucl Acids Res 31: 365-370

Brazma A, Parkinson H, Sarkans U, Shojatalab M, Vilo J, Abeygunawardena N, Holloway E, Kapushesky M, Kemmeren P, Lara GG, Oezcimen A, Rocca-Serra P, Sansone SA (2003) ArrayExpress - a public repository for microarray gene expression data at the EBI. Nucl Acids Res 31(1): 68-71

Brenner S, Johnson M, Bridgham J, Golda G, Lloyd DH, Johnson D, Luo SJ, McCurdy S, Foy M, Ewan M, Roth R, George D, Eletr S, Albrecht G, Vermaas E, Williams SR, Moon K, Burcham T, Pallas M, DuBridge RB, Kirchner J, Fearon K, Mao J, Corcoran K (2000a) Gene expression analysis by massively parallel signature sequencing (MPSS) on microbead arrays. Nat Biotechnol 18: 630-634

Bruford MW, Wayne RK (1993) Microsatellites and their application to population genetic studies. Curr Opin Genet Dev 3: 939-943

Buchanan FC, Friesen MK, Littlejohn RP, Clayton JW (1996) Microsatellites from the beluga whale *Delphinapterus leucas*. Mol Ecol 5: 571-575

Burke J, Davison D, Hide W (1999) d2_cluster: a validated method for clustering EST and full-length cDNA sequences. Genome Res 9: 1135-1142

Camon E, Magrane M, Barrell D, Lee V, Dimmer E, Maslen J, Binns D, Harte N, Lopez R, Apweiler R (2004) The Gene Ontology Annotation (GOA) Database: sharing knowledge in UniProt with Gene Ontology. Nucl Acids Res 32: D262-D266

Cardle L, Ramsey L, Milbourne D, Macaulay M, Marshall D, Waugh R (2000) Computational and experimental characterization of physically clustered simple sequence repeats in plants. Genet Soc Am 156: 847-854

Carollo V, Matthews DE, Lazo GR, Blake TK, Hummel DD, Lui N, Hane DL, Anderson OD (2005) GrainGenes 2.0. An improved resource for the small-grains community. Plant Physiol 139: 643-651

Clark JI, Brooksbank C, Lomax J (2005) It's All GO for Plant Scientists. Plant Physiol 138: 1268-1278

Cochrane G, Aldebert P, Althorpe N, Andersson M, Baker W, Baldwin A, Bates K, Bhattacharyya S, Browne P, van den Broek A, Castro M, Duggan K, Eberhardt R, Faruque N, Gamble J, Kanz C, Kulikova T, Lee C, Leinonen R, Lin Q, Lombard V, Lopez R, McHale M, McWilliam H, Mukherjee G,

Nardone F, Pastor MPG, Sobhany S, Stoehr P, Tzouvara K, Vaughan R, Wu D, Zhu WM, Apweiler R (2006) EMBL nucleotide sequence database: developments in 2005. Nucl Acids Res 34: D10-D15

Cornuet J-M, Piry S, Luikart G, Estoup A, Solignac M (1999) New methods employing multilocus genotypes to select or exclude populations as origins of individuals. Genetics 153: 1989-2105

Craigon DJ, James N, Okyere J, Higgins J, Jotham J, May S (2004) NASCArrays: a repository for microarray data generated by NASC's transcriptomics service. Nucl Acids Res 32(1): D575-577

Cruzan MB (1998) Genetic markers in plant evolutionary ecology. Ecology 79: 400-412

Dawson KJ, Belkhir K (2001) A Bayesian approach to the identification of panmictic populations and the assignment of individuals. Genet Res 78(1): 59-77

Di Rienzo A, Peterson AC, Garza JC, Valdes AM, Slatkin M, Freimer NB (1994). Mutational processes of simple-sequence repeat loci in human populations. Proc Natl Acad Sci USA 91: 3166-3170

Dietrich A, Korn B, Poustka A (1992) Completion of the physical map of XQ28 - The location of the gene for L1CAM on the human X-chromosome. Mam Genom 3: 168-172

Dong Q, Roy L, Freeling M, Walbot V, Brendel V (2003) ZmDB, an integrated database for maize genome research. Nucl Acids Res 31: 244-247

Dong Q, Schlueter SD, Brendel V (2004) PlantGDB, plant genome database and analysis tools. Nucl Acids Res 32(1): D354-359

Dong Q, Lawrence CJ, Schlueter SD, Wilkerson MD, Kurtz S, Lushbough C, Brendel V (2005) Comparative plant genomics resources at plantgdb. Plant Physiol 139: 610-618

Edwards D, Batley J (2004) Plant bioinformatics: from genome to phenome. Trends Biotechnol 22(5): 232-237

Ellegren H (1992) Polymerase-chain-reaction (PCR) analysis of microsatellites - a new approach to studies of genetic relatedness in birds. The Auk 109: 886-895

Engel SR, Linn RA, Taylor JF, Davis SK (1996) Conservation of microsatellite loci across species of artiodactyls: implications for population studies. J Mammol 77: 504-518

Erwin TA, Jewell EG, Love CG, Lim GAC, Li X, Chapman R, Batley J, Stajich J, Mongin E, Stupka E, Ross B, Spangenberg G, Edwards D (2007) BASC: an integrated bioinformatics system for Brassica research. Nucl Acids Res 35: D870-D873

Excoffier L, Laval G, Schneider S (2005) Arlequin ver. 3.0: an integrated software package for population genetics data analysis. Evol Bioinforma Online 1: 47-50

Ferguson A, Taggart JB, Prodöhl PA, McMeel O, Thompson C, Stone C, McGinnity P, Hynes RA (1995) The application of molecular markers to the study and conservation of fish populations, with special reference to *Salmo*. J Fish Biol 47: 103-126

Fizames C, Munos S, Cazettes C, Nacry P, Boucherez J, Gaymard F, Piquemal D, Delorme V, Commes TS, Doumas P, Cooke R, Marti J, Sentenac H, Gojon A (2004) The *Arabidopsis* root transcriptome by serial analysis of gene expression. Gene identification using the genome sequence. Plant Physiol 134: 67-80

Garcia-Hernandez M, Berardini TZ, Chen G, Crist D, Doyle A, Huala E, Knee E, Lambrecht M, Miller N, Mueller LA, Mundodi S, Reiser L, Rhee SY, Scholl R, Tacklind J, Weems DC, Wu Y, Xu I, Yoo D, Yoon J, Zhang P (2002) TAIR: a resource for integrated *Arabidopsis* data. Funct Integr Genom 2: 239-253

Gibbings JG, Cook BP, Dufault MR, Madden SL, Khuri S, Turnbull CJ, Dunwell JM (2003) Global transcript analysis of rice leaf and seed using SAGE technology. Plant Biotechnol J 1: 271-285

Goldstein D, Weale M (2001) Population genomics: Linkage disequilibrium holds the key. Curr Biol 11: R576-R579

Gonzales MD, Archuleta E, Farmer A, Gajendran K, Grant D, Shoemaker R, Beavis WD, Waugh ME (2005) The Legume Information System (LIS): an integrated information resource for comparative legume biology. Nucl Acids Res 33: D660-D665

Grice DE, Leckman JF, Pauls DL, Kurlan R, Kidd KK, Pakstis AJ, Chang FM, Buxbaum JD, Cohen DJ, Gelernter J (1996) Linkage disequilibrium between an allele at the dopamine D4 receptor locus and Tourette syndrome, by the transmission-disequilibrium test. Am J Hum Genet 59(3): 644-652

Gupta PK, Balyan HS, Sharma PC, Ramesh R (1996) Microsatellites in plants: A new class of molecular markers. Curr Sci 70: 45-54

Hardy GH (1908) Mendelian proportions in a mixed population. Science 28: 49-50

Huala E, Dickerman AW, Garcia-Hernandez M, Weems D, Reiser L, LaFond F, Hanley D, Kiphart D, Zhuang MZ, Huang W, Mueller LA, Bhattacharyya D, Bhaya D, Sobral BW, Beavis W, Meinke DW, Town CD, Somerville C, Rhee SY (2001) The *Arabidopsis* Information Resource (TAIR): A comprehensive database and web-based information retrieval, analysis, and visualization system for a model plant. Nucl Acids Res 29: 102-105

Huang X, Madan A (1999) CAP3: A DNA sequence assembly program. Genome Res 9: 868-877

Hubbard T, Andrews D, Caccamo M, Cameron G, Chen Y, Clamp M, Clarke L, Coates G, Cox T, Cunningham F, Curwen V, Cutts T, Down T, Durbin R, Fernandez-Suarez XM, Gilbert J, Hammond M, Herrero J, Hotz H, Howe K, Iyer V, Jekosch K, Kahari A, Kasprzyk A, Keefe D, Keenan S, Kokocinsci F, London D, Longden I, McVicker G, Melsopp C, Meidl P, Potter S, Proctor G, Rae M, Rios D, Schuster M, Searle S, Severin J, Slater G, Smedley D, Smith J, Spooner W, Stabenau A, Stalker J, Storey R, Trevanion S, Ureta-Vidal A, Vogel J, White S, Woodwark C, Birney E (2005) Ensembl 2005. Nucl Acids Res 33: D447-D453

Jaiswal P, Ni J, Yap I, Ware D, Spooner W, Youens-Clark K, Ren L, Liang C, Zhao W, Ratnapu K, Faga B, Canaran P, Fogleman M, Hebbard C, Avraham S, Schmidt S, Casstevens TM, Buckler ES, Stein L, McCouch S (2006) Gramene: a bird's eye view of cereal genomes. Nucl Acids Res 34: D717-D723

Jackson S, Rounsley S, Purugganan M (2006) Comparative sequencing of plant genomes: choices to make. Plant Cell 18(5): 1100-1104

Jewell E, Robinson A, Savage D, Erwin T, Love CG, Lim GAC, Li X, Batley J, Spangenberg GC, Edwards D (2006) SSR Primer and SSR Taxonomy Tree: Biome SSR discovery. Nucl Acids Res 34: W656-W659

Kanehisa M, Goto S, Kawashima S, Okuno Y, Hattori M (2004) The KEGG resource for deciphering the genome. Nucl Acids Res 32: D277-D280

Kanz C, Aldebert P, Althorpe N, Baker W, Baldwin A, Bates K, Browne P, van den Broek A, Castro M, Cochrane G, Duggan K, Eberhardt R, Faruque N, Gamble J, Diez FG, Harte N, Kulikova T, Lin Q, Lombard V, Lopez R, Mancuso R, McHale M, Nardone F, Silventoinen V, Sobhany S, Stoehr P, Tuli MA, Tzouvara K, Vaughan R, Wu D, Zhu WM, Apweiler R (2005) The EMBL Nucleotide Sequence Database. Nucl Acids Res 33: D29-D33

Kichler K, Holder MT, Davis SK, Marquez R, Owens DW (1999) Detection of multiple paternity in the kemp's ridley sea turtle with limited sampling. Mol Ecol 8: 819-830

Krieger CJ, Zhang PA, Mueller LA, Wang A, Paley S, Arnaud M, Pick J, Rhee SY, Karp P (2004) MetaCyc: Recent enhancements to a database of metabolic pathways and enzymes in microorganisms and plants. Nucl Acids Res 32: D438-D442

Lander E, Abrahamson J, Barlow A, Daly M, Lincoln S, Newburg L, Green P (1987) Mapmaker a computer package for constructing genetic-linkage maps. Cytogenet Cell Genet 46(1-4): 642-642

Lawrence CJ, Dong Q, Polacco ML, Seigfried TE, Brendel V (2004) Maizegdb, the community database for maize genetics and genomics. Nucl Acids Res 32: D393-D397

Lee JY, Lee DH (2003) Use of serial analysis of gene expression technology to reveal changes in gene expression in Arabidopsis pollen undergoing cold stress. Plant Physiol 132: 517-529

Lee Y, Tsai J, Sunkara S, Karamycheva S, Pertea G, Sultana R, Antonescu V, Chan A, Cheung F, Quackenbush J (2005) The TIGR Gene Indices: clustering and assembling EST and known genes and integration with eukaryotic genomes. Nucl Acids Res 33: D71-74

Li W, Li J, Liu Z, Li L, Liu J, Qi Q, Liu J, Li L, Li T, Wang X, Lu H, Wu T, Zhu M, Ni P, Han H, Dong W, Ren X, Feng X, Cui P, Li X, Wang H, Xu X, Zhai W, Xu Z, Zhang J, He S, Zhang J, Xu J, Zhang K, Zheng X, Dong J, Zeng W, Tao L, Ye J, Tan J, Ren X, Chen X, He J, Liu D, Tian W, Tian C, Xia H, Bao Q, Li G, Gao H, Cao T, Wang J, Zhao W, Li P, Chen W, Wang X, Zhang Y, Hu J, Wang J, Liu S, Yang J, Zhang G, Xiong Y, Li Z, Mao L, Zhou C, Zhu Z, Chen R, Hao B, Zheng W, Chen S, Guo W, Li G, Liu S, Tao M, Wang J, Zhu L,

Yuan L, Yang H (2002) A draft sequence of the rice genome (*oryza sativa* L. ssp. *indica*). Science: 296: 79-92

Litt M, Luty JA (1989) A hypervariable microsatellite revealed by in vitro amplification of a dinucleotide repeat within the cardiac-muscle actin gene. Am J Hum Genet 44: 397-401

Lomax J (2005) Get ready to GO! A biologist's guide to the Gene Ontology. Brief Bioinform 6: 298-304

Love C, Logan E, Erwin T, Kaur J, Lim GAC, Hopkins C, Batley J, James N, May S, Spangenberg G, Edwards D (2006) Integrating and Interrogating Diverse *Brassica* Data within an EnsEMBL Structured Database. Acta Hort 706: 77-82

Lu C, Tej SS, Luo S, Haudenschild CD, Meyers BC, Green PJ (2005) Elucidation of the Small RNA Component of the Transcriptome. Science 309: 1567-1569

Margulies M, Egholm M, Altman WE, Attiya S, Bader JS, Bemben LA, Berka J, Braverman MS, Chen YJ, Chen ZT, Dewell SB, Du L, Fierro JM, Gomes XV, Godwin BC, He W, Helgesen S, Ho CH, Irzyk GP, Jando SC, Alenquer MLI, Jarvie TP, Jirage KB, Kim JB, Knight JR, Lanza JR, Leamon JH, Lefkowitz SM, Lei M, Li J, Lohman KL, Lu H, Makhijani VB, McDade KE, McKenna MP, Myers EW, Nickerson E, Nobile JR, Plant R, Puc BP, Ronan MT, Roth GT, Sarkis GJ, Simons JF, Simpson JW, Srinivasan M, Tartaro KR, Tomasz A, Vogt KA, Volkmer GA, Wang SH, Wang Y, Weiner MP, Yu PG, Begley RF, Rothberg JM (2005) Genome sequencing in microfabricated high-density picolitre reactors. Nature 437(7057): 376-380

Matthews DE, Carollo VL, Lazo GR, Anderson OD (2003) GrainGenes, the genome database for small-grain crops. Nucl Acids Res 31: 183-186

Matsumura H, Nirasawa S, Terauchi R (1999) Technical advance, transcript profiling in rice (*Oryza sativa* L.) seedlings using serial analysis of gene expression (SAGE). Plant J 20: 719-726

McLaren CG, Bruskiewich RM, Portugal AM, Cosico AB (2005) The international rice information system. A platform for meta-analysis of rice crop data. Plant Physiol 139(2): 637-642

Mewes HW, Amid C, Arnold R, Frishman D, Güldener U, Mannhaupt G, Münsterkötter M, Pagel P, Strack N, Stümpflen V, Warfsmann J, Ruepp A (2004) MIPS: analysis and annotation of proteins from whole genomes. Nucl Acids Res 32: D41-D44

Meyers BC, Tej SS, Vu TH, Haudenschild CD, Agrawal V, Edberg SB, Ghazal H, Decola S (2004a) The use of MPSS for whole-genome transcriptional analysis in Arabidopsis. Genome Res 14: 1641-1653

Moore SS, Sargeant LL, King TJ, Mattick JS, Georges M, Hetzel DJS (1991) The conservation of dinucleotide microsatellites among mammalian genomes allows the use of heterologous PCR primer pairs in closely related species. Genomics 10: 654-660

Moran C (1993) Microsatellite repeats in pig (*Sus domestica*) and chicken (*Gallus domesticus*) genomes. J Hered 84: 274-280

Morgante M, Olivieri AM (1993) PCR-amplified microsatellites as markers in plant genetics. Plant J 3: 175-182

Mortimer J, Batley J, Love C, Logan E, Edwards D (2005) Simple sequence repeat (SSR) and GC distribution in the *Arabidopsis thaliana* genome. J Plant Biotechnol 7: 17-25

Mueller LA, Zhang P, Rhee SY (2003) AraCyc. A biochemical pathway database for *Arabidopsis*. Plant Physiol 132: 453-460

Narvel J, Walker D, Rector B, All J, Parrot W, Boerma R (2001) A retrospective DNA marker assessment of the development of insect resistant soybean. Crop Sci 41: 1931-1939

Nei M (1987) Molecular Evolutionary Genetics. Columbia Univ Press, New York, USA

Ohyanagi H, Tanaka T, Sakai H, Shigemoto Y, Yamaguchi K, Habara T, Fujii Y, Antonio BA, Nagamura Y, Imanishi T, Ikeo K, Itoh T, Gojobori T, Sasaki T (2006) The rice annotation project database (rap-db): hub for *oryza sativa* ssp. *japonica* genome information. Nucl Acids Res 34: D741–D744

Owen PG, Pei M, Karp A, Royle DJ, Edwards KJ (1998) Isolation and characterisation of microsatellite loci in the wheat pathogen *Mycosphaerella graminicola*. Mol Ecol 7: 1611-1612

Page RDM (1996) TreeView: An application to display phylogenetic trees on personal computers. Comp Appl Biosci 12(4): 357-358

Parkinson H, Sarkans U, Shojatalab M, Abeygunawardena N, Contrino S, Coulson R, Farne A, Lara GG, Holloway E, Kapushesky M, Lilja P, Mukherjee G, Oezcimen A, Rayner T, Rocca-Serra P, Sharma A, Sansone S, Brazma A (2005) ArrayExpress—a public repository for microarray gene expression data at the EBI. Nucl Acids Res 33(1): D553-555

Pertea G, Huang X, Liang F, Antonescu V, Sultana R, Karamycheva S, Lee Y, White J, Cheung F, Parvizi B, Tsai J, Quackenbush J (2003) TIGR Gene Indices clustering tools (TGICL): a software system for fast clustering of large EST datasets. Bioinformatics 19(5): 651-652

Plaschke J, Ganal MW, Röder MS (1995) Detection of genetic diversity in closely related bread wheat using microsatellite markers. Theor Appl Genet 91: 1001-1007

Poroyko V, Hejlek LG, Spollen WG, Springer GK, Nguyen HT, Sharp RE, Bohnert HJ (2005) The maize root transcriptome by serial analysis of gene expression. Plant Physiol 138: 1700-1710

Powell W, Machray GC, Provan J (1996) Polymorphism revealed by simple sequence repeats. Trends Plant Sci 1: 215-222

Pritchard JK, Stephens M, Donnelly P (2000) Inference of population structure using multilocus genotype data. Genetics 155: 945-959

Raymond M, Rousset F (1995) Genepop (Version-1.2) - Population-genetics software for exact tests and ecumenicism. J Hered 86(3): 248-249

Reiser L, Rhee SY (2005) Using The *Arabidopsis* Information Resource (TAIR) to find information about *Arabidopsis* genes. In: Baxevanis AD (ed) Current Protocols in Bioinformatics. John Wiley, NY, USA, pp 1.11.1-1.11.45

Rhee SY, Beavis W, Berardini TZ, Chen G, Dixon D, Doyle A, Garcia-Hernandez M, Huala E, Lander G, Montoya M, Miller N, Mueller LA, Mundodi S, Reiser L, Tacklind J, Weems DC, Wu Y, Xu I, Yoo D, Yoon J, Zhang P (2003) The *Arabidopsis* Information Resource (TAIR): a model organism database providing a centralized, curated gateway to *Arabidopsis* biology, research materials and community. Nucl Acids Res 31: 224-228

Rico C, Zadworny D, Kuhnlein U, Fitzgerald GJ (1993) Characterization of hypervariable microsatellite loci in the threespine stickleback *Gasterosteus aculeatus*. Molecular Ecology 2: 271-272.

Robinson AJ, Love CG, Batley J, Barker G, Edwards D (2004a) Simple sequence repeat marker loci discovery using SSR Primer. Bioinfomatics 20: 1475-1476

Robinson SJ, Cram DJ, Lewis CT, Parkin IAP (2004b) Maximizing the efficacy of SAGE analysis identifies novel transcripts in *Arabidopsis*. Plant Physiol 136: 3223-3233

Sasaki T, Matsumoto T, Yamamoto K, Sakata K, Baba T, Katayose Y, Wu J, Niimura Y, Cheng Z, Nagamura Y, Antonio BA, Kanamori H, Hosokawa S, Masukawa M, Arikawa K, Chiden Y, Hayashi M, Okamoto M, Ando T, Aoki H, Arita K, Hamada M, Harada C, Hijishita S, Honda M, Ichikawa Y, Idonuma A, Iijima M, Ikeda M, Ikeno M, Ito S, Ito T, Ito Y, Ito Y, Iwabuchi A, Kamiya K, Karasawa W, Katagiri S, Kikuta A, Kobayashi N, Kono I, Machita K, Maehara T, Mizuno H, Mizubayashi T, Mukai Y, Nagasaki H, Nakashima M, Nakama Y, Nakamichi Y, Nakamura M, Namiki N, Negishi M, Ohta I, Ono N, Saji S, Sakai K, Shibata M, Shimokawa T, Shomura A, Song J, Takazaki Y, Terasawa K, Tsuji K, Waki K, Yamagata H, Yamane H, Yoshiki S, Yoshihara R, Yukawa K, Zhong H, Iwama H, Endo T, Ito H, Hahn JH, Kim HI, Eun MY, Yano M, Jiang J, Gojobori T (2002) The genome sequence and structure of rice chromosome 1. Nature 420: 312-316

Savage D, Batley J, Erwin T, Logan E, Love CG, Lim GAC, Mongin E, Barker G, Spangenberg GC, Edwards D (2005) SNPServer: a real-time SNP discovery tool. Nucl Acids Res 33: W493-W495

Schlötterer C, Amos B, Tautz D (1991) Conservation of polymorphic simple sequence loci in cetacean species. Nature 354: 63-65

Shen L, Gong J, Caldo RA, Nettleton D, Cook D, Wise RP, Dickerson JA (2005) BarleyBase- an expression profiling database for plant genomics. Nucl Acids Res 33(1): D614-618

Stam P (1993) Construction of integrated genetic-linkage maps by means of a new computer package – JOINMAP. Plant J 3(5): 739-744

Stephens M, Sloan JS, Robertson PD, Scheet P, Nickerson DA (2006). Automating sequence-based detection and genotyping of SNPs from diploid samples. Nat Genet 38(3): 375-381

Sumner L, Mendes P, Dixon R (2003) Plant metabolomics: large-scale phytochemistry in the functional genomics era. Phytochemistry 62: 817-836

Swofford DL (1993) Paup - A computer program for phylogenetic inference using maximum parsimony, version 3.1.1 Illinois Natural History Survey, Champaign, Illinois, USA

Syvänen A-C (2001) Accessing genetic variation: Genotyping single nucleotide polymorphisms. Nat Rev Genet 2: 930-942

Tang X, Shen L, Dickerson JA (2005) BarleyExpress: a web-based submission tool for enriched microarray database annotations. Bioinformatics 21(3): 399-401

Tautz D (1989) Hypervariability of simple sequences as a general source for polymorphic DNA markers. Nucl Acids Res 17: 6463-6471

Teulat B, Aldam C, Trehin R, Lebrun P, Barker JHA, Arnold GM, Karp A, Baudouin L, Rognon F (2000) An analysis of genetic diversity in coconut (*Cocos nucifera*) populations from across the geographic range using sequence-tagged microsatellites (SSRs) and AFLPs. Theor Appl Genet 100: 764-771

The Arabidopsis Genome Initiative (2000) Analysis of the genome sequence of the flowering plant *Arabidopsis thaliana*. Nature 408: 796-815

Velculescu VE, Zhang L, Vogelstein B, Kinzler KW (1995) Serial analysis of gene expression. Science 270: 484-487

Ware D, Jaiswal P, Ni JJ, Pan XK, Chang K, Clark K, Teytelman L, Schmidt S, Zhao W, Cartinhour S, McCouch S, Stein L (2002a) Gramene: a resource for comparative grass genomics. Nucl Acids Res 30(1): 103-105

Ware DH, Jaiswal PJ, Ni JJ, Yap I, Pan XK, Clark KY, Teytelman L, Schmidt SC, Zhao W, Chang K, Cartinhour S, Stein LD, McCouch SR (2002b) Gramene, a tool for grass genomics. Plant Physiol 130: 1606-1613

Weems D, Miller N, Garcia-Hernandez M, Huala E, Rhee SY (2004) Design, implementation, and maintenance of a model organism database for *Arabidopsis thaliana*. Comp Funct Genom 5: 362-369

Weinberg W (1908) "Über den Nachweis der Vererbung beim Menschen". Jahreshefte des Vereins für vaterländische Naturkunde in Württemberg 64: 368-382

Weir BS, Cockerham CC (1984) Estimating *F*-Statistics for the analysis of population structure. Evolution 38: 1358-1370

Wright S (1978) Evolution and the Genetics of Populations, Vol 4. Variability within and among Natural Populations. Univ of Chicago Press, Chicago, IL, USA

Zhao WM, Wang J, He XM, Huang XB, Jiao YZ, Dai MT, Wei SL, Fu J, Chen Y, Ren XY, Zhang Y, Ni PX, Zhang JG, Lil SG, Wang J, Wong GKS, Zhao HY, Yu J, Yang HM, Wang J (2004) BGI-RIS: an integrated information resource and comparative analysis workbench for rice genomics. Nucl Acids Res 32: D377-D382

Zhang P, Foerster H, Tissier C, Mueller L, Paley S, Karp P, Rhee SY (2005) MetaCyc and AraCyc. Metabolic pathway databases for plant research. Plant Physiol 138: 27-37

10 An Overview on Plant Genome Initiatives

Christopher A. Cullis

Department of Biology, Case Western Reserve University,
10900 Euclid Avenue, Cleveland, OH 44106-7080, USA;
E-mail: cac5@case.edu

1   INTRODUCTION

1.1   What is a Plant Genome Initiative?

A plant genome initiative is one that develops resources across the genome that is not just tied to a particular gene, pathway or response. Thus genome initiatives do not need to be comprehensive, and may well have a more limited goal in view, namely to understand a complex process that appears to involve hundreds or thousands of genes, but the approach is to accumulate information and biological resources relating to as much of the genome as possible and then determining which parts are of importance for the particular process of interest. By their nature such initiatives also generate resources for a wide community of scientists interested in other aspects of that organism, or in using that organism as a reference or comparison. This broad net of trying to accumulate as much data as possible means that genome initiatives tend to be large, multi-institutional (and multi-national) projects due to their cost and wider community resource aspects of the data generated.

A plant genome initiative cannot be started by just trying to generate sequence data without a certain level of basic information about the particular species of interest. Many of the initiatives have started with the genetic analysis of the plant of interest (such as the making of a molecular map) and then proceed with DNA sequencing efforts to cover the genome to varying degrees, from sample sequencing to a complete genome sequence. Genomic sequence on its own is not necessarily very useful if it cannot be referenced back to other resources for that organism, except when it is being used for specific comparisons with other organisms. One still open question is if the availability of a few complete genomic sequences will be sufficient for the identification of important regions of all other genomes through cross-referencing of conserved regions without the need to get complete genome sequences for many plants. If this does occur then sufficient regions can then be targeted for isolation from many plants that will deliver the appropriate genome sequence information and reduce the need for the complete genome sequences of all plants. However, the C-value paradox is still an issue since the plant nuclear DNA content varies so widely. How deep will random sequencing need to be in order to gain an understanding of a previously unknown genome? Are long tracts of sequence rather than just the sequence of specific regions? The nuclear DNA variation is also an important factor in the development of strategies for generating genome sequences. The larger the genome the greater will be the cost and, possibly, the more marginal the added value of this large amount of sequence. The balance between identifying interesting and useful data and getting a complete genome is important in the development of any genome initiatives.

The definition of a plant genome initiative used in the selection of material to be described here is that it is either an initiative to identify a significant fraction of the genes in a particular plant in order to understand the form and function of that plant (a genome project), or an initiative to develop genome-wide resources for a particular plant or family, such as a comprehensive mutant collection. A number of different stages are involved in a genome project. Initially the basic information, such as a molecular genetic map, is usually necessary as a framework on which the sequence information is hung. Subsequently sequence information is acquired through different avenues, either directly from genomic DNA or from expressed sequences. Ideally the complete genome sequence should be available but this can be cost prohibitive, especially for plants which have large genomes. Alternatives to complete genome sequences are strategies for sampling the genome. The most common fractionations of the genome include identifying the expressed portion (through cDNA sequencing), the un-methylated portion using methyl filtration or methylation-sensitive restriction enzymes, or the sequencing of the low repetition portion of the genome by Cot fraction. Each of these latter approaches generates only partially overlapping information about the genome but when combined can give a reasonably comprehensive view of what are presently identified as the important regions of the genome.

A second form of plant genome initiative would be one where the sequence information has already been identified and the aim is to identify all the proteins and their functions. An example of such an initiative is the Arabidopsis 2010 program which was initiated in 2000 and provided funding to enable researchers to determine the function of all Arabidopsis genes by the year 2010 (Chory et al. 2000). The progress report of this program is available in the mid-course evaluation which was produced in 2005 (*http://www.arabidopsis.org/portals/masc/masc_docs/ AT2010WorkshopFinal.pdf*). It is not certain that such a comprehensive post-genome sequencing effort will be repeated in any other plant species, so that this project may need to provide a sufficient framework on which all others can be compared.

All plant genome initiatives have to be accompanied by an extensive bioinformatics support function. This support has multiple components. Primary objectives are the basic functions of assembling and compiling the data and reporting it out to the community in an accessible fashion. Subsequently, it is important to identify interesting structures contained within the sequence be they expressed regions, regulatory regions or those with novel functions. Interfaces facilitating the interrogation of the information by experts and novices alike need to be developed. The experience from the initial eukaryotic genome projects indicates that this annotation of structure and the functional characterization will be a very time-consuming endeavor. In fact, with the advances in DNA sequencing technology, the initial phase of collecting sequence data will be the shortest and easiest process in the understanding of genomes, while the informatics and functional analyses arising directly or indirectly from the sequence data will continue well beyond the release of the these data and their preliminary characterization.

1.2 Plant Nuclear DNA Content

The amount of DNA in a plant cell varies widely. The DNA content in a reduced gamete (1C) is the value usually given for comparative purposes. These values can be reported in two different ways. The first is as a mass of DNA usually given in picograms per 1C nucleus and the second is in the number of megabase pairs of DNA (the actual equivalence is 1 pg = 965 Mb). The amount of DNA in plants varies by over a 1,000-fold between the largest and smallest genomes so far characterized. For example, one of the smallest genomes belongs to *Arabidopsis thaliana* with 125 Mbp, while the largest to date is *Fritillaria assyriaca* with 124,852 Mbp which is equivalent to 127.4 pg (data from the database maintained by the Royal Botanic Gardens, Kew, *http:// www.rbgkew.org.uk/cval/homepage.html*). However, this range may not represent the true limits since DNA values have been estimated in only about 32% of angiosperm families. This variation does not only occur in comparisons between genera, but it is also true within a genus. Significant intraspecific variation in nuclear DNA amount can also occur, such as that in *Zea mays* ssp. *mays* where Laurie and Bennett (1985) estimated the 1C total nuclear DNA values to range from about 2.45 pg (2,364 Mb) to about 3.35 pg (3,233 Mb). This range in DNA content (both within and between species) does not appear to be associated with variation in the basic number of genes required for growth and development and has led to its being referred to as the C-value paradox. However, knowledge of the nuclear DNA content importantly influences the type of approach that may be appropriate for any genome initiative in at least two different ways – firstly what type of approach would be cost effective to generate the greatest amount of useful data at the least cost and secondly, which particular individual line, or lines, should be used to generate the data.

Following the determination of the total nuclear DNA content is the question as to what proportion of the genome of the plant under consideration is of interest? The plant genome, as with all higher eukaryotic genomes, contains families of sequence of different degrees of repetition. Within the genome the DNA sequences can be divided into a series of classes. There are the sequences that have only one, or a few (< 5) copies, those that occur with hundreds of copies and those that are highly repetitive, although these groups are not discrete with one merging into the next forming a continuum of values of repetition. The low copy number or unique sequences probably represent the genes, the moderately repetitive sequences, many of which may be members of transposable element families, are distributed in a dispersed fashion throughout most of the genome and most of the very highly repetitive sequences are arranged in tandem arrays. The actual physical arrangement of these different classes also affects the approach to genome projects. Obviously the coding regions are important, but are these regions the only ones important for controlling the plant phenotype? With the role of microRNAs becoming more and more prominent, the interest in the non-coding fraction of the genome is becoming more significant. A second consideration is the relative organization of the genome with respect to the organization of the moderately repetitive sequences. These sequences in plant genomes tend to be interspersed frequently with the low copy sequences and the higher the nuclear DNA content the more diluted the gene density. The fact that most of the retrotransposon sequences have a length greater than the average sequencing read results in problems related to the assembly of random sequence reads into long stretches of contiguous sequence.

1.3 Molecular Maps

The genome is organized into chromosomes which are essentially the packets of information. The development of genetic maps, and in particular molecular genetic maps, results in the placement of specific DNA sequences in a spatial relationship, defined by recombination rate, with one another. Many of the genome initiatives aim to fill in the additional sequence information (at varying levels of completion) between these markers to facilitate an understanding of the functional and spatial relationships of the DNA sequences.

The genetic map is an ordering of the regions of the genome according to the rates of recombination that occur along each of the chromosomes. The genetic map is generated by the scoring of numerous polymorphisms in segregating populations, the larger the number of individuals scored the more precise the location of any particular marker becomes. However, the relationship between the genetic distance and the physical distance between two markers is not constant even within a chromosome. Therefore, the conjoining of the genetic and physical maps gives a measure of the relative rates of recombination, per unit number of base pairs, along the chromosomes. The distribution of recombination rates, generally greater at the distal regions of the chromosomes is somewhat mirrored by the spatial distribution of genes, there being a higher gene density at the ends of the chromosomes than nearer the centromere.

The physical map is the linear order of the sequences from one end to the other of each chromosome. The majority of physical maps (even those designated as complete) do not include the complete DNA sequence with overlapping regions of DNA stretching from one end of a chromosome to the other. In generating the physical map, some of the regions will provide special problems. For example, regions that contain long sequences of tandem repeats, such as the genes for the large ribosomal RNAs and the centromeres cannot be completely sequenced in a linear order, since the array is longer than any single sub-fragment that can be sequenced directly, and each repeat is essentially identical. However, knowledge of the number of repeats and the number of chromosomal sites over which these repeats are spread can facilitate an estimation of the length of the region containing the repeats. Any sequences interspersed within this array however, will not be identified in any of the sequencing projects unless they have a specific phenotypic effect and can be mapped genetically and subsequently sequenced. The sequencing through the centromeric region of Arabidopsis identified a number of genes that would have been missed without the sequencing of this region. It also illuminated the structure in and around the centromere which was important especially in the first major plant genome sequencing effort.

Numerous molecular genetic maps for plants are being produced (Table 1; *http://www.ncbi.nlm.nih.gov/genomes/PLANTS/PlantList. html*) as a prelude to both the development of genome sequence characterization and for marker-assisted selection. These mapping efforts will be the scaffold on which additional sequence data, whether for specific crop improvement for evolutionary comparisons, will be attached. Of course, the source of the markers necessary for the development of the molecular map is an important consideration, especially when the species is not fashionable, even if it is an important crop globally.

1.4   Molecular Markers

DNA-based markers revolutionized the whole process of generating genetic maps since, for the first time, a large number of loci could be followed in a single segregating population. The range of genetic markers that are available include restriction fragment length polymorphism (RFLP), random amplified polymorphic DNA (RAPD) (Williams et al. 1990), amplified fragment length polymorphisms (AFLP) (Vos et al. 1995) and simple sequence repeat (SSR) or microsatellites (Senior and Heun 1993). In the context of genome initiatives it is necessary to develop sequence characterized markers as part of the molecular genetic map. The data contained in these marker loci can then be used to anchor shotgun DNA sequence reads in the assembly of such sequence information. Thus, SSRs form the marker system of choice in most cases. RAPDs and AFLPs can still be useful provided that the polymorphisms are converted to sequence characterized amplified regions.

2   GENOME SEQUENCING INITIATIVES

Extensive genome sequence can be acquired in three different types of projects, clone-by-clone (CBC), whole-genome shotgun (WGS), or

Table I Species with molecular genetic map projects. *(http://www.ncbi.nlm.nih.gov/genomes/PLANTS/PlantList_sz.html#EST).*

Aegilops longissima (diploid wheat)
Aegilops tauschi (diploid progenitor of wheat D genome)
Aegilops umbellulata (diploid wheat)
Asparagus officinalis (garden asparagus)
Avena sativa (oat)
Capsicum annuum (pepper)
Eragrostis tef (tef)
Glycine max (soybean)
Hevea brasiliensis (latex rubber tree)
Hordeum bulbosum
Hordeum vulgare (barley)
Linum usitatissimum
Lolium perenne (perennial ryegrass)
Lotus japonicus (lotus)
Manihot esculenta (cassava)
Medicago sativa (alfalfa)
Medicago truncatula (barrel medic)
Oryza sativa (rice)
Phaseolus vulgaris (cultivated bean)
Prunus dulcis (almond)
Quercus robur (pedunculate oak)
Secale cereale (rye)
Setaria italica (finger millet)
Solanum lycopersicoides (nightshade)
Solanum lycopersicum (tomato)
Solanum melongena (eggplant)
Solanum peruvianum (Peruvian tomato)
Solanum tuberosum (potato)
Sorghum bicolor (Sorghum)
Theobroma cacao (cocoa)
Triticum aestivum (bread wheat)
Triticum monococcum (diploid cultivated wheat)
Triticum turgidum (poulard wheat)
Zea mays (corn)

selective sequencing, depending upon the amount of information required and the resources available. The CBC strategy requires large insert (bacterial artificial chromosome, BAC) libraries and detailed clone-based physical maps to develop the minimal tiling paths (MTPs) followed by the large-scale sequencing efforts. A MTP represents the least number of BAC clones that completely cover the region to be sequenced. Therefore the larger the initial BAC library, the smaller the overlaps will be between adjacent clones reducing the total amount of sequencing required. Although this is the most costly approach, CBC produces long contiguous regions of a genome or a region of the genome. Thus it provides the most comprehensive information about the sequence structure of a genome. In contrast, however, WGS uses computing power to produce contiguous regions of sequences (contigs) for a genome relatively quickly by sequencing and assembling small insert libraries (Venter et al. 1996; Adams et al. 2000; Goff et al. 2002). The genome image inferred from such WGS projects tends to be incomplete since the contiguous stretches of sequence are much shorter than those obtained in the CBC projects. The high proportion of repeated sequences in large genomes can pose a major difficulty for WGS in computing capacity, sequence assembly, and financial cost and so has been most useful for small genomes (Adams et al. 2000; Galagan et al. 2003). In addition to the whole-genome approaches, genome filtration strategies, including methylation filtration (MF) (Rabinowicz et al. 1999) and *Cot*-based cloning and sequencing, have been implemented. MF is based on the characteristic of plant genomes in which genes are largely hypomethylated but repeated sequences, especially many of the classes of transposable elements, are highly methylated (Gruenbaum et al. 1981; Bird 1986, 1992; Martienssen 1999). Hypomethylated DNA molecules can be selected in a couple of ways. One is to select short random sheared fragments (about 500 bp) from a sheared DNA preparation that is ligated into a plasmid vector and then propagated in a Mcr^+ *Escherichia coli* strain that develops a methyl filtration library. A second method is to use a methylation sensitive restriction enzyme (such as *Pst*I) to fragment the DNA that is then cloned in a normal Mcr^- *Escherichia coli* strain. Each of these methods should enrich for low copy number sequences and reduce the amount of repetitive family sequencing. The Cot fractionation method is based on the fact that most genic regions are of low copy number and so can be isolated through renaturation kinetics. When sheared, denatured genomic DNA is allowed to renature the repetitive regions reanneal first, and low copy sequences take longer to reanneal. Therefore by choosing the appropriate annealing times and concentrations, those sequences that are present in the genome, for example, at between one and ten copies could be isolated, cloned and sequenced. In both the MF and Cot fractionation approaches, large-scale sequencing of the resulting libraries would have to be followed by the assembly of the sequences into contiguous fragments which should represent a large portion of the genic regions. Since most of the repetitive regions of the genome have been removed, there will be a large number of short contigs. As is the case of the WGS approach, a well developed molecular map is essential in order to assign these contigs to particular chromosomes or chromosome regions. All the fractionation methods result in contigs that are much shorter than the corresponding ones developed through a CBC genome initiative and need additional extraneous information to be available for the genome to be effectively assembled.

Most of the early finished complex genomes, chromosomes, or sub-chromosome regions have been sequenced by a CBC approach (C. *elegans* Sequencing Consortium 1998; Arabidopsis Genome Initiative 2000; International Human Genome Mapping Consortium 2001; Feng et al. 2002; Sasaki et al. 2002; The Rice Chromosome 10 Sequencing Consortium 2003). Although this is the most costly approach, CBC produced long, if not complete, pseudomolecules of a genome and was particularly instructive for the first genomes to be sequenced. Once the framework of how a genome is organized became apparent, following the complete genome sequencing of a few genomes by the CBC approach, the WGS methodology may prove to be more cost effective since the data can be more easily assembled on the frameworks that have been established.

The earliest genome fractionation approach was the development of substantial expressed sequence tag (EST) databases. An EST is a short single read from a cloned cDNA. Initially, these sequencing efforts were costly in the sense that redundant sequencing of the highly expressed cDNAs occurred. The use of normalized cDNA libraries reduced the redundancy in the sequencing effort. As can be seen from Table 2, the EST database from plants is large and includes many species. One unexpected result from such EST databases was the identification of numerous SSR regions within the EST sequences. These SSRs have been used to develop new molecular markers. Drawbacks of the EST approach are that any genes whose expression is either very low, or restricted to tissues that were not sampled in the generation of the ESTs will be missed. Additionally, various members of multigene families that do not differ in the region sequenced will not be recognized as such.

One variant of the CBC approach that only samples part of the genome is one where particular BAC clones enriched for expressed

Table 2 Large scale EST sequencing projects. (*http://www.ncbi.nlm.nih.gov/ genomes/PLANTS/PlantList_sz.html#EST*).

Aegilops speltoides (spelt)
Allium cepa (onion)
Amborella trichopoda
Ananas comosus (pineapple)
Aquilegia formosa x Aquilegia pubescens
Arabidopsis thaliana
Arachis hypogaea (peanut)
Asparagus officinalis (garden asparagus)
Avena sativa (oat)
Beta vulgaris subsp. *vulgaris* (beet)
Betula pendula (European white birch)
Brassica napus (rape)
Brassica rapa spp. *pekinensis* (Chinese cabbage)
Bruguiera gymnorrhiza (Burma mangrove)
Ceratadon purpureus (moss)
Cicer arietinum (chickpea)
Cichorium intybus (chicory)
Citrus aurantium (Seville orange)
Citrus clementina
Citrus jambhiri (jambhiri orange)
Citrus macrophylla (colo)
Citrus reticulata x *Citrus temple*
Citrus reticulata (tangerine)
Citrus sinensis (apfelsine/navel orange)
Citrus unshiu (Satsuma orange)
Citrus x paradisi (grapefruit)
Citrus sinensis x *Poncirus trifoliata* (Carrizo citrange)
Citrus x paradisi x *Poncirus trifoliata*
Coffea arabica (coffee)
Cucumis sativus (cucumber)
Cynodon dactylon (Bermuda grass)
Eleusine coracana (finger millet)
Eragrostis tef (tef)
Eschscholzia californica (California poppy)
Eucalyptus
Euphorbia esula (leafy spurge)

Contd.

Contd.

Euphorbia tirucalli

Festuca arundinacea (tall fescue)

Fragaria x ananassa (strawberry)

Gerbera hybrid cv. 'Terra Regina'

Glycine max (domesticated soybean)

Glycine soja (wild soybean)

Gossypium arboreum (tree cotton)

Gossypium hirsutum (upland cotton)

Gossypium raimondi (cotton)

Hedyotis centrathoides

Hedyotis terminalis

Hevea brasiliensis (Para rubber tree)

Helianthus annuus (sunflower)

Helianthus paradoxus

Hordeum vulgare (barley)

Hordeum vulgare subsp. *spontaneum* (wild barley)

Hordeum vulgare subsp. *vulgare* (two-rowed barley)

Juglans regia (English walnut)

Lactuca sativa (lettuce)

Lactuca serriola (lettuce)

Lilium longiflorum (trumpet lily)

Limonium bicolor

Linum usitatissimum (flax)

Liriodendron tulipifera

Lolium multiflorum (Italian ryegrass)

Lotus japonicus

Lupinus albus (white lupine)

Lycoris longituba

Malus sieboldii (Toringo crab-apple)

Malus x domestica (domesticated apple)

Malus x domestica x Malus sieversii

Manihot esculenta (cassava)

Marchantia polymorpha (liverwort)

Medicago sativa (alfalfa)

Medicago truncatula (barrel medic)

Melaleuca alternifolia (tea tree)

Melaleuca alternifolia (tea tree)

Contd.

Mesembryanthemum crystallinum (common ice plant)

Musa acuminata (Cavendish banana)

Nicotiana benthamiana

Nuphar advena

Oryza sativa (rice)

Panax ginseng (ginseng)

Panicum virgatum (switch grass)

Persea americana (avocado)

Phaseolus coccineus (bean)

Phaseolus vulgaris (kidney bean)

Picea glauca (white spruce)

Picea sitchensis (Sitka spruce)

Pinus radiata (Monterey pine)

Pinus taeda (loblolly pine)

Plumbago zeylanica

Poncirus trifoliata

Populus alba x *Populus tremula* (gray poplar)

Populus deltoides

Populus euphratica

Populus tremula (European aspen)

Populus tremula x *Populus tremuloides* (poplar)

Populus tremuloides (quaking aspen)

Populus trichocarpa (western balsam poplar)

Populus trichocarpa x *Populus deltoides*

Populus trichocarpa x *Populus nigra*

Prunus armeniaca (apricot)

Prunus dulcis (almond)

Prunus persica (peach)

Quercus petraea (sessile oak)

Quercus robur (English oak)

Ricinus communis (castor bean)

Robinia pseudoacacia (black locust)

Rosa chinensis (China rose)

Rosa hybrid cultivar

Saccharum sp. (sugarcane)

Salix viminalis (basket willow)

Saruma henryi

Contd.

Contd.

Secale cereale (rye)
Solanum habrochaites
Solanum lycopersicuum (tomato)
Solanum pennellii
Solanum tuberosum (potato)
Sorghum bicolor (sorghum)
Sorghum halepense (Johnson grass)
Sorghum propinquum
Stevia rebaudiana (stevia)
Tamarix androssowii
Taraxacum kok-saghyz
Theobroma cacao (cacao)
Triticum aestivum (wheat)
Vaccinium spp. (blueberry)
Vitis aestivalis
Vitis hybrid cultivar
Vitis riparia (riverbank grape)
Vitis shuttleworthii (callose grape)
Vitis vinifera (wine grape)
Yucca filamentosa (spoon-leaf yucca)
Zantedeschia aethiopica (arum-lily)
Zea mays (corn)

genes are selected. Rather than sequencing all the BACs that have been identified and placed on the physical map, those BAC contigs enriched for genes can be identified using hybridization to overgo oligonucleotide probes (Han et al. 2000). The probes are designed from non-redundant cDNA or genomic sequences, but still need be masked to eliminate repetitive elements. The overgo probes are then designed from the masked sequence, hybridized to the BACs and the most labeled BACs are placed in the sequencing pipeline. Just sequencing the BACs containing expressed sequences will reduce the absolute amount of sequencing necessary. A possible second benefit to this strategy would arise if there were gene-rich islands, as has been hypothesized, since the BACs selected in this way would have a high density of genes on them. Essentially this strategy depends on identified gene-rich regions in order to reduce the level of repeated sequences in the data.

Another use of the BAC libraries is the generation of BAC-end sequences. These are obtained from sequencing the ends of each BAC clone using primer sequences from the BAC vector. Such paired-end reads give landmarks within the genome that are at known distances apart. If they are in the low copy fraction of the genome they can be used to anchor the BACs and to identify overlapping BACs in the development of a physical map. Obviously, if they are in the repetitive fraction of the genome they do not have much utility. Since BAC-end sequences are expected to be representative of the genome as a whole, then the larger the genome, with the consequential increase in repetitive components, the lower the usefulness of BAC-end sequences.

A final type of genome fractionation involves chromosome isolation. For very large genomes with a variation in chromosome sizes, then the isolation of individual chromosomes is a possible method of genome reduction. This is a possible approach being considered in the case of wheat. However, this approach will only give a snapshot of that chromosome, which, if it is an example of the genome as a whole, will contain the same proportion of repetitive elements as the whole genome. Therefore, this strategy does not reduce the amount of sequencing necessary for identifying genes over that involved in WGS, but does give complete coverage of the chromosome and therefore sheds light on the organization of sequences within chromosomes.

All the sequencing approaches should become more practical as the improvements in sequencing technologies, such as the new 454 process (sequencing by synthesis, SBS) (Margulies et al. 2005; *http://www.454.com/ index.asp*) and the reduction in costs of the traditional Sanger sequencing, are introduced. These developments will make the production of sequence data more rapid and less expensive. For example, a single machine using the SBS methodology can produce up to 20 Mb of sequence in 5 hours. However, a current concern about the new SBS methodology relates to the length of each read. The short length of each read currently available (100 bp) further complicates the assembly of random sequences, but the newer developments of the technology do have longer reads (> 200 bases) which will somewhat alleviate the problem. It does appear that the new sequencing technologies will reduce the sequencing costs and so make more genome sequences readily available. An important area where inexpensive sequencing will be effective is in the sequencing of multiple members of a species to determine the levels of polymorphisms, and which regions of the genome are under active selection. The length of the read in these re-sequencing efforts is not as vital since there is already a genome with which to compare the new sequence. Added to the new sequencing methodology are the efforts of the Joint Genome Institute and the community sequencing program which will provide opportunities for complete WGS sequencing that is identified as important to the plant community. However, without dramatic reduction in sequencing costs, it is still going to be necessary to fractionate the larger genomes in order to make genome sequencing cost effective.

3   EXAMPLES OF GENOME INITIATIVES

The best method to illustrate how the various strategies can be used to generate genome sequence data is to use some examples of current and projected projects. An inspection of the genome sequencing projects in progress or completed is currently skewed towards genomes that are relatively small with only 5 of the 19 having genome sizes greater than 1 Mbp (Table 3). Two of these larger genomes are corn and wheat which will be described here.

Table 3   Large-scale genome sequencing projects. (*http://www.ncbi.nlm.nih. gov/ genomes/PLANTS/PlantList_sz.html#EST*).

Species	Status	Chromosome number	Genome size (bp)
Aquilegia formosa	Pending	7	0.55×10^9
Arabidopsis thaliana (thale cress)	Complete	5	0.154×10^9
Arabidopsis lyrata	In progress	5	0.154×10^9
Brachypodium distachyon (poaceae)	Pending	5	0.34×10^9
Capsella rubella	In progress	5	0.154×10^9
Elaeis guineensis (oil Palm)	In progress	16	0.965×10^9
Gossypium (cotton)	Pending	26	3.2×10^9
Lotus japonicus (lotus)	In progress	6	0.46×10^9
Manihot esculenta (cassava)	In progress	36	0.80×10^9
Medicago truncatula (barrel medic)	Complete	8	0.46×10^9
Mimulus guttatus (monkey flower)	In progress	14	0.43×10^9
Oryza sativa (rice)	Complete	12	0.48×10^9
Populus trichocarpa (black cottonwood)	In progress	19	0.485×10^9
Solanum lycopersicum (tomato)	In progress	12	0.97×10^9
Solanum tuberosum (potato)	In progress	12	2.02×10^9
Sorghum bicolor (sorghum)	In progress	10	1.6×10^9
Triticum aestivum (wheat)	In progress	21	16.7×10^9
Vigna unguiculata (cowpea)	Pending	11	0.58×10^9
Zea mays (corn)	In progress	10	2.6×10^9

3.1 The Maize Sequencing Project

Arabidopsis and rice, the plants chosen for first full genome sequencing projects, both had relatively small genome sizes. In comparison to these two examples, the larger genomes with high levels of dispersed repetitive sequences pose additional problems for generating complete genome sequences both in terms of strategy and cost. However pressure from within the maize (*Zea mays*) genetics and larger plant science community for several years resulted in The Maize Genome Sequencing Project (Chandler and Brendel 2002). Maize was an obvious choice considering the wealth of information that had already been developed and the importance of the crop.

The first consideration when initiating a genome sequence project is the choice of the line or variety to be used. For maize, the inbred B73 was selected as the target for sequencing. One of the advantages of using an inbred line is that it should be homozygous at all its loci. Therefore, any differences identified in the sequencing data should represent different regions of the genome, or rare sequencing errors, and so make assembly of the genomic sequence easier. One of the disadvantages of using an inbred line is that it is homozygous and so no single nucleotide polymorphisms (SNPs) will be identified directly from the sequencing effort.

Important in the choice of maize were the resources that had been developed, including the bacterial artificial chromosome (BAC) libraries (Cone et al. 2002; Tomkins et al. 2002), sample sequences and a very detailed molecular map. The BAC libraries permitted the inclusion of a CBC approach since these libraries had already been used to generate an integrated genetic/physical map using a high-resolution agarose fingerprinting method (Cone et al. 2002; *http://www.genome.arizona. edu/ fpc/maize/*). The maize sequence resources also included a variety of sequence data. Extensive expressed sequence tags (ESTs) from numerous tissues and developmental stages had been deposited in GenBank. Random genome survey sequences (GSSs) were available from numerous sources. These included the sequences around transposon-tagged sites (many transposons insert in and around genes), sequences from random clones, some completely sequenced BAC clones as well as a larger collection of BAC end-sequences, and a sample of sequences selected for hypomethylation or the presence of long open reading frames (Chandler and Brendel 2002). The strategy for developing a genomic sequence of maize needed to be more complex than that for either Arabidopsis or rice since the maize genome is about 20 times larger than that of Arabidopsis and about 6 times larger than that of rice.

Additionally, as mentioned above, the maize genome organization was more complex than either of these two plants. Only about 20% of the maize genome is actually genes which are organized into islands of variable size that are scattered throughout a sea of highly conserved, high-copy retrotransposons and other repetitive sequences (San Miguel et al. 1996). The interspersion of these families of retrotransposons throughout the genome mitigated against a WGS approach as a first pass on the sequence. Since this was the first attempt to sequence one of the larger complex plant genomes, a combination of approaches was proposed so that their relative efficiencies could be evaluated for subsequent implementation of future large genome sequencing projects. Thus a high-resolution, sequence-ready map of the maize genome was developed by the integration of hundreds of thousands of fluorescent-based BAC clone fingerprint reads, end sequences and the shotgun sequence of BACs with known points throughout the genome as the basis of the CBC effort. As an alternative to the CBC approach for this large genome, two gene enrichment technologies, 'methylation-filtering' [Rabinowicz et al. 1999; *http://maize.danforthcenter.org/strategies. html*] and 'high Cot selection' were also followed. All the data that have, and will be, generated are available in a community database at a central web address (*http://www.agcol.arizona.edu/msll/#overview*). The components of The Sequencing the Maize Genome project include creating the high-information-content fingerprinting (HICF) map, sequencing 450,000 BAC ends and 140 BACs (*http://pgir.rutgers.edu/*) (Nelson et al. 2005). The Consortium for Maize Genomics project (*http://maize.danforthcenter.org/*) involves the sequencing and assembling 500,000 reads from Methyl-filtered, High-Cot, and unfiltered libraries, while the Maize Genome Sequencing site represents the collaboration between these two projects (*http://www.maizegenome.org/*). Both methyl-filtration and high-Cot enrichment methods provided a 7- to 8-fold increase in gene discovery rates as compared to random sequencing (Springer et al. 2004). The data from the high-Cot and methyl-filtered libraries are not completely overlapping so they appear to be complementary approaches to identifying interesting regions of the genome. A third gene enrichment strategy using a hypomethylated partial restriction libraries (Emberton et al. 2005) also resulted in about a six-fold enrichment of genes compared to random sequencing, and this approach was particularly efficient at reducing retrotransposon content.

Sequencing the maize genome has used a successful strategy that included a combination of genome sequencing tools. Thus a BAC-based physical map formed the skeleton from which a minimum tiling path was selected. These BACs were sequenced at low redundancy. At the same time a WGS sequence, again at low redundancy was generated along with the genome enrichment strategies described above. Once all three sets of data are merged linking libraries, such as methylation spanning linker libraries can be used to both complete the genic sequences and orient them with respect to each other (Rabinowicz and Bennetzen 2006).

3.2 Wheat Genome Initiative

Maize and Arabidopsis represent a more than ten-fold difference in genome size. Bread wheat is more than five-fold larger than maize and has the added complication of being a hexaploid. Therefore, the strategy to sequence this plant genome needs to be somewhat different from those developed for the earlier genomes sequenced. The International Wheat Genome Sequencing Consortium (IWGSC) has proposed a strategy that includes genome fractionation, sequencing related smaller genomes and sequencing isolated chromosome(s) (Gill et al. 2004). For this endeavor the IWGSC has selected the specific cultivar, Chinese Spring (CS) as the source for the project as it already has available ample genetic and molecular resources (Gill et al. 2004).

Genome fractionation with respect to methyl-filtration does not appear to be as effective with wheat as it is with other genomes. Seventy-four percent of methyl-filtered library clones are repetitive, down from about 90% for random clones. Thus even a good fractionation would still leave a large complexity to sequence. However, a large EST collection has been produced and these do define a gene space in wheat. The use of these ESTs to select from within a BAC library for those BACs that are relatively gene rich can also reduce the level of sequence necessary. However, although the wheat genome is organized into gene-rich and gene-poor regions, even the gene density in the gene-rich regions is much lower than that for smaller genomes. After the sequencing of 66 random BACs the data was analyzed with the use of gene finder programs and resulted in 30 of these BACs having no identifiable genes, and only one of the 66 having as many as 6 genes. Thus low gene density regions have < 1 gene per 160 kb, while highest gene density regions in this sample have 1 gene every 19 kb (Devos 2007).

Since wheat is a hexaploid, possible sources of comparative sequence information are the smaller, diploid relatives. The diploid models for bread wheat include barley, some of the diploid wheats and more recently *Brachypodium*. The genome size of *Brachypodium* is only 0.36 - 0.39 pg and so is in the same range as the rice and Arabidopsis genome sizes and should be amenable for shotgun sequencing. The organization of the International *Brachypodium* Initiative (IBI) (*http:// www.brachypodium.org/IBI/*) was to foster a collaborative framework that promoted and supported using this temperate grass as a reference species for comparative and functional genomics, biological investigation and strategic research in crop improvement and biomass. The success of this international initiative is seen in the fact that this genome has been included in the Community Sequencing Program of the JGI (*http:// www.jgi.doe.gov/sequencing/why/CSP2007/brachypodium.html*).

The wheat chromosomes are large and of sufficiently different size for some of them to be isolated as single chromosome fractions. However, even if a single chromosome such as 3B (Vrana et al. 2000) was isolated in sufficient quantity; this chromosome alone has a genome size greater than the complete rice genome. An additional concern with the sequencing of isolated chromosomes (especially with respect to shotgun sequencing) is that any contamination with other chromosomal material would be problematical for any assembly.

Any successes of the IWGSC will clearly have an impact on the strategies subsequently developed for sequencing large plant genomes in the same way that the successes of the maize sequencing project will impact strategies for medium-sized plant genomes.

4 NON-GENOME SEQUENCING ACTIVITIES

Once the DNA sequence data is available the identification of genes, their functions and control regions needs to be done. There are a number of approaches to these tasks.

4.1 The Identification of Genes

a. **Full-Length cDNAs.** The annotation of the genomic sequence to identify both genes and regulatory motifs can be accomplished in various ways. An important resource in this effort is the generation of sequences that are expressed which has been described above for ESTs. However, EST sequencing only gives parts of the transcribed sequences and so do not necessarily aid in gene prediction. A more useful resource is full-length cDNAs which allow the determination of the beginning and end of transcription. Knowing where genes start and stop is important in identifying regulatory regions around the genes. Therefore, many of the cDNA libraries being developed try immediately for full-length cDNAs, for their added value, rather than simply cloning fragments. An added value of the full-length cDNAs is that they can be used to transform plants to give functional genes and so help in identifying possible functions for novel sequences.

b. **Comparative Genomics**. The regions conserved between species are likely to be functional. Therefore, identifying conserved regions of genomes is likely to also identify genes. Therefore, the sequencing of novel genomes may be important to confirm genic identities as well as finding novel genes themselves.

4.2 The Identification of Gene Function

The use of genome sequence for identifying possible genic sequences is only a first step. Such sequences need to have a function ascribed to them to confirm their status. A method of identifying gene function are by disrupting the gene function and then searching for a phenotype associated with that disruption. Two of the methods for disrupting genes are either by the introduction of mutations into the genes or by silencing the gene families using RNA interference. The former approach, mutation induction, has been carried out in a genome-wide fashion in a number of plants, but RNA interference studies tend to be by each gene family individually.

Mutation libraries have been developed either by chemical mutagenesis or by insertional mutagenesis.

a. **TILLING** (targeting induced local lesions in genomes) has established itself as a powerful tool for functional genomics and reverse genetics (Colbert et al. 2001). This is a high throughput methodology, the conjunction of a high-throughput molecular reverse genetics technique and mutation induction, allowing for the mining of new alleles in a known locus, and can be performed at a very early stage. The tilled population can be produced and then screened as mutant alleles in specific genes are desired. Again, a tilled population is a resource that can be used by the whole community interested in that species.

b. **Insertional Mutagenesis.** Gene knock-outs are where the activity of a gene has been eliminated due to the insertion of a DNA fragment. The two major methods for generating these are either by inserting a T-DNA or a transposon sequence (Azpiroz-Leehan and Feldmann 1997). T-DNA insertion is the most generally applicable method since it can be used for any plant that can be transformed and regenerated. Since each transformant is an independent event with the T-DNA being relatively randomly inserted into the genome a large number of independent transformation events are needed to inactivate every gene. The need for the generation of large numbers of independent transformants therefore limits this technology to plant species or particular lines that are capable of being transformed in a high throughput manner. The two major T-DNA insertion libraries exist in Arabidopsis and rice. The advantage of using transposons is that following a primary transformation they can then be activated and moved into many regions of the genome. Therefore, only a small number of lines with the transposon present need to be initially produced. The transposon can then be launched to move around the genome.

5 FUTURE EXTENDED SEQUENCING OF PLANT GENOMES

The era of large-scale, wide-spread plant genome sequencing opened with the first two plant genomes sequenced (Arabidopsis and rice). The rapid expansion of sequencing efforts has been facilitated by the reduction in the costs of sequencing and the organization of communities advocating particular species models. With this flurry of activity, many of the obvious candidates for genome sequencing, namely model species with small genomes, species of economic importance or those with a coherent community effort, have either already been completed or are underway. Therefore, a strategy needs to be developed for selecting the next round of choices. Should the choices be made as part of a coherent strategy based on a mixture of scientific and economic needs while still recognizing the value of including phylogenetic position as a selection criterion (Jackson et al. 2006). The first two species with completed genome sequences, *A. thaliana* and *O. sativa*, last shared a common ancestor 150 to 200 million years ago. Subsequent species that have been chosen for sequencing have been selected either for their small genomes or because of the interests of a particular research community. It has been proposed that a portion of the of plant genome sequencing efforts should be directed to selected genomes in a systematic fashion with respect to the relationships of plant groups, so that the number and distribution of sequenced genomes would allow investigators the examination of the evolution of genome structure and function within the context of a robust, well-defined phylogeny (Jackson et al. 2006). The selection of 350 plant lineages based on this line of reasoning would provide sequenced genomes that are at most 20 million years divergent from any other angiosperm species (Jackson et al. 2006). Considering the number of 'complete' sequences produced thus far, either a greatly increased level of funding or an even more dramatic reduction in the costs of sequencing will be needed to complete this catalogue.

6 CONCLUDING REMARKS

Large-scale plant genome initiatives have generally been developed through international consortia (e.g., Arabidopsis, rice, maize and wheat). These consortia efforts have been supported through both public and private funding sources. The rationale for these projects has ranged from potential crop improvement to the identification of evolutionary relationships. Since there has been a utility basis for many of these projects, crops important to the developing countries have not been as prominent in the list. The major underutilized crops for Africa (Lost crops of Africa, Volume 2 Vegetables) are poorly represented in the current list although some have become the topics of collaborations and interest, examples being cowpea and cassava where sequencing is already in progress (Kirkhouse Trust, JGI).

Many of the proposals for large-scale sequencing projects are aimed at the potential coding sequences. This focus on the genes may miss many of the small RNAs that appear to be essential elements in the control of gene function. The finding that short sequences present in the non-coding, otherwise frequently labeled 'junk', DNA are present in coding sequences (Rigoutsos 2007) opens the possibility that many regulatory regions are to be found in this fraction. Therefore, there still may be an importance in obtaining complete genome sequences even if that might still entail a great deal of 'unnecessary' sequencing.

The genome sequencing projects, in particular, the pilot studies on maize, have indicated that the new gene-enrichment, gene-finishing and gene-orientation technologies are efficient, robust and comprehensive. A combination of these strategies should succeed in sequencing the gene-space of large genome plants, and in locating most of these genes and adjacent sequences on the genetic and physical maps, and thus facilitate the sequencing of the gene space of any large plant genome.

As has been noted above, the identification of the DNA sequence, either through genome sequence or ESTs is only the start of the process of understanding the control of plant growth and development. The next stages in plant genome initiatives will include characterizations of the natural variations in plant genomes and the way they affect growth and development. Following or simultaneous with, DNA analysis will be the characterization of the proteome and metabolome with more systems biology approaches being necessary. An overarching theme will continue to be the informatic analysis of the data and the development of databases that are accessible to all the research and development community.

References

Adams MD, Celniker SE, Holt RA, Evans CA, Gocayne JD, Amanatides PG, Scherer SE, Li PW, Hoskins RA, Galle RF, George RA, Lewis SE, Richards S, Ashburner M, Henderson SN, Sutton GG, Wortman JR, Yandell MD, Zhang Q, Chen LX, Brandon RC, Rogers YH, Blazej RG, Champe M, Pfeiffer BD, Wan KH, Doyle C, Baxter EG, Helt G, Nelson CR, Gabor GL, Abril JF, Agbayani A, An HJ, Andrews-Pfannkoch C, Baldwin D, Ballew RM, Basu A, Baxendale J, Bayraktaroglu L, Beasley EM, Beeson KY, Benos PV, Berman BP, Bhandari D, Bolshakov S, Borkova D, Botchan MR, Bouck J, Brokstein P, Brottier P, Burtis KC, Busam DA, Butler H, Cadieu E, Center A, Chandra I, Cherry JM, Cawley S, Dahlke C, Davenport LB, Davies P, de Pablos B, Delcher A, Deng Z, Mays AD, Dew I, Dietz SM, Dodson K, Doup LE, Downes M, Dugan-Rocha S, Dunkov BC, Dunn P, Durbin KJ, Evangelista CC, Ferraz C, Ferriera S, Fleischmann W, Fosler C, Gabrielian AE, Garg NS, Gelbart WM, Glasser K, Glodek A, Gong F, Gorrell JH, Gu Z, Guan P, Harris M, Harris NL, Harvey D, Heiman TJ, Hernandez JR, Houck J, Hostin D, Houston KA, Howland TJ, Wei MH, Ibegwam C, Jalali M, Kalush F, Karpen GH, Ke Z, Kennison JA, Ketchum KA, Kimmel BE, Kodira CD, Kraft C, Kravitz S, Kulp D, Lai Z, Lasko P, Lei Y, Levitsky AA, Li J, Li Z, Liang Y, Lin X, Liu X, Mattei B, McIntosh TC, McLeod MP, McPherson D, Merkulov G, Milshina NV, Mobarry C, Morris J, Moshrefi A, Mount SM, Moy M, Murphy B, Murphy L, Muzny DM, Nelson DL, Nelson DR, Nelson KA, Nixon K, Nusskern DR, Pacleb JM, Palazzolo M, Pittman GS, Pan S, Pollard J, Puri V, Reese MG, Reinert K, Remington K, Saunders D, Scheeler F, Shen H, Shue BC, Siden-Kiamos I, Simpson M, Skupski MP, Smith T, Spier E, Spradling AC, Stapleton M, Strong R, Sun E, Svirskas R, Tector C, Turner R, Venter E, Wang H, Wang X, Wang ZY, Wassarman DA, Weinstock GM, Weissenbach J, Williams SM, Woodage T, Worley KC, Wu D, Yang S, Yao QA, Ye J, Yeh RF, Zaveri JS, Zhan M, Zhang G, Zhao Q, Zheng L, Zheng XH, Zhong FN, Zhong W, Zhou X, Zhu S, Zhu X, Smith HO, Gibbs RA, Myers EW, Rubin GM, Venter JC (2000) The genome sequence of *Drosophila melanogaster*. Science 287: 2185-2195

Azpiroz-Leehan R, Feldmann KA (1997) T-DNA insertion mutagenesis in Arabidopsis: going back and forth. Trends Genet 13: 152-156

C. *elegans* Sequencing Consortium (1998) Genome sequence of the nematode C. *elegans*: a platform for investigating biology. Science 282: 2012-2018

Chandler VL, Brendel V (2002) The Maize Genome Sequencing Project. Plant Physiol 130: 1594-1597

Chory J, Ecker JR, Briggs S, Caboche M, Coruzzi GM, Cook D, Dangl J, Grant S, Guerinot ML, Henikoff S, Martienssen R, Okada K, Raikhel NV, Somerville

CR, Weigel D (2000) National Science Foundation-sponsored workshop report: "The 2010 Project" functional genomics and the virtual plant. A blueprint for understanding how plants are built and how to improve them. Plant Physiol 123: 423-426

Colbert T, Till BJ, Tompa R, Reynolds S, Steine MN, Yeung AT, McCallum CM, Comai L, Henikoff S (2001) High-throughput screening for induced point mutations. Plant Physiol 126: 480-484

Cone K, McMullen M, Bi IV, Davis G, Yim Y-S, Gardiner J, Polacco M, Sanchez-Villeda H, Fang Z, Schroeder S, Havermann SA, Bowers JE, Paterson AH, Soderlund CA, Engler FW, Wing RA, Coe EH Jr (2002) Genetic, physical and informatics resources for maize: on the road to an integrated map. Plant Physiol 130: 1598-1605

Devos K (2007) Unbiased characterization of the hexaploid wheat genome. In: Plant and Animal Genome XIV Conf, San Diego, CA, USA, p 25

Emberton J, Ma J, Yuan Y, SanMiguel P, Bennetzen JL (2005) Gene enrichment in maize with hypomethylated partial restriction (HMPR) libraries. Genome Res 15: 1441-1446

Feng Q, Zhang Y, Hao P, Wang S, Fu G, Huang Y, Li Y, Zhu J, Liu Y, Hu X, Jia P, Zhang Y, Zhao Q, Ying K, Yu S, Tang Y, Weng Q, Zhang L, Lu Y, Mu J, Lu Y, Zhang LS, Yu Z, Fan D, Liu X, Lu T, Li C, Wu Y, Sun T, Lei H, Li T, Hu H, Guan J, Wu M, Zhang R, Zhou B, Chen L, Jin Z, Wang R, Yin H, Cai Z, Ren S, Lv G, Gu W, Zhu G, Tu Y, Jia J, Zhang Y, Chen J, Kang H, Chen X, Shao C, Sun Y, Hu Q, Zhang X, Zhang W, Wang L, Ding C, Sheng H, Gu J, Chen S, Ni L, Zhu F, Chen W, Lan L, Lai Y, Cheng Z, Gu M, Jiang J, Li J, Hong G, Xue Y, Han B (2002) Sequence and analysis of rice chromosome 4. Nature 420: 316-320

Galagan JE, Calvo SE, Borkovich KA, Selker EU, Read ND, Jaffe D, FitzHugh W, Ma L-J, Smirnov S, Purcell S, Rehman B, Elkins T, Engels R, Wang S, Nielsen CB, Butler J, Endrizzi M, Qui D, Ianakiev P, Bell-Pedersen D, Nelson MA, Werner-Washburne M, Selitrennikoff CP, Kinsey JA, Braun EL, Zelter A, Schulte U, Kothe GO, Jedd G, Mewes W, Staben C, Marcotte E, Greenberg D, Roy A, Foley K, Naylor J, Stange-Thomann N, Barrett R, Gnerre S, Kamal M, Kamvysselis M, Mauceli E, Bielke C, Rudd R, Frishman D, Krystofova S, Rasmussen C, Metzenberg R, Perkins DD, Kroken S, Cogoni C, Macino G, Catcheside D, Li W, Pratt RJ, Osmani SA, DeSouza CPC, Glass L, Orbach MJ, Berglund JA, Voelker R, Yarden O, Plamann M, Seiler S, Dunlap J, Radford A, Aramayo R, Natvig DO, Alex LA, Mannhaupt G, Ebbole DJ, Freitag M, Paulsen I, Sachs MS, Lander ES, Nusbaum C, Birren B (2003) The genome sequence of the filamentous fungus *Neurospora crassa*. Nature 422: 859-868

Gill BS, Appels R, Botha-Oberholster A-M, Buell CR, Jeffrey L, Bennetzen JL, Chalhoub B, Chumley F, Dvorak J, Iwanaga M, Keller B, Li W, McCombie WR, Ogihara Y, Quetier F, Sasaki T (2004) A workshop report on wheat genome sequencing: International Genome Research on Wheat Consortium. Genetics 168: 1087-1096

Goff SA, Ricke D, Lan TH, Presting G, Wang RL, Dunn M, Glazebrook J, Sessions A, Oeller P, Varma H, Hadley H, Hutchinson D, Martin C, Katagiri F, Lange BM, Moughamer T, Xia Y, Budworth P, Zhong JP, Miguel T, Paszkowski U, Zhang SP, Colbert M, Sun WL, Chen LL, Cooper B, Park S, Wood TC, Mao L, Quail P, Wing R, Dean R, Yu YS, Zharkikh A, Shen R, Sahasrabudhe S, Thomas A, Cannings R, Gutin A, Pruss D, Reid J, Tavtigian S, Mitchell J, Eldredge G, Scholl T, Miller RM, Bhatnagar S, Adey N, Rubano T, Tusneem N, Robinson R, Feldhaus J, Macalma T, Oliphant A, Briggs S (2002) A draft sequence of the rice genome (*Oryza sativa* L. ssp. *japonica*). Science 296: 92-100

Gruenbaum Y, Naveh-Many T, Cedar H, Razin A (1981) Sequence specificity of methylation in higher plant DNA. Nature 292: 860-862

Han CS, Sutherland RD, Jewett PB, Campbell ML, Meincke LJ, Tesmer JG, Mundt MO, Fawcett JJ, Kim UJ, Deaven LL, Doggett NA (2000) Construction of a BAC contig map of chromosome 16q by two-dimensional overgo hybridization. Genome Res 10: 714-721

International Human Genome Mapping Consortium (2001) Initial sequencing and analysis of the human genome. Nature 409: 860-921

Jackson S, Rounsley S, Purugganan M (2006) Comparative sequencing of plant genomes: Choices to make. Plant Cell 18: 1100-1104

Kirkhouse Trust (*http://www.kirkhousetrust.org/index.html*)

Laurie DA, Bennett MD (1985) Nuclear DNA content in the genera *Zea* and *Sorghum*. Intergeneric, interspecific and intraspecific variation. Heredity 55: 307-313

Martienssen RA (1999) Differential methylation of genes and retrotransposons facilitates shotgun sequencing of the corn genome. Nat Genet 23: 305-308

Margulies M, Egholm M, Altman WE, Attiya S, Bader JS, Bemben LA, Berka J, Braverman MS, Chen Y-J, Chen Z, Dewell SB, Du L, Fierro1 JM, Gomes XV, Godwin BC, He W, Helgesen S, Ho CH, Irzyk GP, Jando SC, Alenquer MLI, Jarvie TP, Jirage KB, Kim J-B, Knight JR, Lanza JR, Leamon JH, Lefkowitz SM, Lei M, Li J, Lohman KL, Lu H, Makhijani VB, McDade KE, McKenna MP, Myers EW, Nickerson E, Nobile JR, Plant R, Puc BP, Ronan MT, Roth GT, Sarkis GJ, Simons JF, Simpson JW, Srinivasan M, Tartaro KR, Tomasz A, Vogt KA, Volkmer GA, Wang SH, Wang Y, Weiner MP, Yu1 P, Begley RF, Rothberg JM (2005) Genome sequencing in microfabricated high-density picolitre reactors. Nature 437: 376-380

Nelson WM, Bharti AK, Butler E, Wei F, Fuks G, Kim H, Wing RA, Messing J, Soderlund C (2005) Whole-genome validation of high-information-content fingerprinting. Plant Physiol 139: 27-38

Rabinowicz PD, Bennetzen JL (2006) The maize genome as a model for efficient sequence analysis of large plant genomes. Curr Opin Plant Biol 9: 149-156

Rabinowicz PD, Schutz K, Dedhia N, Yordan C, Parnell LD, Stein L, McCombie WR, Martienssen RA (1999) Differential methylation of genes and retrotransposons facilitates shotgun sequencing of the maize genome. Nat Genet 23: 305-308

Rigoutsos I (2007) Evidence for the existence of organism-specific regulatory elements that are linked to RNAi. In: Plant and Animal Genome XIV Conf, San Diego, CA, USA, S4

San Miguel P, Tikhonov A, Jin YK, Motchoulskaia N, Zakharov D, Melake-Berhan A, Springer PS, Edwards KJ, Lee M, Avramova Z, Bennetzen JL (1996) Nested retrotransposons in the intergenic regions of the maize genome. Science 274: 765-768

Sasaki T, Matsumoto T, Yamamoto K, Sakata K, Baba T, Katayose Y, Wu J, Niimura Y, Cheng Z, Nagamura Y, Antonio BA, Kanamori H, Hosokawa S, Masukawa M, Arikawa K, Chiden Y, Hayashi M, Okamoto M, Ando T, Aoki H, Arita K, Hamada M, Harada C, Hijishita S, Honda M, Ichikawa Y, Idonuma A, Iijima M, Ikeda M, Ikeno M, Ito S, Ito T, Ito Y, Ito Y, Iwabuchi A, Kamiya K, Karasawa W, Katagiri S, Kikuta A, Kobayashi N, Kono I, Machita K, Maehara T, Mizuno H, Mizubayashi T, Mukai Y, Nagasaki H, Nakashima M, Nakama Y, Nakamichi Y, Nakamura M, Namiki N, Negishi M, Ohta I, Ono N, Saji S, Sakai K, Shibata M, Shimokawa T, Shomura A, Song J, Takazaki Y, Terasawa K, Tsuji K, Waki K, Yamagata H, Yamane H, Yoshiki S, Yoshihara R, Yukawa K, Zhong H, Iwama H, Endo T, Ito H, Hahn JH, Kim HI, Eun MY, Yano M, Jiang J, Gojobori T (2002) The genome sequence and structure of rice chromosome 1. Nature 420: 312-316

Senior ML, Heun M (1993) Mapping maize microsatellites and polymerase chain-reaction confirmation of the targeted repeats using a CT primer. Genome 36: 884-889

Springer NM, Xu X, Barbazuk WB (2004) Utility of different gene enrichment approaches toward identifying and sequencing the maize gene space. Plant Physiol 136: 3023-3033

The Arabidopsis Genome Initiative (2000) Analysis of the genome sequence of the flowering plant *Arabidopsis thaliana*. Nature 408: 796-815

The Rice Chromosome 10 Sequencing Consortium (2003) In-depth view of structure, activity, and evolution of rice chromosome 10. Science 300: 1566-1569

Tomkins JP, Davis G, Main D, Yim Y, Duru N, Musket T, Goicoechea JL, Frisch DA, Coe EH Jr, Wing RA (2002) Construction and characterization of a deep-coverage bacterial artificial chromosome library for maize. Crop Sci 42: 928-933

Venter JC, Smith HO, Hood L (1996) A new strategy for genome sequencing. Nature 381: 364-366

Vos P, Hogers R, Bleeker M, Rijans M, van de Lee T, Hornes M, Frijters A, Pot J, Peleman J, Kuiper M, Zabeau M (1995) AFLP - A new technique for DNA-fingerprinting. Nucl Acids Res 23: 4407-4414

Vrana J, Kubalakova M, Simkova H, Cíhalíkova J, Lysak MA, Dolezel J (2000) Flow sorting of mitotic chromosomes in common wheat (*Triticum aestivum* L.). Genetics 156: 2033-2041

Williams JGK, Kubelik AR, Livak J, Rafalski AJA, Tingey SV (1990) DNA polymorphisms amplified by arbitrary primers are useful as genetic-markers Nucl Acids Res 18: 531-6535

11 Computing Strategies and Software for Gene Mapping

Brian S. Yandell[1] **and Peter Bradbury**[2]

[1] University of Wisconsin, 1675, Observatory Drive, Biometry Program, Madison, WI 53706, USA;
E-mail: byandell@wisc.edu

[2] USDA-ARS, 741 Rhodes Hall, Ithaca, NY 14853, USA

1 INTRODUCTION

This chapter focuses on computing strategies and software for gene mapping. We separately address software strategies for experimental crosses, known as quantitative trait loci (QTL) mapping, from those used in natural populations for association analysis. Both of these approaches look for correlations between genotypes and phenotypes. For most of the development in this chapter, we focus on a single phenotype, but we briefly note strategies that can examine multiple correlated phenotypes.

The goal of gene mapping is model selection for the genetic architecture of a phenotypic trait. That is, we wish to infer what genomic regions, or genetic loci, are associated with a phenotype, and what mode of gene action is involved. Genetic loci typically cover several megabases of DNA containing many closely linked genes. Gene action is often interpreted in terms of the additive and/or dominance effect of single loci, and epistatic interaction among two or more loci. A well estimated genetic architecture for a trait of interest can be used in disease prognosis, marker-assisted selection or studies of the evolution of the trait.

Depending on the method of analysis, the genomic region for a locus may cover a single genetic marker or a set of markers. Genetic markers are short segments of DNA, or in some cases qualitative traits, that can be scored to partially or fully identify the inheritance of parental alleles. The markers are ideally arranged in a genetic linkage map corresponding to the physical map or sequence of the genome under study. While these previously determined maps are now routinely available for model organisms, such as Arabidopsis and rice, there are many taxa with no marker map, or only with maps based on the marker data in hand.

This chapter does not address the problem of building marker maps, except to point out that there are standard tools to build maps for experimental crosses, including MapMaker (Lander and Green 1987; Lander et al. 1987), R/qtl (Broman et al. 2003), MultiPoint (Mester et al. 2004) and JoinMap (Jansen 1993). Genetic maps are typically in units of centi-Morgan (cM), with two loci separated by 1 cM having an expected recombination frequency of 1%. Roughly speaking, 1 cM is about 1 megabase of DNA, depending on taxa. However, recombination frequency is not linear with genetic distance. The choice of recombination model, defining the relationship between genetic distance and recombination fraction, plays a minor role in gene mapping for experimental crosses once a linkage map is built. Most gene mapping methods for experimental crosses assume no crossover interference. Typically there are further assumptions of independent crossovers with equal likelihood across the genome (Haldane map function), although most packages allow other options. More complicated experimental cross designs, particularly in outbred populations, may require multipoint mapping. There are many subtle issues about the relationship of recombination to distance that are beyond the scope of this chapter.

Gene mapping involves multiple tests for correlation between a set of genetic markers and a trait of interest. Typically for an individual experiment, a large number of markers are tested against a phenotype, which leads to multiple testing issues when each marker-trait combination is tested individually. We will address these issues in context of specific software strategies.

QTL mapping in experimental crosses usually begins with two inbred lines. The individuals in such an experimental cross, are created, nurtured and measured under uniform conditions so that the primary differences among individuals are due to their genetic 'treatment'. In contrast, association analysis considers the relationship between genotype and phenotype in a natural population. It uses the extent of linkage disequilibrium (LD) between a trait and markers to infer the location of QTLs.

The existence of an association between a marker and a trait in an experimental cross implies that either the marker itself is the cause of the observed phenotypic variation or that it is linked to a causal polymorphism. There are two problems with interpreting correlation as causation in an association study of a natural population: (1) a linked marker might not be in LD with the trait; and (2) a marker that is in LD with a trait might not be linked to it.

Several distinctions between QTL mapping with experimental crosses and association mapping with natural populations are worth mentioning beyond the issue of causal inference. The resolution of QTL mapping in experimental crosses is typically fairly coarse, on the order of 5-20 cM, whereas association mapping can lead to much finer resolution maps. Experimental crosses from two inbred lines have at most two alleles at any genomic locus, while natural populations may have multiple alleles. As a result, heterozygosity is fairly uniform in experimental crosses, and markers are usually informative, or not, of parentage for the whole population. However, in natural populations, markers may have quite different heterozygosity and may only be informative for a certain subset of individuals. All individuals in an experimental cross have equal genetic correlation on average, but this does not hold in natural populations. Finally, missing data presents much more difficult problems in natural populations, where one may need to infer the phase of inheritance to estimate haplotypes.

2   QTL ANALYSIS WITH INBRED LINES

Modern methods for QTL analysis in experimental crosses derived from inbred lines largely employ the interval mapping framework developed by Lander and Botstein (1989). This fundamental paper viewed the relationship between phenotype and genotype as a genomic question, providing a visual LOD score map to profile evidence for association across the genome. The linkage map inference inherent in this work (Lander and Green 1987) provided an algorithmic approach to model missing genotype information between markers.

Good expositions of QTL methods for experimental crosses can be found in Broman (2001) or Hackett (2003). Doerge et al. (1997) reviewed the statistical issues, while Mackay (2001) placed QTL mapping in the context of identifying underlying mechanisms. Several recent papers have addressed the difficult issue of moving from QTL intervals to confirmed genes (Guo and Lange 2000; Nadeau and Frankel 2000; Glazier et al. 2002; Korstanje and Paigen 2002; Page et al. 2003; Darvasi 2005).

This section gives an overview of QTL analysis for inbred lines, contrasting the algorithms commonly used. Rather than show screen shots of individual packages, we primarily use graphs developed in R (R Core Development Team 2006) to illustrate concepts. We point out which packages use what methods as we go along. We begin with a detailed investigation for mapping a single QTL. This leads to questions of assessing significance (thresholds, support intervals), which leads to model selection for multiple QTL. We finish with a brief overview of packages.

2.1 Single QTL Mapping

QTL mapping models the relationship between phenotype and genotype at each location across the genome. We briefly review methods employed to assess this relationship profiled across the genome. We begin with a complete data situation with 100 individuals and 201 markers spaced every 1 cM, with a single QTL at 100 cM. Here all methods agree exactly with each other.

Marker regression (MR) examines the association between each marker and a phenotype. This was the available method until 1989, performed marker by marker using *t*-tests for backcross or ANOVA for intercross. When markers are arranged in a linkage map, *p*-value summaries provide a crude genome-wide profile. However, it was widely recognized that regression with a single marker confounds the allele substitution effect and linkage between each marker and the pertinent genetic locus. Further, any missing data reduces power, as those individuals must be dropped for that marker. This method is still sometimes used as a quick initial examination, particularly in genomes with no linkage map yet available.

Simple interval mapping (SIM) models the relationship between phenotype and genotype by testing for a QTL at each location across the genome (Lander and Botstein 1989). This involves a likelihood ratio test, rescaled as a familiar LOD score. That is, interval mapping (IM) states that the phenotype has a normal, or bell-shaped, histogram for any given genotype, with a different mean depending on the genotype. When the genotype is not known, IM assigns probabilities to missing genotype data based on informative flanking markers. The full likelihood mixes over all possible missing genotype values. That is, the distribution for an individual is a weighted average of normal distributions, with the weights being the probabilities for QTL genotypes given flanking markers. Individuals with the same flanking marker genotypes would have the same mixture distribution. The profile likelihood is the product of these distributions at the phenotype values maximized for unknown effects using the expectation-maximization (EM) algorithm (see Lander and Botstein 1989). The log odds (LOD) is the log base 10 of this likelihood profile divided by the null likelihood for the no QTL model.

Regression mapping, also known as Haley-Knott (HK) regression (Haley and Knott 1992; Martinez and Curnow 1992), considers regression of the phenotype on the expected genotype, which approximates the more correct mixture detailed above. That is, they agree in mean value, but differ in variance and in distribution shape. This is not a serious problem if markers are closely spaced and there are only a few missing marker genotypes. However, with selective genotyping, this Haley-Knott regression can be seriously biased (Kao 2000). Xu (1995) showed that this method can overestimate residual variance when QTL effects are large or markers are widely spaced.

When there are no missing data, interval mapping and Haley-Knott regression are exactly identical. Marker regression agrees as well at every fully informative marker with these curves once it is rescaled in terms of LOD scores. That is, there is no approximation involved when we have complete data at markers, and all methods agree. Consider the following simple example with 100 individuals in a backcross fully genotyped at 201 markers spaced every 1 cM. The QTL is located at 100 cM, with a substitution effect of 2 relative to a standard deviation of 1. The LOD curve peaks near 100 cM, but slowly trails off (Figure 1a). The attenuated substitution effect at a locus that has a recombination rate of r with the QTL is $(1-2r)a$, where a is the substitution effect at the QTL (Figure 1b). The long attenuation of the LOD curve is due to this confounding of QTL substitution effect and linkage of nearby markers (Wright and Kong 1997). The expected LOD at the QTL is $(n/2) * \log10(1 + a^2/2)$, which is attenuated to $(n/2) * \log10((1 + a^2/2)/(1 + 2a^2r(1 - r)))$ at the linked marker. The attenuation can only be relieved by increasing sample size or considering multiple QTL models.

2.1.1 Missing Genotypes

Linkage maps for some time have had markers every 5 to 20 cM, requiring some way to fill in for missing genotypes between markers. Thus genotype information between markers is completely missing. Further, some markers may have missing values due to technical reasons unrelated to phenotype. In other situations, genotype data may be missing in a pattern associated with the phenotype, either by design or chance. This subsection examines the impact of various types of missing data on QTL mapping.

Marker data missing at random does not introduce any appreciable bias to QTL mapping. However, missing data must be replaced by assumptions about what that data might have been. This introduces

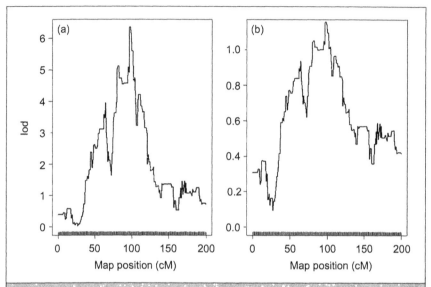

Fig. I Simulation with I QTL at I00 cM and complete marker information. Markers every I cM. (a) LOD profile scan for simple interval mapping. Black line is interval mapping; gray dashed line is expected LOD profile. Note long-term attenuation of peak away from I00 cM. (b) Substitution effect attenuated by linkage (I-2r) to nearby markers. Estimated effect from interval mapping in black; idealized effect in gray dotted.

uncertainty into LOD maps, and the impact of that uncertainty depends on the algorithm employed. Several ways to address missing genotype data have been implemented in QTL software. It is important to understand these because some methods have known bias when data are not missing at random.

It is useful here to introduce a third method of QTL mapping known as multiple imputation (IMP). The basic idea is to fill in the missing genotype at each 1 cM step using the assumed map function. This is done multiple times, recognizing that any particular realization is flawed. These multiple imputations are then averaged in a careful way to produce a log posterior density (LPD) that is very close to the LOD score (Sen and Churchill 2001).

What is the impact of marker spacing on LOD profiles? Simple IM fills in missing data between markers using the EM algorithm and the map function. This leads to a smooth parabola shape for the LOD between partially informative markers. HK regression tends to dampen those parabolas slightly. Multiple imputation averages as well, but may occasionally wiggle due to sampling variation. Figure 2a shows the effect

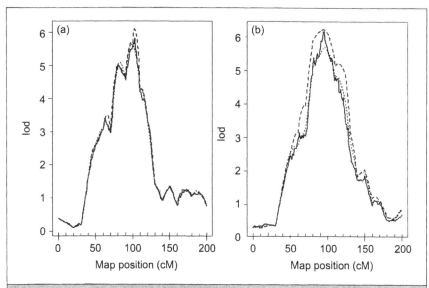

Fig. 2 Impact of marker spacing and genotype data missing at random. Markers every 10 cM, with profile evaluated every 1 cM. (a) Complete marker data. (b) Half of marker data missing at random. Dashed line = interval mapping, dotted = Haley-Knott regression, solid = multiple imputation; gray dotted is expected substitution effect.

of 10 cM spacing on the single QTL example introduced earlier. When we in addition remove half the marker data at random, the LOD curves keep the same basic shape (Figure 2b).

Markers may have missing data due to a design decision, such as selective genotyping. Selective genotyping implies a biased pattern of missing marker information. Typically, some fraction (10-25%) of extreme high and extreme low phenotype individuals are fully genotyped, while those in the middle range are not genotyped. All phenotype data are used for analysis, even those with no genotype information. Figure 3a shows how HK regression greatly overestimates the strength of the QTL signal when only extreme quartiles are genotyped. The other two methods tend to have reduced peaks relative to the figures seen earlier, which is not surprising since much data has been lost, leading to a reduction in power. However, they capture the same essential strength of relationship between phenotype and genotype and are not inflated by the pattern of missing genotype data.

The impact of selective genotyping on Haley-Knott regression can be more or less ameliorated by having a framework map of fully informative markers. Figure 3b shows a situation with fully informative markers every 20 cM starting at 10 cM. The peak for Haley-Knott regression is still biased upwards. Note that the addition of these fully informative framework markers drops the proportion of missing data per marker from 50% to at most 20% in this situation. For more information on designing experiments with selective genotyping (see Sen et al. 2005).

It should be pointed out that the EM method can introduce artifacts in the presence of some patterns of missing data. Sometimes we see an unusual spike in the LOD map between partially informative markers. This indicates a region where the EM method is achieving 'too good' a fit to the data. Multiple imputation and HK regression tend to dampen such effects. The bottom line is that missing data is 'filled in' by assumptions in one way or another. No way is perfect, and artifacts can emerge. Always use caution interpreting QTL analyses in the presence of much missing data.

2.1.2 Detection of QTL

LOD maps as shown above provide a sense of where strong correlation between phenotype and genotype lie. But how large is large enough to say a QTL is detected with confidence? Theoretical guidelines based on high-powered math suggest a LOD threshold of about 3 (Lander and Botstein 1989; Lander and Kruglyak 1995), with some adjustment for design. In practice, it is wise to use resampling methods to assess the strength of the LOD signal.

A permutation threshold can be computed with most packages. The idea is to permute, or shuffle, the phenotypes independent of the genotypes. For each permutation, construct the LOD profile and record the maximum. The distribution of maximum LOD under the 'null' model of no QTL is approximated by a histogram of such values. Typically, 1,000 permutations are recommended for genome-wide purposes (Churchill and Doerge 1994). For the 10 cM data spacing shown in Figure 3, the EM thresholds at 1%, 5% and 10% are, respectively, 2.61, 1.81, and 1.53. Similar thresholds are found for the Haley-Knott and multiple imputation methods (not shown). Figure 4 shows these thresholds superimposed on the LOD maps. There is strong evidence to support a QTL, but there is also a wide region of the LOD curve that exceeds the 1% threshold. Note that permutation thresholds for the sex chromosome may need to constructed separately (Broman et al. 2006).

Common practice involves constructing a LOD support interval that spans a genomic region where LOD values are within 1 to 2 LOD of the peak. Strictly speaking a LOD support interval is not analogous to a

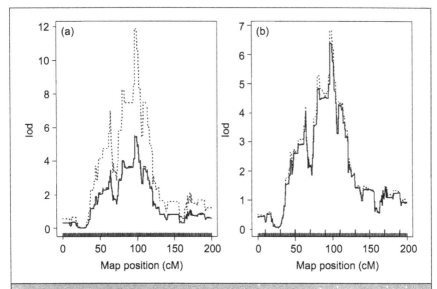

Fig. 3 Impact of selective genotyping. (a) Only extreme quartiles are genotyped while the genotypes for the middle 50% of individuals are missing. (b) Complete genotyping at framework map (markers every 20 cM indicated by larger tics); selective genotyping at all other markers. Note different vertical scales on the two figures.

confidence interval, although it is commonly misinterpreted as such. Its properties, in terms of percent of times it covers the true QTL location, depend on the marker spacings, pattern of missing data, and presence of other linked and epistatic QTLs. Still, it is a useful guide to the uncertainty in the QTL location. The 1.5 LOD support interval shown in Figure 4 is 76 to 109 cM for 10 cM marker spacing, 80 to 101 cM for 1 cM spacing. Permutation thresholds for single QTL scans are available in most QTL packages, although some only offer them for single QTL scans. Permutation thresholds for more complicated models, with multiple QTL and/or covariates, require more care (Doerge and Churchill 1996). Recently, some faster methods for permutation thresholds have emerged (Zou et al. 2004; Jin et al. 2007) and are gradually being incorporated into QTL packages.

Another resampling approach, bootstrapping, has been applied to QTL mapping. Several packages (Seaton et al. 2002; Mester et al. 2004; Wang et al. 2003) offer bootstrap of the distribution of the probability of the presence of a QTL across intervals. This tool is not well studied in this context, can be misleading, and should be approached with caution (Manichaikul et al. 2006).

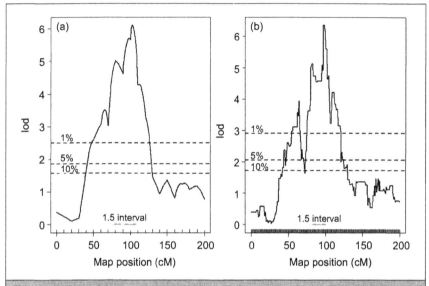

Fig. 4 Permutation-based thresholds for maximum LOD. Dotted horizontal lines at 1%, 5% and 10% threshold using EM method. Gray horizontal line is 1.5 LOD support interval from 76 to 109 cM. (a) Markers every 10 cM; (b) markers every 1 cM. Note slightly higher thresholds for 1 cM spacing.

2.1.3 Model Selection for Multiple QTL

Detection of a single QTL is an important first step. However, most complex traits are likely influenced by several, if not hundreds of genetic loci. We cannot hope to uncover the 'true model' in any given experimental cross, but we can infer major aspects of the genetic architecture that are supported by the data. Extension from a single QTL to multiple QTLs has been implemented in several distinct ways. We illustrate this with the one QTL example above and with another simulation having several QTLs. We focus on R/qtl (Broman et al. 2003), R/qtlbim (Yandell et al. 2007) and QTLCart (Basten et al. 1999) in this demonstration, as they have the key features of the methods found in most packages. Further, their graphics can be readily annotated and prepared for publication. We show graphs of one- and two-dimensional scans, as well as model selection tools. We briefly discuss model selection criteria, though that is beyond the scope of this chapter.

We need to briefly state that epistasis herein refers to the effect on a phenotype of statistical interaction among distinct genetic loci. Model-based epistasis is related to biological epistasis, which W. Bateson defined in 1907: "The allelic state at one locus can mask or uncover the effects of allelic variation at another" (*cf.* Hollander 1955). We model epistatic interaction for complex traits with QTL as we do in ANOVA. Thus marker regression (MR) can be extended to multiple QTL using standard statistical packages. This quick and dirty method should only be considered a first pass, and epistasis uncoverd with MR is in general not to be trusted.

The process of model selection typically involves finding a balance between models that are too simple, missing key features and introducing bias, models that are overly complicated, inflating the variance of parameter estimates. A more complicated model with more QTLs will fit better and have a higher likelihood. But we pay a price for this: interpretation of a larger model is more complicated, and its components—including loci and genotypic effects—are each less precisely estimated. Further, the utility of a model is in its ability to predict effects of genoytpe on phenotype in a new experiment. An overly simple model will give biased predictions, missing linked QTLs and important epistasis. However, a model that is too complicated can be biased in other ways, being constrained to particulars of the current experiment. Thus, apparent evidence of subtle QTLs and epistasis may be artifacts of the data at hand, and may not generalize to other settings. These problems are not new to QTLs, and they have been well studied in stepwise regression. The basic idea in comparing models of different sizes, varying by the number of QTLs and the degree of epistasis, is to use an information criterion that equals the likelihood less some penalty that measures model 'complexity'. No one criterion is 'best', as each involves reducing a complicated, multidimensional comparison to a single number. We often compare models based on their maximum LOD scores. This criterion has no penalty for complexity, and is most appropriate when doing a few comparisons, say one vs. two QTLs, with or without epistasis.

Broman and Speed (2002) compare various methods of model selection for multiple QTLs that are located only at markers spaced every 10 cM. This is one of the only simulation studies to date comparing multiple QTL strategies. We refer the reader to this paper for a discussion of information criteria that measure the bias/variance tradeoff.

2.1.4 Multiple QTL Estimation Approaches

This subsection reviews the approaches to multiple QTL model fitting commonly used in software. These include regression mapping approximations, maximum likelihood, and Bayesian posteriors. A hybrid between regression and maximum likelihood is also highlighted, as it is found in several packages. Model fitting is typically coupled with a model selection procedure, as we indicate in particular for the recommended maximum likelihood and Bayesian approaches.

As pointed out above, HK regression mapping is challenged by missing genotype data. This problem is exacerbated with multiple QTL (Xu 1995; Kao 2000). Still, the method is fast and works well when there are few missing data subject to selective genotyping. QTL Express (Seaton et al. 2002) is a popular package for this method. Tools for HK regression with multiple interacting QTLs are available as well in R/qtl (Broman et al. 2003).

Hybrid methods, known as composite interval mapping (CIM) (Zeng 1994) or multiple QTL mapping (Jansen 1993) were proposed as ways to scan the genome with interval mapping while approximately adjusting for other QTLs using nearby markers, or co-factors. This method is fast and fairly easy to implement, hence it has been incorporated into several packages, including PLABQTL (Utz and Melchinger 1996), MapQTL (van Ooijen and Maliepaard 1996) and MapManager/QTX (Meer et al. 2004). However, the approximation can be problematic, and its properties depend on the minimal spacing window to linked co-factors as well as the number of co-factors. An early analysis of morphological shape (Liu et al. 1996) using CIM was later revised using MIM described below (Zeng et al. 2000). Broman and Speed (2002) showed that CIM is effective when there are 'enough' co-factors, but misses linked QTL if there are too few co-factors. Returning to the fully informative simulation of a single QTL, Figure 5 shows how CIM can narrow the support interval for a QTL, but at the same time the peak is slightly elevated as the variance is artificially deflated by the co-factors. The 5% permutation threshold for CIM is estimated at 2.04 (300 permutations, default settings), compared to 2.09 for IM (1000 permutations). CIM should be used with caution as an exploratory tool, in conjunction with other methods described below.

Methods that estimate all QTLs together began emerging in the late 1990s and into this century. These methods include an extension of EM for maximum likelihood (Kao et al. 1999; Kao and Zeng 2002), known as multiple interval mapping (MIM). MIM is available in QTLCart/WinQTLCart (Basten et al. 1999) and in MultiQTL (Mester et al. 2004). It turns out there are a number of technical issues that arise, making this a difficult problem. Basically, it is hard to know when you have actually reached maximum for a model with many QTLs! On top of that, there is the issue of deciding among models. Again, Broman and Speed (2002) provide simulation studies of the stepwise regression model selection strategy adapted to multiple QTL mapping. MIM applied to the earlier

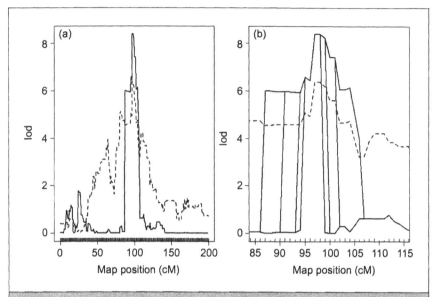

Fig. 5 Composite interval mapping (CIM) LOD profile. IM profile is dashed, CIM solid. Right panel zooms in on peak area, with CIM curves for 5 co-factors and window widths 10 (default), 5, 2 and 1 cM.

example definitively concludes there is exactly one QTL, thus reducing to simple interval mapping in this instance.

Two primary Bayesian interval mapping (BIM) methods have emerged, and they have been incorporated into available packages: multiple imputation and Markov chain Monte Carlo (MCMC). The Bayesian approach focuses on studying random samples from the posterior distribution, which is basically the likelihood weighted by a prior distribution on unknowns, rescaled to have area 1 to make it a distribution. The posterior is a useful device to examine the entire likelihood, rather than focusing only on the maximum peak. Priors play the role of formally incorporating uncertainty about unknowns into our models. These include positions of QTLs, their genotypic effects and epistatic effects, and even the complexity of the genetic architecture.

Multiple imputation (IMP), introduced above, is available in R/qtl (Broman et al. 2003) and the Matlab application, Pseudomarker (Sen and Churchill 2001). This method profiles the log posterior density (LPD), yielding curves markedly similar to the LOD curves, as shown in Figures 2-4 above. The LOD maximizes the genotypic effects at each locus, while the LPD averages over the effects; these are essentially the same when the assumed phenotype model is normal. The advantage of the multiple imputation method lies in its simplicity. We fill in (impute) missing genotype data based on flanking markers, then compute the LPD; repeat this several times and (carefully) average the results. Thus, it can readily be extended to multiple QTL using standard linear model tools for model building and model selection. Many useful tools along this line are incorporated into R/qtl (Broman et al. 2003).

A second BIM method was developed using Markov chain Monte Carlo (MCMC). The MCMC method in itself is not Bayesian—it has been used for maximum likelihood in human QTL studies (Heath 1997)—but it can be used to obtain random samples from the posterior. MCMC methods have proven quite useful for complex models in a variety of settings where a Bayesian perspective, modeling the uncertainty about many relationships, is important (*cf.* Gelman et al. 2003). Initially handling a fixed number of QTLs, (Satagopan et al. 1996) MCMC methods for QTL now incorporate uncertainty about the genetic architecture (Satagopan and Yandell 1996; Sillanpää and Arjas E 1998; Stephens and Fisch 1998; Gaffney 2001; Yi 2004; Wang et al. 2005; Yi et al. 2005; Yandell et al. 2007). This allows us to use the Bayesian approach for model selection, in which we allow the number, position and genotypic effects of QTL to be unknown.

Figure 6 shows MCMC applied to the fully informative one QTL example used to this point. The peaks for LOD and LPD are nearly the same, but the MCMC curve drops off more quickly. This is also apparent with the substitution effect. The dropoff is more dramatic with larger sample sizes (here we have 100) and/or larger substitution effects. The reason for this drop-off is that the question has changed somewhat. Up until now, the LOD or LPD profiles compared a model with one QTL at the locus under consideration against the null model of no QTL. The MCMC samples allow us to compare models with and without a particular locus while allowing other QTL to be present. Thus, away from the peak, the question is about a second QTL allowing for a major QTL found near 100 cM. This type of comparison is analogous to type III ANOVA, in which we test for a second predictor adjusting for the effect of the first being predictor. Simple IM profiles shown earlier are analogous to type I ANOVA, asking about one predictor at a time. CIM approximates this type III ANOVA idea by using co-factors (Figure 5).

MIM provides formal inference on genetic architecture, following a stepwise approach to model building that compares simpler models to more complicated models by adding or dropping main QTLs and/or epistatic QTLs. Thus, at each point, we consider one model at a time. Multiple imputation fits a model with a set number of QTL, again to be

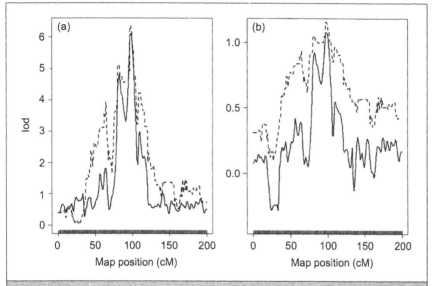

Fig. 6 Bayesian LPD allowing for 2 QTL. Solid curve is from Bayesian posterior, simple interval mapping is dashed. (a) LOD and LPD; (b) Substitution effect.

compared with other genetic architectures through profiles or other summaries. The MCMC approach allows us to sample all possible models, or more exactly the more probably models of the genetic architecture. Thus we have information immediately about a variety of plausible models and can construct summaries to explore these in detail.

2.1.5 Detailed Analysis of Multiple QTL Simulated Cross

We now consider a simulation with four QTLs on three chromosomes, including two pairs of epistatic loci (Table 1). We draw a sample of 100 individuals from a backcross, with markers spaced roughly every 10 cM on chromosomes that are 60 cM in length. There is a small amount of data missing at random. The goal is to recover the genetic architecture as much as possible. Our strategy is to consider models with one, two or an arbitrary number of QTLs and examine how strong the data are to support them. We give detailed analysis using IM in conjunction with multiple imputation, MIM and BIM via MCMC.

Simple IM picks up strong evidence for a QTL on chromosome 1 and weak evidence on chromosome 2 (Figure 7a and Table 2). The effects are underestimated (Figure 7b). None of this is surprising, since IM only considers one effect at a time. The QTL on chromosome 2 is suggestive

Table 1 Backcross simulation with 4 QTLs on 3 chromosomes, and two pairs of epistatic loci, both with QTLs 2. Standard deviation was 1; effects were supplied while heritabilities were estimated from simulated sample of 100 individuals.

qtl	chr	pos	effect	herit	qtl2	effect2	herit2
1	1	15	1.5	25.6%			
2	1	45	0.0	0.0%			
3	2	12	-1.0	11.4%	2	-2.0	11.4%
4	3	15	0.0	0.0%	2	3.0	26.5%

Table 2 IM one-dimensional summary. LOD scores for single main QTL at best position on each chromosome. The notation "c2.loc15" means chromosome 2, location 15 cM.

	chr	pos	lod
C1M2	1	15.9	4.156
c2.loc15	2	29.7	2.298
C3M5	3	45.9	0.816

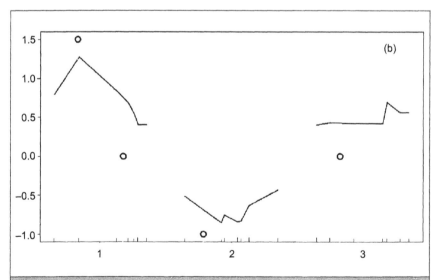

Fig. 7 Simple interval mapping on multiple QTL example. (a) LOD scan with 1,000 permutation-based thresholds constructed with method HK; (b) effect scans (line = estimate, circle = true). Vertical gray dashed lines at true location of QTLs.

(1% and 5% critical values based on 1000 HK permutations are 2.49 and 1.89, respectively).

A two-dimensional EM scan of the genome shows some of the epistatic effects (Figure 8). Similar scans are obtained with HK and IMP for this multiple QTL simulation. Note the strong evidence for epistasis between chromosomes 1 and 3, and little apparent evidence for any other QTL. All methods pick up the epistasis between chromosomes 1 and 3, but they show little indication of the other epistatic pair (1 and 2). Normally, a 'zscale' would appear to the right, but that was suppressed so that the red lines at true values could be added. Summaries shown in

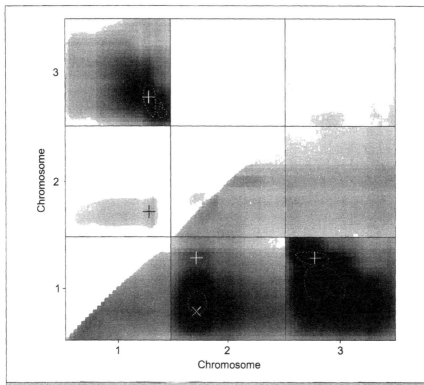

Fig. 8 Two-dimensional profile of simulated data with multiple QTL. Scan is based on EM, but HK and IMP are similar. Contours are 1.5 down from peak. Lower triangle shows LOD comparing full model with epistasis to no QTL. Upper triangle showing LOD of full model to additive model assesses epistasis. Diagonal has LOD for single QTL model. Notice that chromosome 1 and 2 epistasis is barely visible in upper triangle, but appears to contribute in lower triangle. Crosses at true loci pairs; X at mistaken locus pair.

Table 3 support one of the two epistatic pairs (LODs of 8.06 and 1.95, respectively). The 1% critical value based on 1,000 permutations (HK method) is 3.27; the *p*-value for the lesser epistatic pair is 0.17. Thus there is strong evidence for the 1 by 3 epistasis, but very weak evidence for the 1 by 2 epistasis. The latter would be kept for exploratory purposes, but probably not believed to be real.

Given the one-dimensional and two-dimensional scans, we now have a hypothetical model. An ad-hoc approach available in R/qtl involves ANOVA averaged over imputed samples (Table 4). This strongly supports a three QTL model with one epistatic pair, with the suggestion for a fourth QTL, on chromosome 2, possibly epistatic to the distal QTL on chromosome 1.

Now let's consider a strategy for building the genetic architecture using MIM, to be performed after these initial one QTL and two QTL investigations. We use WinQTLCart (Basten et al. 1999), as it has a good graphical interface and is free. We fit a new model using the MIM

Table 3 IM two-dimensional summaries. LOD scores for "full" model with two QTLs (lod.full), "additive" model with two QTLs (lod.add) or epistatic pair adjusted (type II) for main QTL effects (lod.int). Comparisons of full vs. best single QTL (lod.fvl) and additive vs. best single QTL (lod.avl) are also provided. Only entries with LOD > I shown.

2-QTL "best" summary evaluated at best full model per pair:

	pos1f	pos2f	lod.full	lod.fvl	lod.int	pos1a	pos2a	lod.add	lod.avl
c1:c1	15.93	40.6	4.65	0.497	0.247	6.83	11.4	4.41	0.249
c1:c2	20.05	12.8	**8.71**	**4.550**	1.497	17.99	15.0	7.21	3.053
c1:c3	48.12	13.0	**9.78**	**5.620**	4.249	15.93	52.1	5.53	1.371
c2:c2	29.70	36.1	3.45	1.148	0.541	31.80	33.9	2.91	0.607
c2:c3	29.70	60.0	3.04	0.742	0.000	29.70	60.0	3.04	0.741
c3:c3	2.12	24.3	1.53	0.709	0.536	42.88	45.9	0.99	0.173

2-QTL "int" summary evaluated at best epistasis per pair:

	pos1	pos2	lod.full	lod.fvl	lod.int	lod.add	lod.avl	
c1:c1	40.60	51.75	2.73	−1.4263	0.863	1.867	−2.289	
c1:c2	48.12	12.82	5.45	1.2936	**1.954**	3.495	−0.661	
c1:c3	48.12	13.00	9.78	5.6200	**8.060**	1.716	−2.440	
c2:c2	12.82	21.36	3.24	0.9394	1.325	1.913	−0.385	
c2:c3	6.41	4.23	2.25	−0.0508	0.511	1.737	−0.562	
c3:c3	2.12	24.35	1.53	0.7092	0.956	0.569	−0.247	

Table 4 ANOVA on best main QTL and epistatic pairs inferred from 1-QTL and 2-QTL scans. ANOVA table and Type III tests of effects based on 128 imputations. Performed using sim.geno, makeqtl and fitqtl in R/qtl (Broman et al. 2003).

	df	SS	MS	LOD	%var	Pvalue (F)
Model	6	115.64912	19.2748536	17.54648	56.90287	1.909584e-14
Error	89	87.59038	0.9841616			
Total	95	203.23951				

	df	Type III SS	LOD	%var	F value	Pvalue(F)	
Chr1@15.9	1	26.155	5.447	12.869	26.576	1.52e-06	***
Chr1@48.1	3	61.268	11.055	30.146	20.751	2.79e-10	***
Chr2@12.8	2	16.559	3.610	8.147	8.413	0.000450	***
Chr3@13	2	51.000	9.565	25.094	25.911	1.36e-09	***
Chr1@48.1:Chr2@12.8	1	6.266	1.440	3.083	6.367	0.013403	*
Chr1@48.1:Chr3@13	1	50.220	9.448	24.710	51.028	2.39e-10	***

forward search method, which is similar to the forward selection from regression applied to markers followed by CIM. We then refine the model by alternatively optimizing QTL positions, searching to add or testing to delete main QTLs and/or QTL interactions between pairs of QTLs. This is somewhat an art form, and can take many steps. There is no guarantee that different paths will lead to the same final model. The model achieved actually included two closely linked QTLs on chromosome 1 near 45 cM with opposite main effects. The BIC criterion (see WinQTLCart or Broman and Speed 2002) accepted all five QTLs, but the two closely linked QTLs were not really believable. For instance, they had high negative correlation of effects, and the one without epistatic effects had a very modest LOD. Therefore, we dropped this fifth QTL and obtained a model with four QTLs and two epistatic pairs, which is quite close to the truth (Table 5). Preserving hierarchy, the least significant effect is the epistasis between chromosomes 1 and 2, which agrees with the design of this simulation. In short, MIM using the BIC criterion almost recovered the correct model, with some subjective intervention during model search and model selection.

Figure 9 shows some one-dimensional summaries of the MIM fit. Each LOD profile is the added contribution of a QTL conditional on the maximum likelihood estimates of the three other QTLs in the model. The red curves are for the three QTLs picked up by MIM on its own. The blue

Table 5 MIM inferred QTL model. Compare estimates and heritabilities with values in Table 1. LOD scores test main QTL or epistatic pair adjusted (type III) for other effects.

qtl	chr	pos	effect	herit	LOD	qtl2	effect2	herit2	LOD
1	1	16	1.16	15.9%	3.26				
2	1	48	−0.05	0.0%	0.01				
3	2	18	−0.96	10.9%	2.42	2	−1.43	6.0%	1.46
4	3	10	−0.09	0.0%	0.03	2	3.00	26.5%	5.39

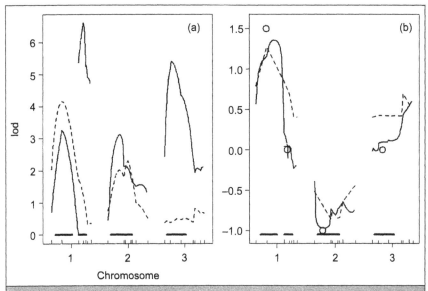

Fig. 9 MIM one-dimensional profiles. Detected QTL as solid lines. IM profile as dashed for comparison. True QTL indicated by gray vertical lines. Horizontal lines at base of figures specify 1.5 LOD support intervals. (a) LOD profiles for contribution of QTL in presence of other QTL (adjusted type III test from MIM, solid) or on its own (single QTL type I, IM, dashed); (b) main effects profile, with true values added as circles.

curve corresponds to the second QTL on chromosome 1. The IM profile is included for comparison (black). Estimates of main effects of QTL are shown in Figure 9b, with true values in purple. Currently, there is no two-dimensional graphic for MIM fit.

The strategy for model selection with BIM using MCMC samples is somewhat different. We first draw many samples (default is 120,000, saving every 40th) from the more probably genetic architectures. We then use Bayes factors and Bayesian model averaging to uncover evidence for the better models. The one-dimensional and two-dimensional summaries are different as well. They measure the contribution of a particular locus (or pair of loci for 2-D scans) after adjusting for all other possible *QTL*. That is, these are adjusted LPD, allowing for multiple QTL and averaging over all possible genetic architectures. This is distinct from IM scans, which have no adjustment for other QTLs; it is also distinct from MIM scans, which fix the genetic architecture for a set number of QTLs when scanning a particular QTL. The LPD peaks on Figure 10a are higher than for MIM, because other QTL effects are adjusted by model averaging rather than conditioning on the maximum likelihood estimates. The properties of these BIM scans are an area of active research.

The R/qtlbim software can separate linked effects, although effects for linked QTLs are averaged together in the 1-D projections of Figure 10. The estimates of genotypic effects for the main QTL (Figure 9b) are again close to the true values. Estimates of epistatic effects projected onto each QTL in Figure 10b should be interpreted with caution – better to view them with a 2-dimensional scan (not shown). For instance, the epistatic

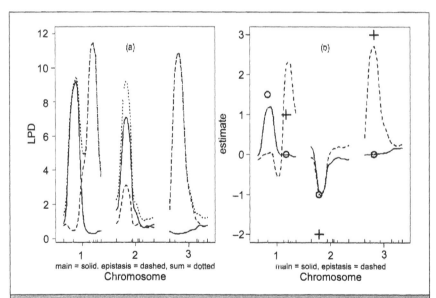

Fig. 10 BIM one dimensional profiles. (a) log posterior density (LPD) for combined effects (dotted), main QTL (solid) and epistatic effects (dashed); (b) main QTL (solid line = estimate, circle = true) and projected epistatic effects (dashed = estimate, cross = true). Contributions of epistatic pairs are shown at both loci, although it is not possible with this representation to determine which loci are paired.

effect on chromosome one distal end is an average of the two pairs of epistatic effects.

Table 6 shows one dimensional BIM summaries. The expected number of QTLs (n.qtl) is estimated from the posterior distribution of the number of QTLs (Yandell et al. 2007). That is, MCMC samples had on average 2.66 QTLs on chromosome 1 and almost two on chromosomes 2 and 3. It appears from Figure 10 that these extra QTLs are not major contributors, as the peaks for contributions from main QTL and epistatic QTLs are unimodal.

Figure 11 and Table 7 show a two-dimensional summary of LPD. Note how adjustment for other QTLs leads to a tightening of peaks relative to Figure 8. Once again, the contribution of the epistatic interaction between chromosomes 1 and 2 is not very strong, although it is more evident in Figure 11 than in Figure 8.

Other summaries can be useful. Figure 12 profiles Bayes factors, rescaled as 2log(BF), and means by genotype. Values of 2log(BF) above 2.1 are considered significant; values below zero are truncated. Other possibilities include the posterior intensities, variance estimates, or heritabilities.

Figure 13 shows posteriors and Bayes factor ratios for several important summaries. The posterior mode for number of QTLs is 7, but there is only weak to moderate evidence for more than 4 QTLs (BF ratios of ~3 comparing 4 to 5-9). Several chromosome patterns are equally

Table 6 BIM one dimensional summaries. Separate tables for log posterior density (LPD) and estimate of genotype effects. n.qtl is expected number of QTL on chromosome; pos are positions of the peak total effects per chromosome; m.pos are positions of main peaks in posterior; e.pos is position of epistatic peak.

LPD of pheno.normal for main, epistasis, sum

	n.qtl	pos	m.pos	e.pos	main	epistasis	sum
c1	2.66	48.1	15.9	48.1	8.86	11.78	11.81
c2	1.93	15.0	15.0	15.0	7.14	3.15	9.27
c3	1.99	10.7	60.0	10.7	0.79	10.70	10.72

estimate of pheno.normal for main, epistasis

	n.qtl	pos	m.pos	e.pos	main	epistasis	
c1	2.66	48.1	15.9	48.1	1.173	2.36	
c2	1.93	15.0	15.0	15.0	−0.977	−1.04	
c3	1.99	10.7	60.0	10.7	0.165	2.65	

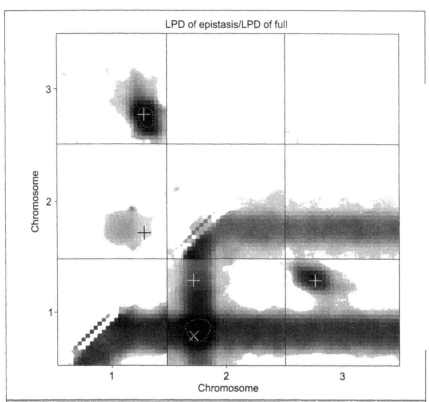

LPD of epistasis/LPD of full

Fig. 11 BIM two-dimensional summary. Lower triangle has LPD contribution of both QTL plus epistasis; upper triangle has contribution of epistasis only. Crosses indicate locations of true loci pairs; X at mistaken loci pair. Gray contours are 1.5 LPD support interval.

Table 7 BIM two-dimensional summaries. n.qtl is expected number of QTL on chromosome; l.pos are positions of main peaks (lower triangle) in posterior; u.pos are positions of epistatic peaks (upper triangle).

uupper: LPD of pheno.normal for epistasis
lower: LPD of pheno.normal for full

	n.qtl	l.pos 1	l.pos2	lower	u.pos 1	u.pos2	upper
c1:c1	2.34	24.2	32.38	10.91	6.83	57.19	1.398
c1:c2	5.18	20.0	17.09	14.51	38.55	25.50	4.598
c1:c3	5.37	48.1	13.00	11.32	48.12	13.00	11.292
c2:c2	1.23	0.0	10.68	11.89	4.27	36.13	2.846
c2:c3	3.97	15.0	60.00	8.19	6.41	4.23	0.856
c3:c3	1.32	0.0	6.35	2.17	0.00	6.35	1.840

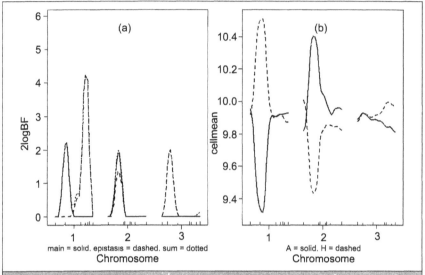

Fig. 12 Further BIM one dimensional profiles. (a) Bayes factors as 2log(BF), with main QTL (solid), epistatic effect (dashed) and combined effects (dotted, often hidden behind main or epistatic). Contributions of epistatic pairs are shown at both loci, although it is not possible with this representation to determine which loci are paired. (b) Means by genotype (solid = AA genotype; dashed = AB genotype).

favored, but the simplest has two on chromosome 1, one on chromosome 2, and one on chromosome 3 (coded as 2*1,2,3). Finally, the epistatic pairs 1.3 and 1.2 have the highest posteriors and the highest Bayes factors relative to any other epistatic pairs. In short, these summaries support the true model.

2.1.6 Covariates and Gene-Environment Interactions

Rarely is an experimental cross conducted in a single environment with all individuals handled identically, leading to measurements of just one phenotype. Complications arise, planting times vary, or there are broader scientific questions about differences across environments. We often do not have an ideal measurement of the characteristic we are most interested in studying. Instead, we measure multiple traits that are correlated, hoping that one of them will show strong heritability. It can be useful to think of such multiple correlated traits as covariates, measurements that covary with each other. Covariates are very important in understanding the effect of genotype on phenotype. First, including covariates can reduce residual variation and potentially enhance the power to detect QTL. Second, there may be important

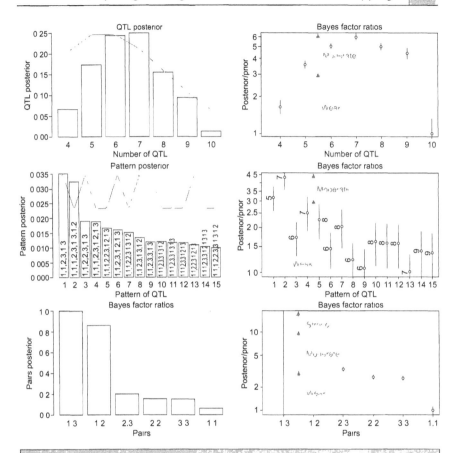

Fig. 13 Bayes factors for (a-b) number of QTL, (c-d) chromosome pattern of QTL and (e-f) epistatic pairs. Prior is rescaled and overlaid on posterior in left panels; vertical arrows in right panels indicate 3/10/30-fold ratios for Bayes factor comparisons.

differences in genotypic effects that depend on covariates or the environment. It is beyond the scope of this chapter to examine GxE interactions and multiple trait analysis in detail. Instead, we highlight a few important issues concerning adjustment for covariates and genotype by environment (GxE) interaction, indicating how they may be assessed with current software.

The most important covariates for plant breeding typically involve aspects of the environment. That is, different genotypes might perform best (in terms of the phenotype) in different environments. There are basically two types of experiments with crosses over different environment (Paterson et al. 1991; Stuber et al. 1992). Design I has individuals from one cross evaluated in two or more environments, while Design II has different genetic material in each environment. Many experiments have multiple traits measured on each individual. We might view these multiple traits in some sense as evaluating the same genetic material in different environments. That is, measurements of multiple traits per individual and measurements of a single trait in multiple environments for an individual (or clones or progeny) can be analyzed in much the same way as a Design I experiment.

Design I experiments with measurements on two or more environments, or correlated traits, for each individual in a cross can be examined in a variety of ways. For instance, we can first scan each trait on its own, noticing where peaks appear in common. However, we should not be surprised to find the same genomic regions cropping up, as we probably started with correlated traits. We can subsequently consider one trait as a covariate for the other, particularly when traits are over time or there is some other suggestion of causation. A dramatic reduction of peaks after adjusting for a covariate provides evidence that the QTL has an indirect effect through that covariate (although recall that statistical evidence in itself does not imply causation!). Alternatively, additional QTL can be detected after adjusting for covariates that reduce residual variation (*cf.* Li et al. 2006; Stylianou et al. 2006;). The primary advantage of joint analysis of multiple phenotypes is the ability to distinguish between pleiotropy, where one gene affects many traits, and close linkage of QTLs that independently affects separate traits (Jiang and Zeng 1995; Vieira et al. 2000; Li et al. 2006). Another approach is to combine correlated traits using some multivariate approach such as principal components (Liu et al. 1996).

Any category that divides a cross up into groups can lead to a Design II situation. Sex in animals and dioecious plants can be viewed as an example of Design II. Stratifying by age (young, old) or on experiments done over time are other ways. The key for Design II is that different, independent individuals are evaluated. While it is helpful to examine each 'environment' separately, this leads to a reduction in power to detect QTL: no evidence of QTL is inconclusive regarding GxE interaction. We must conduct a combined analysis adjusting for environment using all the data to properly assess GxE interaction. Be sure to adjust for the interaction of genotype and environment (known as GxE, or interacting covariates) rather than just for the main effect of environment. An excellent example of this is found in the work of Solberg et al. (2004, 2006).

Thus, the effect of QTL may depend on the value of the covariate for both Design I and Design II. The formal LOD (or LPD) score for a QTL allowing for GxE assesses both the effect of QTL on phenotype and the interaction of QTL with the covariate. Separately we can test if the GxE effect adds anything to assess the influence of the covariate. The papers cited below give examples using available software, primarily Pseudomarker or R/qtl (Solberg et al. 2004, 2006; Li et al. 2006; Stylianou et al. 2006), WinQTLCart (Jiang and Zeng 1995; Vieira et al. 2000) and R/ qtlbim (Yi et al. 2005). Many packages back to MapMaker/QTL (Lander et al. 1987; Lander and Botstein 1989) have allowed some form of adjustment for covariates. However, only a few packages appear to have full interacting covariate GxE adjustments. Pseudomarker (Sen and Churchill 2001), R/qtl (intcov option to scans; Broman et al. 2003) and R/ qtlbim (intcov for MCMC sampling; Yandell et al. 2007) allow adjustment for covariates to individual phenotypes. WinQTLCart (Basten et al. 1999) and MultiQTL (Mester et al. 2004) conduct multiple trait mapping of a modest number of correlated traits, providing joint LODs that formally test a QTL for any considered trait.

Expression QTL (eQTL) studies are now appearing with thousands of traits per individual in an experimental cross. WebQTL (Wang et al. 2003) is a handy, intuitive tool designed for expression traits, although it is largely focused on HK regression and correlation among traits. See Lan et al. (2006) for one way to extend this approach.

2.1.7 Overview of Available Packages

A number of packages were created for QTL analysis of inbred lines in the late 1980s and early 1990s. A handful of those survive today, some of them static and some of them under continual development. We focus attention here on the more current packages, with occasional reference to historical packages. Terms used here to describe methods and properties are explained in more detail above.

Marker regression can be used in any statistical package, and in the QTL packages QTLCart (Basten et al. 1999) and R/qtl (Broman et al. 2003).

The package MapMaker/QTL (Lander et al. 1987; Lander and Botstein 1989) greatly modified the conceptual framework for gene mapping, and its ideas are central to all other packages found today. MapMaker/QTL is still available as the original source (the Windows exe file does not appear to work under Windows/XP). Most users seeking to conduct simple interval mapping via the EM algorithm now rely on currently maintained packages that have incorporated the interval mapping algorithm, with slight variations, including

WinQTLCart or QTL Cartographer (Basten et al. 1999), R/qtl (Broman et al. 2003), MapQTL (van Ooijen and Maliepaard 1996) and MultiQTL (Mester et al. 2004).

Haley-Knott regression is used in QTL Express (Seaton et al. 2002), PLABQTL (Utz and Melchinger 1996), MapManager/QTX (Meer et al. 2004), and WebQTL (Wang et al. 2003). It is also available as an option in R/qtl (Broman et al. 2003) and R/qtlbim (Yandell et al. 2007).

Table 8 shows a summary of key features in packages for inbred lines. All except MultQTL and MapQTL are free. Most of the free packages have an open source, which means you can examine and

Table 8 Comparison of packages for inbred lines. Platform: W=Windows, L=Linux, M=MacOSX. Analysis method: B =Bayesian interval mapping, EM = expectation-maximation of likelihood, HK = Haley-Knott regression (marker regression only for R/bqtl). Platform is standalone (solo), R statistical system, Matlab, the Web (Java), or another package on the list. Most packages have a graphics user interface (GUI), typically coming from Windows, the Web (Java), or a platform application (R or Matlab). Out indicates capability to handle some outbred populations; * use blocking factors or a limited outbred option (e.g. 4-way cross in R/qtl). Most software is Free; some providing source, others providing applications only. GxE for interacting covariates or multiple trait mapping. X for presence, O for absence, * for limited ability, ? for unknown.

Package	W	L	M	B	EM	HK	Platform	GUI	Out	Free	GxE
MapMaker/ QTL	X	X	X	O	X	O	solo	O	O	X	*
Pseudomarker	X	X	X	X	O	O	Matlab	X	*	X	X
R/qtl	X	X	X	O	X	X	R	J/qtl	*	X	X
R/qtlbim	X	X	X	X	O	X	R/qtl	O	*	X	X
R/bim	X	X	X	X	O	O	R/qtl	O	*	X	O
R/bqtl	X	X	X	X	O	*	R	O	O	X	O
QTLCart	X	X	X	O	X	O	solo	WinQTL	O	X	X
WinQTL	X	O	O	X	X	O	QTLCart	X	O	X	X
Webqtl	X	X	X	O	O	X	Java	X	O	X	O
PLABQTL	X	O	O	O	O	X	solo	O	O	X	*
MapManager/ QTX	X	O	O	O	O	X	solo	X	O	X	O
MultiQTL	X	O	X	O	X	?	solo	X	*	O	X
MapQTL	X	O	O	O	X	?	solo	X	X	O	*
QTLExpress	X	X	X	O	O	X	Java	X	X	X	O
QTLCafe	X	X	X	O	O	X	QTLExpress	X	X	X	O

modify the code used for the QTL analysis. To our knowledge, only R/qtl properly handles the X chromosome (Broman et al. 2006). Some packages have extensive manuals, with screen shots available at their web sites. See the references for current URLs to individual packages.

Some software only available in source form is not included in Table 8 Multimapper (Sillanpää and Arjas 1998) conducts Bayesian interval mapping and model selection, with summaries in terms of the posterior intensity per locus. DIRECT (Ljungberg et al. 2004) is a very fast algorithm for solving linear models, and it has been incorporated into R/qtl and WebQTL. QTLNetwork (Yang et al. 2005) is a recent package with appealing graphics, but it is poorly documented to date, and the underlying methods of analysis are unclear.

Of the packages listed in Table 8 only QTLCart, MultiQTL, and R/qtlbim fully consider model selection with an arbitrary number of QTLs and epistasis. The former two use MIM while the latter uses Bayesian model averaging over possible genetic architectures. The R/qtl and Pseudomarker packages have some tools for arbitrarily large genetic architectures, but primarily focuses on two QTL with epistasis. R/bim allows for multiple QTL in a Bayesian model averaging framework, but cannot handle epistasis. QTLExpress does a limited investigation of epistasis for pairs of linked QTLs. PLABQTL and MapQTL employ CIM to adjust for other QTL when conducting a 1-QTL profile. MapManager/QTX and WebQTL rely on user-supplied markers to manually adjust for other QTL, somewhat analogous to CIM.

Several packages now employ Bayesian methods for interval mapping, most built on the R system (R Core Development Team 2006). R/qtl (Broman et al. 2003) includes multiple imputation, in addition to classical methods mentioned above. R/bqtl (Borevitz et al. 2002) was an early entrant, using marker regression in a Bayesian framework. The packages R/bim (Satagopan et al. 1996; Satagopan and Yandell 1996; Gaffney 2001) and R/qtlbim (Yi et al. 2005: Yandell et al. 2007) estimate the full posterior for models involving an arbitrary number of loci that may be in intervals between markers. R/qtlbim allows for epistasis and gene by environment interaction (see section 1.4). Pseudomarker and R/qtl both incorporate multiple imputation (Sen and Churchill 2001), with graphics for two QTLs and some tools for examining more that two QTLs.

The packages MultiQTL, R/qtl, Pseudomarker, and R/qtlbim all handle gene by environment (GxE) interactions, also known as interacting covariates. Other packages such as PLABQTL and QTLExpress handle covariates in a limited way.

Each package for QTL analysis of inbred lines has its own data input format. Several packages allow multiple importing formats, or can export data in a few different forms. This step is the biggest headache about using packages—figuring how to get your data in. Fortunately, most packages include well-documented examples. Further, most authors are open to email questions about package use. The MapMaker/QTL format is widely used, but it is not easy to set up the first time. The CSV format used in R/qtl allows one to build data in a spreadsheet, in a format that can be opened by Excel or by R. The R/qtl package has several other input and output formats, and it is not that difficult (with the help of a programmer) to customize output from R to most other packages.

Nothing has been said yet about non-normal phenotypes. There are some papers in this area, but only modest availability in packages to date. Nonparametric analysis (Kruglyak and Lander 1995) basically involves replacing trait values by their ranks; it is available in several packages. R/qtl includes binary traits, and R/qtlbim can handle ordinal traits (qualitative rankings such as poor/fair/good/excellent). Semi-parametric methods have been developed but are not broadly available yet (see Jin et al. 2007). Other approaches, such as Poisson regression, have been used in specialized software that, to our knowledge, has not been released.

2.1.8 QTL Analysis with Outbred Lines

This section is quite brief. Some packages such as QTLExpress, MapQTL and PLABQTL handle certain types of outbred crosses, including full sibs, half sibs and other relatedness designs. Some of these can combine different inbred crosses (*cf.* QTLExpress). SOLAR (Almasy and Blangero 1998) is a general purpose linkage and QTL mapping package using identity by descent (IBD) that works well for modest sized pedigrees. HAPPY (Mott et al. 2000) is specifically designed for heterogeneous stocks created from known founders.

In a way, QTL mapping for inbred lines can be adapted to experimental crosses (e.g. backcross or intercross) based on outbred founders. The easiest way is to use markers that distinguish among the founders. There is some loss of precision, as the QTL genotype can have more than two alleles. If your experiment is an F1 resulting from crossing two outbred founders, and the phases (haplotypes) of the founders are known, one can use the '4way' cross type in R/qtl for analysis and follow methods detailed above.

The QTL analysis methods with inbred lines detailed in this section are in theory extendable to outbred population, taking care of IBD and multiple alleles. All the problems and subtleties encountered above carry over and become harder. Further investigation of outbred crosses is beyond the scope of this chapter.

3 ASSOCIATION ANALYSIS

While an in-depth description of association analysis is beyond the scope of this chapter, several recent reviews describe association analysis in some detail (Flint-Garcia et al. 2003; Gupta et al. 2005; Hirschhorn and Daly 2005; Breseghello and Sorrells 2006; Yu and Buckler 2006). The review by Gupta et al. also provides a list of software as supplemental electronic material.

Association analysis is also called linkage disequilibrium (LD) mapping because it uses the extent of LD between a trait and markers to identify and find the location of QTLs. Such an association might imply that either the marker or some polymorphism linked to it is the cause of the observed phenotypic variation. However, two problems exist with this reasoning. First, a linked marker might not be in LD with the trait. Second, a marker that is in LD with a trait might not be linked to it. The story of the development and refinement of association analysis, and of the software for performing it, is largely a story of how these two problems have been addressed.

In the first case, LD between a linked marker and a quantitative trait loci (QTL) will be difficult to detect when the frequency of the marker is very different from that of the QTL. Presumably, sequence polymorphisms of some type underlie both markers and QTLs. A marker is simply a polymorphism that can be detected by an assay. A QTL is a polymorphism that causes a measurable change in phenotype. In the ideal situation, the marker and QTL are the same polymorphism and consequently have the same frequency. Otherwise, a closely linked marker may have a very different frequency in a population and as a result not be very useful for detecting a QTL.

This problem can be addressed by increasing marker density to make it more likely that at least one linked marker will be in LD with the QTL or be the QTL itself. For example, using the candidate gene approach to association mapping, an entire gene may be resequenced and all the sequence polymorphisms identified. For genome-wide scans, of course, that approach is not feasible. Another approach has been to use haplotypes instead of single markers to look for associations. Since the number of haplotypes will generally be greater than the number of individual markers, there may be more opportunities to match the frequency of the QTL.

The second problem with association analysis is that unlinked markers may be in LD with a trait of interest. LD structure can be affected by a number of factors including mutation, recombination, selection, mating patterns, and population admixture. Strategies for dealing with these problems include genomic control (GC), structured association (SA), family-based studies, and the use of marker data to correct for polygenic background effects.

In spite of these drawbacks, association analysis has been gaining interest rapidly in the plant genetics community. The key advantages of association analysis that are driving interest in this approach include the ability to use existing germplasm without the need to develop special populations, the ability to survey the diversity of alleles present in a broad-based population instead of being restricted to those present in the two parents of a mapping population, and the ability to map with high resolution. Resolution is determined by how rapidly LD decays in the population being sampled and can vary greatly depending on the species being sampled. For example, LD has been found to span just 1 kb in a diverse maize population, over 100 kb in a population of US elite maize inbred lines, and about 10cM in sugarcane, a vegetatively propagated species (Flint-Garcia et al. 2003).

As the use of association analysis, especially in plants, is a relatively recent development, the methodology and software is still undergoing development. No standard, accepted methodology exists and, consequently, standard software does not exist either. Nonetheless, software has been released that is useful for association analysis and related tasks, though knowledge is required on the part of the user to be sure that the analysis is appropriate. Related tasks include inference of population structure, derivation of measures of relatedness between individuals, haplotype inference, and analysis and visualization of LD structure.

Most of the software developed to date for association analysis has been developed for human genetics. Often that software is not directly useful for plant genetics. First, two tools widely used by plant geneticists, planned crosses and inbreeding, are not available in human genetics. As a result, family structures in human studies tend to be quite different from those in plant studies. The result is that the methods of analysis best suited for human genetics studies are often not optimal for plant genetics. Second, a lot of human genetics studies involve case-control studies for diseases. Phenotype data from case control studies is binary, affected versus unaffected. Relatively little plant phenotype data is binary.

While some very useful software has been developed for association analysis of case-control studies, it has been left out of this review since it is not likely to find much use by plant geneticists. On the other hand, some of the family-based methods could be used for plant species that are both naturally cross-pollinated and difficult to inbreed. As a result some software that uses family-based methods has been included.

Association analysis without some form of correction for population structure is straightforward and can be run using a variety of general statistical software packages. For example, Czika and Yu (2004) describe how to use SAS to perform marker-trait association tests for unrelated individuals or for populations with known family structure. In addition, freely available software has been written for association analysis that uses various strategies to cope with population structure. This software includes TASSEL, Powermarker, QTDT, MTDFREML, GC, BAMA, and TreeLD.

TASSEL (Yu et al. 2006) is written using Java and, as a result, runs on most computing platforms. It has an elaborate graphic user interface with a large number of functions for data management, analysis, and visualization. It accepts input either as text files or by way of the GDPC (Genomic Diversity and Phenotype Connection) browser (Casstevens and Buckler 2004), middleware providing a web connection to databases that have been made available through a GDPC server. Data imported independently from different sources can be combined for analysis. In addition, TASSEL will extract SNPs and indels from aligned sequence using flexible filtering criteria.

TASSEL has a number of analysis routines. Most notably, it implements a mixed model approach to association analysis (Yu et al. 2006) that uses both large-scale population structure and pair-wise kinship coefficients derived from marker data to correct for population stratification. The mixed model function, called MLM, requires a matrix of kinship coefficients to correct for population substructure and can optionally incorporate a population structure matrix or Q-matrix. TASSEL can calculate kinship coefficients for homozygous inbred lines. For heterozygotes, the kinship matrix can be calculated from marker data using the program SpaGeDi (Hardy and Vekemans 2002). The Q-matrix can be calculated using the program STRUCTURE (Pritchard et al. 2000). Structured association analysis, which uses the Q-matrix but not the kinship matrix, can be carried out using a fixed-effect linear model using the GLM function or logistic regression. In addition, TASSEL can be used to calculate and display LD measures for pairs of markers, calculate an evolutionary tree, or cladogram, and calculate population diversity statistics.

Powermarker (Liu and Muse 2005) is another example of multi-function software with a well-designed graphic user interface. Written in MS Visual C++ under the Microsoft .NET framework, Powermarker requires the MS Windows operating system. Data can be input from either text or Excel files. Powermarker provides data management functions and a number of descriptive genetic statistics, including measures of LD, population heterozygosity, inbreeding coefficients, tests of Hardy-Weinberg equilibrium, and Wright's F-statistics. It calculates genotype and allele frequencies and estimates haplotype frequencies. For association analysis, Powermarker will calculate an F-test for association between individual markers and traits and will perform Zaykin's haplotype trend regression (HTR) (Zaykin et al. 2002). Both of the association analyses assume a homogenous population without underlying structure or stratification.

QTDT (Abecasis et al. 2000) is software that implements the quantitative transmission disequilibrium test, which uses family structure to correct for population stratification. The analysis method uses a maximum likelihood approach to partition the genetic effects into within and between family components. The within family component provides an estimate of the genetic effect of a marker that is free of any population structure effect. As it makes use of family structure, QTDT, in effect, combines association and linkage analysis. QTDT analyzes data from nuclear families. At a minimum, it requires trios of parents plus one offspring or full sib pairs. Larger sibships can be analyzed as well. Other software, such as Merlin (Abecasis et al. 2002), must be used to calculate the IBD matrix required by the analysis. It also requires map positions of the genetic markers. QTDT has a command line interface, uses text files for input and output and has been compiled for Windows, Linux, and SunOS.

MTDFREML (Boldman et al. 1995) is software designed to set up and solve mixed model equations with individuals treated as random effects and an additive genetic relationship matrix used to define the covariance between individuals, similar to the model used in TASSEL's MLM function. The original program calculates the inverse of the additive genetic relationship matrix directly from pedigrees. It was recently modified (Zhang et al. 2006) to use a relationship matrix calculated directly from markers. MTDFREML can be used to estimate variance components, predict breeding values, and estimate associations between markers and phenotypes. The software is quite flexible, can incorporate covariates, and analyze multivariate models. Written in FORTRAN, it must be compiled by the user. In addition, it requires either FSPAK or SPARSPAK, which are sparse matrix libraries available from other sources. The interface is command-line. Text files are used for both input and output. The paper by Zhang et al. (2006) gives an example using data for canine hip dysplasia that uses MTDFREML to compute marker effects and other genetic parameters.

Genomic control (Bacanu et al. 2002; Devlin and Roeder 1999) is a method of correcting for unknown population structure by using bi-allelic markers distributed across the genome, that are not expected to be linked to QTLs. This method calculates a correction factor, λ, based on the values of a test statistic at the null loci. The test statistics for the loci being tested for association with a trait are then divided by λ. The programs, GC and GCF, are implemented in the R statistical programming language. Both programs estimate λ, then use that to construct a test of either a single locus or a pair of loci plus interaction. Devlin et al. 2004 recommend using GCF when testing a large number of candidate loci or when the required a-level is small.

BAMA provides a Bayesian solution which tests multiple loci simultaneously (Kilpikari and Sillanpaa 2003). As such, it avoids problems with multiple testing and over-estimation of the effect size when using the same data for detection and estimation. It assumes that the population being investigated has no substructure. The program is distributed as C source code, which must be compiled by the user, is designed for Linux or Unix operating systems, uses a command-line interface and text files for input and output.

TreeLD (Zollner and Pritchard 2005) provides an approach to association analysis that is different from the other programs mentioned. First, the method uses marker data to model the ancestry as a set of coalescent trees. Next, association between markers and phenotypes is evaluated by looking at the distribution of phenotypes among the tips of the trees. The authors demonstrate the effectiveness of the method using both simulated and real datasets. The method should not be used when population structure issues exist. The software requires phased genotypes and marker positions as input, though PHASE, described below, has been used to infer haplotypes from diplotype data with unknown phase. TreeLD is very computationally intensive. Documentation at the TreeLD website (Pritchard) notes that a thorough analysis of a data set with 250 individuals and 130 markers required 48 hours on a 10 processor Linux cluster.

As suggested above, haplotypes may have advantages over individual SNPs for conducting association analyses. In fact, haplotypes could be regarded as multi-allelic markers. Consequently, they have greater information content than individual SNPs. Buntjer et al. 2005 discuss the use of haplotypes in plant association analysis. For inbred lines, haplotypes can be identified directly. For homozygous individuals, TASSEL has a function that derives haplotypes from SNPs using a sliding window. Those haplotypes can then be used as markers in association analysis. For heterzygotes, however, haplotypes must be inferred. Separate software can be used to infer haplotypes, and the resulting haplotypes can be used as input for another analysis platform. Powermarker, mentioned earlier, uses an EM algorithm for inferring haplotypes. PHASE and Haplotyper use two different Bayesian methods to identify haplotypes.

The LD viewers, a related category of software, can be helpful for visualizing haplotypes or for interpreting the results of association analysis. Examples of LD viewers include Haploview (Barrett et al. 2005;Wu et al. 2006), MIDAS (Gaunt et al. 2006), JLIN (Carter et al. 2006), LDA (Ding et al. 2003), and GOLD (Abecasis and Cookson 2000; Ding et al. 2003). Haploview is easily the most versatile of these programs. It accepts phased chromosomes or unphased diplotypes as input. Family structure information can be incorporated but is optional. It will calculate and graph several pairwise LD measures, including D' and r^2. The user can select groups of markers for haplotype analysis or have Haploview automatically generate haplotypes. Haplotypes and haplotype frequencies can be output, but the user will have to recode genotypes using that information for further analysis. As Haploview's association testing is restricted to case-control and TDT trios, it will be of limited value for analyzing plant genetics data. GOLD provides nice graphics but is designed for use with human genetics data and may be challenging to adapt to other types of data. Both LDA and JLIN take SNP data with unknown phase as input and calculate and graph pairwise LD measures but provide no other functionality. MIDAS (Multiallelic Interallelic Disequilibrium Analysis Software) is unique in a couple of respects. First, it is designed to evaluate multiallelic markers rather than SNPS only. Second, it is a Python program and uses Tkinter for its graphic user interface. Haploview, JLIN, and LDA are all java applications. As mentioned earlier, both TASSEL and Powermarker calculate pairwise LD statistics and display the results graphically.

A final category of software which is required for SA assigns members of a population to subpopulations or mixtures of them. By far the most widely used of these is STRUCTURE (Pritchard et al. 2000). It uses a Bayesian model to construct subpopulations that minimize LD within a set of unlinked markers. While it requires the user to choose the number of subpopulations, the software can be used to estimate the number of subpopulations by running the analysis multiple times with different numbers of subpopulations then choosing the number that produces the best fit. PSMIX (Wu et al. 2006) is an R package that uses a maximum likelihood approach that is solved using an EM algorithm. It is computationally less intensive than STRUCTURE and, the authors find that it gives comparable results. The PSMIX article provides a good review of some additional programs for finding population substructure.

3.1 Association Analysis Examples

We illustrate the use of some of the association analysis software and highlight some of the issues that arise when doing this type of analysis with examples taken from studies of maize. In the following examples, the trait being analyzed is days to silk for maize inbred lines. The lines used were chosen to represent as much of the diversity present in the species as possible while restricting them to a manageable range of flowering dates. The trait 'days to silk' was chosen because flowering time often associates strongly with population structure. As plants must flower at roughly the same time in order to be cross-pollinated, populations tend to become stratified into flowering time or maturity groups.

Measurements were taken in 1999, 2000, and 2001 in Clayton, NC. The days to silk data and sequence for the dwarf8 gene was downloaded from the Panzea database (www.panzea.org) using the middleware application GDPC (Casstevens and Buckler 2004). The matrix of kinship coefficients and the population parameters used are part of the TASSEL tutorial. The population parameters were derived from SSR data using the program STRUCTURE (Pritchard et al. 2000). The kinship coefficient matrix was calculated using the program SPAGeDi (Hardy and Vekemans 2002) using data for 553 random SNPs. The random SNP data is also available as part of the TASSEL tutorial. Statistical analyses were run using TASSEL.

Using good quality phenotypic data with reasonably high heritability is critical to association analysis. While beyond the scope of this chapter, using principles of experiment design, checking for outliers and unreliable data, making certain that important assumptions are not violated, and following accepted statistical procedures are important steps in producing good phenotype data. For example, in each year that the days to silk data were taken, two or three replicates of data were taken but in some cases much data were missing from some of the replicates. Consequently, using least square means as estimates of days to silk within each year or across years was better than using simple averages.

The effectiveness of the analysis depends in part on how the kinship coefficients are derived. For this data, using SPAGeDi to calculate kinship coefficients is relatively straightforward. Information describing the data must be included in the input dataset. The data file specified that there were zero categories, zero spatial coordinates, and that the ploidy level was 1 since the data was for inbreds. Ritland's method was used to create a matrix of kinship coefficients. Negative values in the output matrix obtained from SPAGeDi were set to zero before running the subsequent association analysis.

As the results described below show, this method of calculating a kinship (K) matrix works well for this data. In part, this may be a result of the fact that maize is a natural outcrosser and lacks strongly differentiated subpopulations. Applying the K matrix method to self-pollinated species with major population substructure, such as rice, may require a modified approach.

The first set of examples uses the candidate gene approach to association analysis. This method entails using different lines of evidence to develop a list of genes which could be important in controlling the expression of a trait of interest. The genes identified are then resequenced for each of the individual taxa in the study. The resulting sequences are aligned, and sequence polymorphisms identified. The results below use data previously analyzed and reported by Thornsberry et al. (2001). In this study, the dwarf8 gene was chosen because QTL studies and mutagenesis had suggested that it affects maize flowering time and plant height. For simplicity, the results below only look at SNPs. As the published analysis shows, including indels is critical for proper interpretation.

The software TASSEL was used to extract SNPs from the aligned sequence and to perform the analyses. Three related analytical methods were examined. In each case, each SNP was analyzed individually. First, a fixed effect linear model (GLM) was solved using the SNP as a classification variable. Second, the population parameters (Q-matrix) derived from STRUCTURE were added to the model as covariates (GLM + Q). As STRUCTURE was used to assign each line to three populations, each line had three parameters which added to 1. As a result, the three sets of parameters were linearly dependent and only two of the three needed to be included in the model as covariates. The F-test of SNP is the same regardless of which two are actually used. Third, a mixed linear model (MLM) solution (Yu et al. 2006) was used which treated the SNP and the population parameters as fixed effects and taxa as a random effect with the kinship matrix (K-matrix) defining the additive genetic covariance structure.

An important difference between the models is the way in which taxa (the inbred lines) are treated. In the case of the two GLM analyses, taxa do not enter the model explicitly. Each line can be thought of as a random sample from two underlying populations, that differ only by the value of the SNP. When the model contains only the SNP main effect, it is equivalent to a t-test of the difference between the means of the two populations. If additional terms are added to the model, such as replications or environments, the SNP effect must be tested using the taxa nested within SNP mean square, not the model residual. In the case of MLM, taxa enter the model as random effects with a known covariance structure. This structure is specified by the K-matrix. The K-matrix supplies more information about relationships between lines than the Q-matrix and does a better job of removing spurious effects due to population substructure (Yu et al. 2006).

The probability values derived from the F-tests of selected SNPs for association with the days to silk data for each of the analysis methods is shown in Table 9. As expected, the different analysis methods give similar results. The p-values are generally lowest for GLM, especially in the case of site 2625. As several hypotheses are being tested, some form of multiple test correction should be used to evaluate the results. While

Table 9 Probability values for F-tests of selected SNP sites in the dwarf8 gene using different analysis methods.

Site	GLM	GLM + Q	MLM
184	0.3921	0.5684	0.4528
677	0.0012	0.0019	0.0007
680	0.0157	0.057	0.0298
695	0.8967	0.3757	0.5568
699	0.0006	0.0013	0.0004
713	0.7852	0.3555	0.4713
736	0.8501	0.4321	0.5638
741	0.0031	0.0422	0.0185
756	0.0101	0.0841	0.0746
1663	0.0000	0.0004	0.0003
2511	0.6236	0.5742	0.5632
2625	0.0007	0.0515	0.051
2880	0.5412	0.4678	0.511
3000	0.6979	0.254	0.213
3459	0.0014	0.0017	0.007
3490	0.0001	0.0003	0.0002
3570	0.0001	0.0003	0.0002

conservative, a Bonferroni correction is easy to use. If an α-level of 0.01 is desired for rejection of the null hypothesis of no association, the Bonferroni corrected α-level is $0.01/17 = .0006$. Using that cut-off would lead us to reject the null hypothesis for sites 1663, 3490, and 3570 for all three methods and 699 for GLM and MLM. However, the choice of α-level is clearly arbitrary and should only serve as a guide to our interpretation of the results. Sites 677 in the MLM results and 2625 in the GLM results are close to the cutoff and could be considered as well. Even in this small example, the results of the three analyses vary, but without additional information, there is no way to decide which is best.

The same days to silk data used to generate the results in Table 9 were also tested for association with 553 random unlinked SNPs from maize genes. As linkage disequilibrium in maize decays very rapidly with distance, randomly chosen SNPs are not expected to be linked to any individual trait and should not be associated with it. The SNPs with the lowest p-values from the GLM analysis are shown in Table 10. For

Table 10 Probability values for F-tests of a subset of 553 random SNPs from maize genes. All 553 SNPs were tested. Those with the smallest p-values for the GLM method are shown here.

SNP	GLM	GLM+Q	MLM
514	0.00000	0.0212	0.0025
10	0.00000	0.0447	0.2464
429	0.00000	0.0031	0.0148
469	0.00000	0.0111	0.0103
319	0.00002	0.0308	0.076
318	0.00002	0.0000	0.0004
368	0.00003	0.0502	0.0954
45	0.00003	0.2761	0.6529
203	0.00003	0.0498	0.0909
46	0.00004	0.0431	0.1814
464	0.00005	0.3304	0.3111
478	0.00005	0.1382	0.175
388	0.00007	0.0527	0.2574
398	0.00014	0.6201	0.2859
307	0.00018	0.1304	0.2586
157	0.00020	0.0324	0.0827
526	0.00022	0.0001	0.0008
173	0.00022	0.2148	0.0969
1	0.00027	0.0061	0.0072

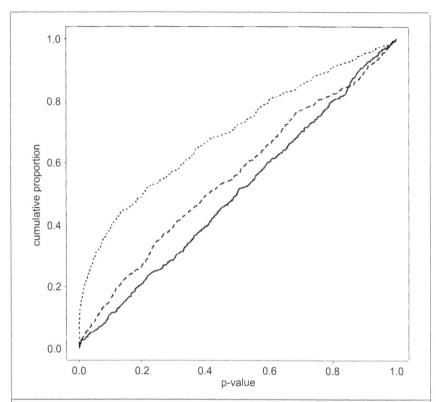

Fig. 14 Cumulative probability curve from F-tests of the association of 553 random SNPs with days to silk for three analysis methods. A straight line with a slope of 1 is expected under the null hypothesis of no association. Solid = MLM, dashed = GLM + covariate, dotted = GLM.

this set the Bonferroni corrected α-level corresponding to an overall desired α-level of 0.01 is $0.01/553 \cong$ 2E-5. At that level, using GLM several SNPs appear to be associated with flowering date. Only two of those associations remain using GLM+Q. None of them are close to our chosen significance level using MLM.

A more effective way to summarize the data from this example is shown in Figure 14, a graph of the cumulative distribution functions. To generate the graph, the p-values from the F-tests for SNP were sorted in ascending order individually for each method. An order statistic was assigned to each value with 1 for the lowest p-value and 553 for the highest. The order statistic divided by 553 is plotted on the y-axis and the actual p-value on the x-axis. Under the null hypothesis of no association,

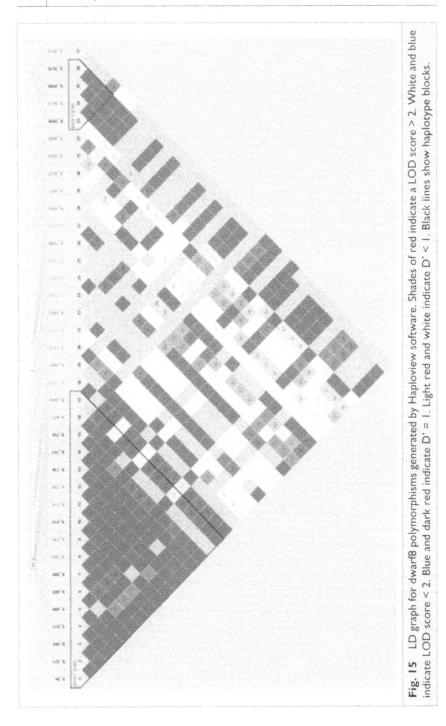

Fig. 15 LD graph for dwarf8 polymorphisms generated by Haploview software. Shades of red indicate a LOD score > 2. White and blue indicate LOD score < 2. Blue and dark red indicate D' = I. Light red and white indicate D' < I. Black lines show haplotype blocks.

the points are expected to lie on a straight line with a slope of 1. A deviation from that line is an indication of false positives probably resulting from underlying population substructure. As was obvious from Table 10, the GLM method has a number of false positives. Adding population covariates to the model helps, but the cumulative probability curve still deviates from a straight line. The MLM results lie almost exactly on the line expected under the null hypothesis, indicating that in this example the MLM method effectively eliminated association due to population substructure alone. Other traits show the same trend, though the strength of that trend varies depending on how strongly a trait is associated with population structure (Yu et al. 2006).

Graphs showing the extent of LD between markers can help to interpret the output from an association analysis. As described earlier, a number of software packages can be used to visualize the pattern of linkage disequilibrium between markers. The LD graph shown in Figure 15 was generated by Haploview. This figure graphs the relationships between SNPs and indels from the dwarf8 sequence alignment used in the first example. It reveals two haplotype blocks defined by high internal levels of LD. In addition, those two blocks are seen to be in LD with each other. The graph shows why, using this dataset, several polymorphisms are almost equally likely to be the cause of phenotypic variation. Relative positions of the polymorphisms are shown in the bar above the graph. Gray lines indicate SNPs and green lines indicate indels because the indels were identified in the input file of marker names. The color coding is based on values of D' and its associated LOD scores or tests of significance. Values of D' are displayed in the sqaures. In addition, Haploview provides other color schemes, can display r^2 values as well, has a display which shows haplotype values, but only analyzes data for bi-allelic markers.

Acknowledgements

Yandell was supported in part by National Institutes of Health (NIH) Grants R01 GM069430, NIDDK 5803701 and NIDDK 66369-01. The authors wish to thank Karl Broman, Natalia de Leon and Nengjun Yi for helpful comments on early drafts of this chapter.

References

Abecasis GR, Cardon LR, Cookson WOC (2000) A general test of association for quantitative traits in nuclear families. Am J Hum Genet 66: 279-292

Abecasis GR, Cookson WO (2000) GOLD—Graphical Overview of Linkage Disequilibrium. Bioinformatics 16: 182-183: www.sph.umich.edu/csg/abecasis/GOLD

Abecasis GR, Cherny SS, Cookson WO, Cardon LR (2002) Merlin—rapid analysis of dense genetic maps using sparse gene flow trees. Nat Genet 30: 97-101: www.sph.umich.edu/csg/abecasis/Merlin

Almasy L, Blangero J (1998) Multipoint quantitative trait linkage analysis in general pedigrees. Am J Hum Genet 62: 1198-1211: www.sfbr.org/solar

Bacanu S, Devlin B, Roeder K (2002) Association studies for quantitative traits in structured populations. Genet Epidemiol 22: 78-93

Barrett JC, Fry B, Maller J, Daly MJ (2005) Haploview: analysis and visualization of LD and haplotype maps. Bioinformatics 21: 263-265: www.broad.mit.edu/mpg/haploview

Basten CJ, Weir BS, Zeng ZB (1999) QTL Cartographer: A Reference Manual and Tutorial for QTL Mapping. Center for Quantitative Genetics, NCSU: statgen.ncsu.edu/qtlcart

Boldman KG, Kriese LA, VanVleck LD, Van Tassell CP, and Kachman SD (1995) A manual for use of MTDFREML. A set of programs to obtain estimates of variance and covariance. USDA, Agriculture Research Service, Clay Center, NE: aipl.arsusda.gov/curtvt/mtdfreml.html

Borevitz JO, Maloof JN, Lutes J, Dabi T, Redfern JL, Trainer GT, Werner JD, Asami T, Berry CC, Weigel D, Chory J (2002) Quantitative trait loci controlling light and hormone response in two accessions of *Arabidopsis thaliana*. Genetics 160(2): 683-96: hacuna.ucsd.edu/bqtl

Breseghello F, Sorrells ME (2006) Association analysis as a strategy for improvement of quantitative traits in plants. Crop Sci 46: 1323-1330

Broman KW (2001) Review of statistical methods for QTL mapping in experimental crosses. Lab Anim 30(7): 44-52

Broman KW, Sen Ś, Owens SE, Manichaikul A, Southard-Smith EM, Churchill GA (2006) The X chromosome in quantitative trait locus mapping. Genetics 174: 2151-2158

Broman KW, Speed TP (2002) A model selection approach for the identification of quantitative trait loci in experimental crosses (with discussion). J Roy Stat Soc B 64: 641-656, 731-775

Broman KW, Wu H, Sen Ś, Churchill GA (2003) R/qtl: QTL mapping in experimental crosses. Bioinformatics 19: 889-890: www.rqtl.org

Buntjer JB, Sorensen AP, Peleman JD (2005) Haplotype diversity: the link between statistical and biological association. Trends Plant Sci 10: 466-471

Carter K, McCaskie P, Palmer L (2006) JLIN: A java based linkage disequilibrium plotter. BMC Bioinformatics 7: 60: www.genepi.org.au/jlin

Casstevens TM, Buckler ES (2004) GDPC: connecting researchers with multiple integrated data sources. Bioinformatics 20: 2839-2840: www.maizegenetics.net/gdpc

Churchill GA, Doerge RW (1994) Empirical threshold values for quantitative trait mapping. Genetics 138: 963-971

Czika W, Yu X (2004) Gene frequencies and linkage disequilibrium. In: AM Saxton (ed) Genetic Analysis of Complex Traits Using SAS. Cary, NC, USA, SAS Institute Inc: 179-200

Darvasi A (2005) Dissecting complex traits: the geneticists' 'Around the world in 80 days'. Trends Genet 21: 373-376

Devlin B, Roeder K (1999) Genomic control for association studies. Bioinformatics 55: 997-1004

Devlin B, Bacanu S, Roeder K (2004) Genomic Control to the extreme. Nat Genet 36: 1129-1130

Doerge RW, Churchill GA (1996) Permutation tests for multiple loci affecting a quantitative character. Genetics 142: 285-294

Doerge RW, Zeng ZB, Weir BS (1997) Statistical issues in the search for genes affecting quantitative traits in experimental populations. Statist Sci 12: 195-219

Ding K, Zhou K, He F, Shen Y (2003) LDA – a java-based linkage disequilibrium analyzer. Bioinformatics 19: 2147-2148: www.chgb.org.cn/lda/lda.htm

Flint-Garcia SA, Thornsberry JM, Buckler ES (2003) Structure of linkage disequilibrium in plants. Annu Rev Plant Biol 54: 357-374

Gaffney PJ (2001) An efficient reversible jump Markov chain Monte Carlo approach to detect multiple loci and their effects in inbred crosses. PhD Dissertation, Dept of Statistics, UW-Madison, WI, USA

Gaunt TR, Rodriguez S, Zapata C, Day INM (2006) MIDAS: software for analysis and visualisation of interallelic disequilibrium between multiallelic markers. BMC Bioinformatics 7: 227-237: www.genes.org.uk/software/midas

Gelman A, Carlin JB, Stern HS, Rubin DB (2003) Bayesian Data Analysis. 2nd edn. CRC Press, London, UK

Glazier AM, Nadeau JH, Aitman TJ (2002) Finding genes that underlie complex traits. Science 298: 2345-2349

Guo SW, Lange K (2000) Genetic mapping of complex traits: promises, problems, and prospects. Theor Pop Biol 57: 1-11

Gupta P, Rustgi S, Kulwal P (2005) Linkage disequilibrium and association studies in higher plants: Present status and future prospects. Plant Mol Biol 57: 461-485

Hackett CA (2003) Statistical methods for QTL mapping in cereals. Plant Mol Biol 48: 585-599

Haley C, Knott S (1992) A simple regression method for mapping quantitative trait loci of linked factors. J Genet 8: 299-309

Hardy OJ, Vekemans X (2002) Spagedi: a versatile computer program to analyse spatial genetic structure at the individual or population levels. Mol Ecol Notes 2: 618-620: www.ulb.ac.be/sciences/lagev/spagedi.html

Heath SC (1997) Markov chain Monte Carlo segregation and linkage analysis for oligogenic models. Am J Hum Genet 61: 748-760

Hirschhorn JN, Daly MJ (2005) Genome-wide association studies for common diseases and complex traits. Nat Rev Genet 6: 95-108

Hollander WF (1955) Epistasis and hypostasis. J Hered 46: 222-225

Jansen RC (1993) Interval mapping of multiple quantitative trait loci. Genetics 135: 205-211

Jiang C, Zeng ZB (1995) Multiple trait analysis of genetic mapping for quantitative trait loci. Genetics 140: 1111-1127

Jin C, Fine JP, Yandell BS (2007) A unified semiparametric framework for QTL analyses, with application to spike phenotypes. J Am Statist Assoc 102: 56-67

Kao CH (2000) On the differences between maximum likelihood and regression interval mapping in the analysis of quantitative trait loci. Genetics 156: 855-865

Kao CH, Zeng ZB (2002) Modeling epistasis of quantitative trait loci using Cockerham's model. Genetics 160: 1243-1261

Kao CH, Zeng ZB, Teasdale RD (1999) Multiple interval mapping for quantitative trait loci. Genetics 152: 1203-1216

Kilpikari R, Sillanpaa MJ (2003) Bayesian analysis of multilocus association in quantitative and qualitative traits. Genet Epidemiol 25: 122-135

Korstanje R, Paigen B (2002) From QTL to gene: the harvest begins. Nat Genet 31: 235-236

Kruglyak L, Lander ES (1995) A nonparametric approach for mapping quantitative trait loci. Genetics 139: 1421-1428

Lan H, Chen M, Byers JE, Yandell BS, Stapleton DS, Mata CM, Mui ETK, Flowers MT, Schueler KL, Manly KF, Williams RW, Kendziorski C, Attie AD (2006) Combined expression trait correlations and expression quantitative trait locus mapping. PLoS Genetics 2: e6

Lander E, Abrahamson J, Barlow A, Daly M, Lincoln S, Newburg L, Green P (1987) MapMaker: A computer package for constructing genetic linkage maps. Cytogenet Cell Genet 46: 642-642

Lander ES, Botstein D (1989) Mapping Mendelian factors underlying quantitative traits using RFLP linkage maps. Genetics 121: 185-199

Lander ES, Green P (1987) Construction of multilocus genetic linkage maps in humans. Proc Natl Acad Sci USA 84: 2363-2367: www.broad.mit.edu/genome_software

Lander ES, Kruglyak L (1995) Genetic dissection of complex traits- guidelines for interpreting and reporting linkage results. Nat Genet 11: 241-247

Li R, Tsaih SW, Shockley K, Stylianou IM, Wergedal J, Paigen B, Churchill GA (2006) Structural model analysis of multiple quantitative traits. PLoS Genetics 2: e114

Liu J, Mercer JM, Stam LF, Gibson GC, Zeng ZB, Laurie CC (1996) Genetic analysis of a morphological shape difference in the male genitalia of *Drosophila simulans* and *D. mauritiana*. Genetics 142: 1129-1145

Liu K, Muse SV (2005) PowerMarker: an integrated analysis environment for genetic marker analysis. Bioinformatics 21: 2128-2129: www.powermarker.net

Ljungberg K, Holmgren S, Carlborg Ö (2004) Simultaneous search for multiple QTL using the global optimization algorithm DIRECT. Bioinformatics 20: 1887-1895: user.it.uu.se/~kl

Mackay TFC (2001) The genetic architecture of quantitative traits. Annu Rev Genet 35: 303-339

Manichaikul A, Dupuis J, Sen Ś, Broman KW (2006) Poor performance of bootstrap confidence intervals for the location of a quantitative trait locus. Genetics 174:481-489

Martinez O, Curnow RN (1992) Estimation the locations and the sizes of the effects of quantitative trait loci using flanking markers. Theor Appl Genet 85: 480-488

Meer JM, Cudmore Jr RH, Manly KF (2004) MapManager/QTX: software for complex trait analysis. www.mapmanager.org/mmQTX.html

Mester DI, Ronin YI, Nevo E, Korol AB (2004) Fast and high precision algorithms for optimization in large-scale genomic problems. Comp Biol Chem 28: 281-290 www.multiqtl.com

Mott R, Talbot CJ, Turri MG, Collins AC, Flint J (2000) A new method for fine-mapping quantitative trait loci in outbred animal stocks. Proc Natl Acad Sci USA 97(23):12649-12654: www.well.ox.ac.uk/~rmott/happy.html

Nadeau JH, Frankel WN (2000) The roads from phenotypic variation to gene discovery: mutagenesis versus QTLs. Nat Genet 25: 381-384

Page GF, George V, Go RC, Page PZ, Allison DB (2003) "Are we there yet?": Deciding when one has demonstrated specific genetic causation in complex diseases and quantitative traits. Am J Hum Genet 73: 711-719

Paterson AH, Damon S, Hewitt JD, Zamir D, Rabinowitch HD, Lincoln SE, Lander ES, Tanksley SD (1991) Mendelian factors underlying quantitative traits in tomato: comparison across species, generations and environments. Genetics 127: 181-197

Pritchard JK, Stephens M, Donnelly P (2000) Inference of population structure using multilocus genotype data. Genetics 155: 945-959: pritch.bsd.uchicago.edu/software.html

Satagopan JM, Yandell BS (1996) Estimating the number of quantitative trait loci via Bayesian model determination. Special Contributed Paper Session on Genetic Analysis of Quantitative Traits and Complex Diseases, Biometric Section, Joint Statistical Meetings, Chicago, IL, USA

Satagopan JM, Yandell BS, Newton MA, Osborn TC (1996) Markov chain Monte Carlo approach to detect polygene loci for complex traits. Genetics 144: 805-816

Seaton G, Haley CS, Knott SA, Kearsey M, Visscher PM (2002) QTL Express: mapping quantitative trait loci in simple and complex pedigrees. Bioinformatics 18: 339-340: qtl.cap.ed.ac.uk

Sen Ś, Churchill GA (2001) A statistical framework for quantitative trait mapping. Genetics 159: 371-387: www.jax.org/staff/churchill/labsite/software/pseudomarker

Sen Ś, Satagopan JM, and Churchill GA (2005) Quantitative trait locus study design from an information perspective. Genetics 170: 447-464: www.jax.org/staff/churchill/labsite/software/pseudomarker

Sillanpää MJ, Arjas E (1998) Bayesian mapping of multiple quantitative trait loci from incomplete inbred line cross data. Genetics 148: 1373-1388: www.rni.helsinki.fi/~mjs

Solberg LC, Baum AE, Ahmadiyeh N, Shimomura K, Li R, Turek FW, Churchill GA, Takahashi JS, Redei EE (2004) Sex- and lineage-specific inheritance of depression-like behavior in the rat. Mamm Genom 15: 648-662

Solberg LC, Baum AE, Ahmadiyeh N, Shimomura K, Li R, Turek FW, Takahashi JS, Churchill GA, Redei EE (2006) Genetic analysis of the stress-responsive adrenocortical axis. Physiol Genom 00: 000-000: dx.doi.org/10.1152/physiolgenomics.00052.2006

Stephens DA, Fisch RD (1998) Bayesian analysis of quantitative trait locus data using reversible jump Markov chain Monte Carlo. Biometrics 54: 1334-1347

Stuber CW, Lincoln SE, Wolff DW, Helentjaris T, Lander ES (1992) Identification of genetic factors contributing to heterosis in a hybrid from two elite maize inbred lines using molecular markers. Genetics 132: 823-839

Stylianou IM, Tsai SW, diPetrillo K, Ishimori N, Li R, Paigen B, Churchill G (2006) Complex genetic architecture revealed by analysis of HDL in chromosome substitution strains in F2 crosses. Genetics 174: 999-1007

Thornsberry JM, Goodman MM, Doebley J, Kresovich S, Nielsen D, Buckler ES 4th (2001) Dwarf8 polymorphisms associate with variation in flowering time. Nat Genet 28: 286-289

Utz HF, Melchinger AE (1996) PLABQTL: a program for composite interval mapping of QTL. J Quant Trait Loci 2 : www.uni-hohenheim.de/ipspwww/soft.html

van Ooijen JW, Maliepaard C (1996) MapQTL version 3.0: software for the calculation of QTL positions on genetic maps. CPRO-DLO, Wageningen, The Netherlands: www.kyazma.nl

Vieira C, Pasyukova EG, Zeng ZB, Hackett JB, Lyman RF, Mackay TFC (2000) Genotype-environment interaction for quantitative trait loci affecting life span in *Drosophila melanogaster*. Genetics 154: 213-227

Wang H, Zhang YM, Li X, Masinde GL, Mohan S, Baylink DJ, Xu S (2005) Bayesian shrinkage estimation of quantitative trait loci parameters. Genetics 170: 465-480

Wang J, Williams RW, Manly KF (2003) WebQTL: Web-based complex trait analysis. Neuroinformatics 1: 299-308: www.genenetwork.org

Wright FA, Kong A (1997) Linkage mapping in experimental crosses: the robustness of single-gene models. Genetics 146: 417-425

Wu B, Nianjun L, Hongyu Z (2006) PSMIX: an R package for population structure inference via maximum likelihood method. BMC Bioinformatics 7: 317: bioinformatics.med.yale.edu/PSMIX

Xu S (1995) A comment on the simple regression method for interval mapping. Genetics 141: 1657-1659

Yandell BS, Mehta T, Banerjee S, Shriner D, Venkataraman R, Moon JY, Neely WW, Wu H, von Smith R, Yi N (2007) R/qtlbim: QTL with Bayesian interval mapping in experimental crosses. Bioinformatics 23: 641-643: www.rqtlbim.org

Yang J, Hu CC, Ye XZ, Zhu J (2005) QTLNetwork 2.0. Institute of Bioinformatics, Zhejiang University, Hangzhou, China: ibi.zju.edu.cn/software/qtlnetwork

Yi N (2004) A unified Markov chain Monte Carlo framework for mapping multiple quantitative trait loci. Genetics 167: 967-975

Yi N, Yandell BS, Churchill GA, Allison DB, Eisen EJ, Pomp D (2005) Bayesian model selection for genome-wide epistatic quantitative trait loci analysis. Genetics 170: 1333-1344

Yu J, Buckler ES (2006) Genetic association mapping and genome organization of maize. Curr Opin Biotechnol 17: 155-160

Yu J, Pressoir G, Briggs WH, Vroh Bi I, Yamasaki M, Doebley JF, McMullen MD, Gaut BS, Nielsen DM, Holland JB, Kresovich S, Buckler ES (2006) A unified mixed-model method for association mapping that accounts for multiple levels of relatedness. Nat Genet 38: 203-208: www.maizegenetics.net/bioinformatics/tassel

Zaykin DV, Westfall PH, Young SS, Karnoub MA, Wagner MJ, Ehm MG (2002) Testing association of statistically inferred haplotypes with discrete and continuous traits in samples of unrelated individuals. Hum Hered 53: 79-91

Zeng ZB (1994) Precision mapping of quantitative trait loci. Genetics 136: 1457-1468

Zeng ZB, Liu J, Stam LF, Kao CH, Mercer JM, Laurie CC (2000) Genetic architecture of a morphological shape difference between two drosophila species. Genetics 154: 299-310

Zhang Z, Todhunter RJ, Buckler ES, van Vleck LD (2006) Technical note: use of marker based relationships with multiple-trait derivative-free restricted maximal likelihood. J Anim Sci 85: 881-885

Zollner S, Pritchard JK (2005) Coalescent-based association mapping and fine mapping of complex trait loci. Genetics 169: 1071-1092

Zou F, Fine JP, Hu J, Lin DY (2004) An efficient resampling method for assessing genome-wide statistical significance in mapping quantitative trait loci. Genetics 168: 2307-2316

Index